Superconducting Quantum Electronics

Superconducting Quantum Electronics

Edited by V. Kose

Foreword by Werner Buckel

With Contributions by
M. Albrecht H. Bachmair G. Brunk
K. H. Gundlach P. Gutmann C. Heiden
J. Hinken R. P. Huebener W. Kessel
H. Koch H. Lübbig J. Niemeyer
R. Pöpel H. Rogalla

With 180 Figures

Springer-Verlag Berlin Heidelberg New York
London Paris Tokyo Hong Kong

Director and Professor Dr. *Volkmar Kose*

Physikalisch-Technische Bundesanstalt, Bundesallee 100
D-3300 Braunschweig, Fed. Rep. of Germany

ISBN 3-540-51176-8 Springer-Verlag Berlin Heidelberg New York
ISBN 0-387-51176-8 Springer-Verlag New York Berlin Heidelberg

Library of Congress Cataloging-in-Publication Data. Superconducting quantum electronics / edited by Volkmar Kose; with contributions by M. Albrecht ... [et al.]. p. cm. Includes index. 1. Superconductivity. 2. Quantum electronics. 3. Josephson effect. I. Kose, Volkmar. QC611.97.T46S87 1989 537.6'23–dc20 89-11379

Printing: Zechnersche Buchdruckerei, 6720 Speyer
Binding: J. Schäffer GmbH & Co. KG., 6718 Grünstadt
2157/3150–543210 – Printed on acid-free paper

Foreword

With the surprising discovery of superconductivity at temperatures above 100 K, this field was not only brought into the public eye, but also stimulated research in universities, scientific institutions and industry, thus continuing the fascinating development which began with the discovery of the Josephson effect in the sixties. Cryoelectronics has become a special branch of cryophysics and cryotechnics and today plays a prominent role whenever high resolution and precision measurements are required.

Motivated by this development, seven years ago scientists working in cryoelectronics in the Federal Republic of Germany felt the necessity for regular meetings allowing a free exchange of ideas and results achieved. Seminars under the title of "Kryoelektronische Bauelemente" were held for the first time at the Physikalisch-Technische Bundesanstalt in Braunschweig in 1982 on the occasion of the 100th anniversary of the birth of Walther Meißner, a pioneer in superconductivity. Since then, meetings have been held every year at different venues in Germany.

It is now felt that the status of this field necessitates a review of the results of the past, a description of the current state of the art, and a discussion of future perspectives. This book, entitled SUPERCONDUCTING QUANTUM ELECTRONICS is a collection of invited lectures and contributions which will inform the reader on the most interesting problems involving fundamentals, sensitive detectors and precision metrology being studied by different groups.

There is no doubt that the new superconductors open up interesting possibilities for almost all measuring devices: for example, integrated circuits in microelectronics will be developed which combine classic components such as semiconductors with the new superconductors. Computers working at 77 K are already on the market. The integration of the new superconductors into these computers does not involve any cryogenic problem, but the tailoring of suitable films will require much research work.

This book summarizes the successful development of classic superconductors and discusses the outlook for the future. We hope that this volume will stimulate further discussion, collaboration among scientists in different laboratories, new developments and applications of superconductivity in science and technology. More rapid progress in this field can be expected. Superconductivity is still as fascinating as it was in the past.

Werner Buckel

Preface

With the discovery of macroscopic quantum phenomena such as superconductivity, the Josephson effect, flux quantization and superconducting quantum interference it was possible for the first time to describe properties of macroscopic objects directly in terms of quantum physics. Thus our knowledge of physical nature was extended beyond the well known classical physics. For scientists and engineers, it was a novel experience to study and realize quantum physics and quantum systems on a macroscopic scale.

This volume entitled SUPERCONDUCTING QUANTUM ELECTRONICS will give a contribution to this most interesting subject. It presents and discusses in three parts the state of the art in our understanding of the *fundamentals,* describes *sensitive detectors* and outstanding applications in *precision metrology.* New doors are opened by this unique development in numerous scientific and technical areas.

Today, in medical diagnosis this makes possible the detection of biomagnetic signals from human hearts and brains, in electrical metrology the generation of dc voltages on a 1 volt level with a reproducibility of that of atomic clocks ($10^{-12}/\sqrt{\mathrm{Hz}}$), and in astronomy a profounder knowledge of the universe by extending observations of the interstellar medium into the millimeter and submillimeter wavelength region.

The authors of this book have tried in their particular fields of activity to respond to the new challenge of high critical temperature superconductivity. Important questions will arise in the future on whether experiments can be prepared in such a way that also at higher operating temperatures, the striking characteristics of quantum mechanical systems can be verified, utilized, and maintained.

In the part on *fundamentals* Heinz Lübbig presents in a unique way the nonlinear dynamics of the Josephson effect in junctions and SQUIDs (Superconducting Quantum Interference Devices). By introducing a new state variable he outlines the linear response theory and discusses the corresponding quantum limitations. This makes the tunnel junction a favored object for nonlinear analysis and circuit theory, in particular if dissipation is conceptually involved.

On this basis, in his article Gerd Brunk describes the principle of mapping the essential physical junction properties within a network synthesis and derives electrical equivalent circuits, which are extremely useful for the correct modelling of superconducting electronic devices.

Ralf Pöpel shows in his contribution of basic interest that the exact solution of the Mattis-Bardeen equations for bulk materials and thin films perfectly describe the electromagnetic properties of superconductors for all temperatures, frequencies and mean free paths, also including strong coupled superconductors. Only five material parameters have to be known to compare the theoretical results without any fitting parameters with the various experimental ones. The agreement is extraordinarily good and experimentalists should be encouraged to make full use of the exact solutions of the Mattis-Bardeen equations, particularly for measurements in the energy gap frequency region.

The second part covers articles on *sensitive detectors and diagnosis*. The relevant technology, fabrication, and operation of single and multilayer high-T_c Josephson junctions, SQUIDs and other devices are described by Christoph Heiden and Horst Rogalla. This work is motivated chiefly by the expectation that their use will be widespread if such devices can be operated well above liquid helium temperatures. The extremely high resolution of SQUID magnetometers compared with any conventional magnetometer permits a slight decrease of the sensitivity on most applications if it is operated at higher temperatures, e.g. at 77 K. However, the small coherence length in high-T_c superconductors makes any reliable and reproducible production in this field extremely difficult. It is all the more astonishing to see the promising results already achieved.

Magnetocardiography and magnetoencephalography characterize two novel methods for obtaining information on a patient's condition. These modern diagnostic tools in medicine are not conceivable without the application of high-precision SQUID magnetometry. Hans Koch summarizes the various constraints on achieving an optimum design. However, there is no doubt today that the standard diagnostic procedures will be supplemented in the near future by these valuable new methods.

The unique functions of a Josephson junction as a self-oscillating mixer and as a spectral detector enables its amplitude and its frequency information to be determined for an unknown electromagnetic signal. Johann Hinken describes this remarkable broad band spectrum analyzer which needs no separate local oscillator and is characterized by its high amplitude resolution and ability to scan the whole frequency range from about 1 to 1000 GHz in about 10 seconds.

The probing of astronomical objects in the interstellar medium in the millimeter and submillimeter wavelength region has resulted in exciting discoveries. Besides an optimum antenna area, a prerequisite for obtaining such information is to have extremely sensistive mixers which operate close to their quantum limit. Karl Heinz Gundlach reports on this subject, which it is interesting to note, requires the Josephson current to be suppressed in order to take full advantage of the quasiparticle tunneling.

Besides aesthetics, the cover illustration of this book also demonstrates one aspect of the quantum nature of cryoelectronics. Two magnetic flux quanta can be seen trapped in a Josephson tunnel junction of an area which could be covered by two crossed human hairs - a visible manifestation of a macroscopic quantum phenomenon. Rudolf P. Huebener reports on the fascinating diagnosis of thin films and the relevant devices. Low-temperature scanning electron microscopy is a powerful tool for directly probing the activated or nonactivated cryoelectronic device with its inherent high spatial resolution.

The third part on *precision metrology* deals with quantum measures which are superior to any conventional measure of classical physics.

Jürgen Niemeyer focusses on recent achievements in the integration of about 15 000 all-refractory metal Josephson tunnel junctions in a series array. This makes it possible to generate dc voltages up to 10 volts with extremely high reproducibility, by irradiating microwaves. This is a metrological milestone in two respects: First is its importance in the worldwide, uniform dissemination of the unit of voltage as of 01.01.1990. Second, the series arrays are essential parts of novel Josephson potentiometers working at levels up to 10 volts with which arbitrary voltage or resistance ratios can be traced back to well known microwave frequency ratios.

In addition to the generation and measurement of precision voltage ratios using Josephson potentiometry, the cryogenic current comparator as reviewed by Peter Gutmann and Hans Bachmair completes the metrological scene with precision dc current and resistance ratios. This cryogenic instrument will be an indispensable tool in the worldwide uniform dissemination of the unit of electrical resistance starting 01.01.1990. It also allows a resistance scale with an accuracy never before attained to be established.

In the last contribution to this volume, Martin Albrecht and Wolfgang Kessel present a novel way to produce calculable noise power spectra in the microwave region which could be used for diverse noise metrology applications. The synthetic generation of random noise signals is based on fast superconducting SQUID shift registers in a feedback loop with clock rates on the order of 10 GHz. Conventional thermal noise standards deteriorate with time when operating at the high temperatures required, which also have to be stabilized and measured. By way of contrast, the pseudo random noise mentioned is simply calculated by means of purely electrical quantities.

On behalf of the authors, the editor wishes to express his gratitude to the Stiftung Volkswagenwerk which sponsored several projects reported in this volume and kindly gave financial assistance to the seminars on "Kryoelektronische Bauelemente" held at the various venues in the Federal Republic of Germany.

The editor also wishes to thank all the authors, and Dr. Hefter of the Springer Verlag, for their fruitful and efficient cooperation, and Shirley Patricia Helm for her perusal of some of the English translations. I am exceptionally grateful to Inge Bode of the Physikalisch-Technische Bundesanstalt for completely processing the text with great skill in a very short time.

Volkmar Kose

Contents

Part I Fundamentals

Part II Sensitive Detectors

Low-Temperature Scanning Electron Microscopy of Superconducting Thin Films and Tunnel Junctions
R. P. Huebener

Part III Precision Metrology

Josephson Series Array Potentiometer
J. Niemeyer

Cryogenic Current Comparator Metrology
P. Gutmann, H. Bachmair

Fast SQUID Pseudo Random Generators
M. Albrecht, W. Kessel

Part I

Fundamentals

Classical Dynamics of Josephson Tunnelling and Its Quantum Limitations

H. Lübbig
Physikalisch-Technische Bundesanstalt, Institut Berlin, D-1000 Berlin

1. Introduction

In the dynamics of Josephson tunnel junctions the quantum phase shift of the superconducting order parameter is the important degree of freedom. The highly non-linear stationary response to external excitations and the resistivity associated with the transport of superconducting current carriers through the junction under non-stationary conditions make the Josephson junction a favoured object for fundamental physical questions and for important quantum electronic applications as well. This article is intended to present a comprehensive description of the classical dynamics of the quantum phase shift occurring in superconducting Josephson tunnel junctions in the framework of *linear response theory*. This description is equivalent to the Kubo formalism known from irreversible thermodynamics. Firstly, it allows the dynamical properties of all kinds of superconductors to be emphasized, connected by tunnel junctions. Secondly, it permits an equivalent circuit of the junction to be set up. This quantum network description turns out to be extremely useful for modelling, simulating, and designing superconducting electronic circuits (cf. article by G. BRUNK). The quantum limitations of the classical Josephson effect will be discussed and the present state of macroscopic quantum phenomena will be indicated.

The specific properties of superconductors are the result of the phenomenon of macroscopic coherence of the current carriers in the superconducting material as originated by a pairing mechanism of the electrons with opposite momenta and spins. This is the basic postulate in the conception of superconductivity. Conventional superconductivity is well defined by the microscopic Bardeen-Cooper-Schrieffer (BCS) theory, or appropriate extensions, which completely describe the macroscopic properties observed in superconductors with 'low' critical temperatures $T_c < 20$ K [1]. For non-conventional superconductors, 'high-T_c superconductors', which were discovered barely two years ago, different and conceptually quite new microscopic models are under investigation, but there is no common and consistent theoretical picture at present [2]. It has been demonstrated experimentally that pairs are also responsible for their macroscopic properties.

Besides the spatial coherence of the charge carriers, another type of dynamical coherence plays a similar fundamental role in understanding the nature of superconductivity. This coherence phenomenon occurs when a (stationary or non-stationary) current flows through a weak junction linking two bulk superconductors. If the intersection is weak enough to build up a finite phase gradient, i.e. a finite difference φ of the quantum phases of the wave functions on both sides of the junction, the current is an explicit periodic function of the quantum phase shift φ. On the other hand, the dynamical behaviour of the quantum phase shift is completely determined by the electromagnetic field localized in the junction area. This type of phase coherent transport of pair and single electrons through a weakly coupling junction between two bulk superconductors is known as the *Josephson effect* [3 - 5]. There are two different lines of interest in the physics of Josephson junctions: Firstly, Josephson junctions can be used to probe dynamical properties of the superconductors coupled to each other. Secondly, the dynamical properties of the junctions can be applied to construct quantum electronic systems for practical purposes. The mutual interdependence of both these main lines of interest has considerably promoted the understanding of this macroscopic coherence effect since the prediction by BRIAN D. JOSEPHSON in 1962 [6] and its experimental demonstration by P.W. ANDERSON and J.M. ROWELL in 1963. Concerning the term Josephson junctions, first of all the diversity of relevant junction configurations must be taken into consideration: tunnel junctions, proximity effect junctions, and microbridge junctions. All these types of junctions are usually referred to as Josephson junctions although these structures differ in their geometry and material constituents, and in the basic physical processes associated with the phase coherent phenomena which are observed when a voltage or a current is applied to the contact. Besides this, a distinction must be made between junctions fabricated artificially and those junction configurations which form naturally at grain boundaries within the bulk material or on its surfaces. The latter junction type is a morphological constituent of ceramic materials of high-T_c superconductors. Among the different types of junctions the tunnel junction has a particular position for several reasons:

1) Tunnelling assisted phase coherence was the basic process originally studied in Josephson's prediction.
2) The dynamics of coherent tunnelling of pairs and single electrons became clear to a large extent in the middle of the sixties as a consequence of the microscopic analysis by WERTHAMER, and by LARKIN and OVCHINNIKOV based upon the BCS theory [7]. This dynamical description remains the only complete theory, and allows a quantum network description to be constructed.
3) The dynamics of tunnel junctions provides a guiding principle in explaining the more highly varied behaviour of other weak-link types.

The dynamics of tunnel junctions has been used to demonstrate the *quantum limitations* of the 'classical' Josephson effect: Despite the quantum nature of its origin, the quantum phase shift φ across the tunnel junction under most experimental conditions is a continuously varying classical variable of motion. Consider, for example, a superconducting ring which is interrupted by a tunnel junction and coupled inductively to the external signal, a so-called single-junction SQUID (Superconducting Quantum Interference Device). The quantum phase shift across the junction under this condition equals the (normalized) magnetic flux enclosed in the superconducting loop $\varphi = 2\pi\Phi/\Phi_0$, $\Phi_0 = h/2e$. In usual applications Φ is a continuously varying macroscopic variable with well defined and 'sharp' values irrelevant to the state of the set of observables which completely determine the dynamical state of the system. This means that all these variables can be measured simultaneously without any fundamental uncertainty, whereas quantum-mechanical variables are restricted by definition to satisfy Heisenberg's uncertainty principle.- The current I_x flowing through the junction and the quantum phase shift equivalent flux Φ threading the superconducting loop play the role of canonical conjugated variables in the sense of classical Hamilton dynamics.

Quantum dynamical behaviour of the quantum phase shift φ has been demonstrated both for extremely low and extremely high damping conditions. These conditions are realized in 'small capacitance' tunnel junctions (geometrical capacitance $C < 1$ pF, critical current $I_c \approx 1$ μA) at very low temperatures ($T < 50$ mK), and the quantum phase shift has to be considered as a quantized variable of motion. The *'secondary quantization'* leads to the dynamics of the conjugated pair of non-commuting variables φ and Q, where Q denotes the charge operator of the capacitance of the tunnel junction, and is given by $Q = (2e/i)\partial/\partial\varphi$. The corresponding equation of motion is similar to the Schrödinger equation for one electron, and defines by this analogy a 'quantum particle'. However, dissipation is involved here conceptionally and that makes a distinction to ordinary quantum-mechanics. Since tunnel junctions are macroscopic objects this kind of quantum dynamics is referred to as macroscopic or sometimes mesoscopic quantum dynamics. The concept of dissipative macroscopic quantum dynamics has been inspired mainly by A.J. LEGGETT [8]. Two types of macroscopic quantum phenomena have been observed recently, namely 'Macroscopic' Quantum Tunnelling (MQT), and effects which reflect properties of 'macroscopic' energy bands like Bloch oscillations and Single-Electron-Tunneling (SET). The quantum-mechanical equation of motion is derived from the classical equation of motion in the usual way leaving the term accounting for the resistivity of the junction unchanged. It is of primary interest to understand the connection between the internal dissipative degrees of freedom of the junction and those of the environment, i.e. the thermal bath.

The first section of this article deals with basic properties of Josephson junctions and important features are described which result from the nonlinearity of the junction dynamics directly. Neither techniques to obtain solutions nor special phenomena are discussed. In the second section an introduction to linear response theory is given based on the Werthamer theory. In addition, the basic features of the classical dynamics are discussed. The final section is devoted to a discussion of the corresponding quantum limitations.

2. Basic Properties of Josephson Junctions

2.1 The DC and the AC Josephson Effect

In its simplest version the equation of motion of a Josephson junction is analogous to that one of a damped plane mechanical pendulum - with angle variable φ, and with $C(\Phi_0/2\pi)$ and $(\Phi_0/2\pi R)$ being equivalent to the inertia and the damping coefficient, respectively - moving in a uniform gravity field and driven by the external torque I_x:

$$\left(C\Phi_0/2\pi\right)\ddot{\varphi} + \left(\Phi_0/2\pi R\right)\dot{\varphi} + I_c \sin\varphi = I_x \ . \tag{2.1}$$

In the present context I_x denote the current flowing through the junction, I_c the critical current of the junction, C its (geometrical) capacitance, and R the effective resistance acting on the charge transfer across the junction. This is the so-called *resistively shunted junction (RSJ) model*. The quantum phase shift φ is directly related to the electromagnetic fields localized in the junction area. The postulate of gauge invariance of the quantum state with respect to the electromagnetic fields leads directly to

$$\hbar\dot{\varphi}(t) = 2e\, v(t) \ , \tag{2.2}$$

where v denotes the electrical voltage across the junction. This fundamental relation is called the Josephson relation. (Here for the sake of simplicity we applied the gauge with vanishing vector potential without losing generality.) The supercurrent term $I_c\sin\varphi$ in Eq. (2.1) directly records the quantum phase shift, which is determined by the time integral of the voltage, Eq. (2.2), i.e. the equivalent magnetic flux Φ, namely

$$\varphi = 2\pi\, \Phi/\Phi_0 \ . \tag{2.3}$$

Φ is called the quantum phase shift equivalent Aharonov-Bohm flux.

If no voltage appears across the junction, the system is in the stationary state, $I_x = I(\varphi) = I_c\sin\varphi$. The coupling energy

$$F_J(\varphi) = (\Phi_o/2\pi)\int_o^{\varphi} I(\varphi')\,d\varphi' = E_P(1-\cos\varphi)\,, \quad E_P = \Phi_o I_c/2\pi\ , \tag{2.4}$$

which is necessary to establish this non-dissipative state of phase coherence between the reservoirs in the electrodes, is an important quantity for several reasons:

1.) The quantum phase shift is state coordinate, and therefore the stationary current $I(\varphi)$ is related to the φ-derivative of the free energy F_J due to this equilibrium state:

$$I(\varphi) = (2\pi/\Phi_o)\, dF_J/d\Phi\ . \tag{2.5}$$

$I(\varphi)$ has the following symmetry: $I(\varphi) = -I(-\varphi) = I(\varphi + 2\pi)$. The condition of stability, $d^2 F_J > 0$, i.e. $|\varphi| < \pi/2$, shows that a stationary supercurrent $I_c\,|\sin\varphi| < I_c$ can flow dissipation-free through the intersection between the two superconductors. This describes the *DC Josephson effect*. 2.) As a consequence of the fact that the coupling energy explicitly depends upon the equivalent magnetic flux Φ, Eqs. (2.3, 2.4), the Josephson junction under stationary conditions acts like a non-linear inductance

$$\tilde{L}(\Phi) := \left[dI(\Phi)/d\Phi\right]^{-1} = L_o/\cos\left(2\pi\Phi/\Phi_o\right),\ L_o = \Phi_o/2\pi I_c\ , \tag{2.6}$$

within the range of stability $|\Phi| < \Phi_o/4$. It is helpfull to insert numbers: for $I_x/I_c = 0.5$, $I_c = 1\ \mu A$ one obtains $\tilde{L} = 0.4$ nH; a number which is relevant to the inductance-coefficient of the external circuit if the junction is excited inductively, which is the case in SQUID's [9]. A single-junction SQUID, sometimes called RF SQUID, consists of a superconducting loop which is interrupted by a Josephson junction. The internal dynamics of the junction controls the dynamics of the system. If the inductance-coefficient L of the superconducting ring is smaller than the intrinsic junction inductance-coefficient L_o, the SQUID is in the stationary state, i.e. $\dot{\Phi} = 0$, and the corresponding equation of motion reads

$$2\pi\Phi/\Phi_o + (L/L_o)\sin\left(2\pi\Phi/\Phi_o\right) = 2\pi\Phi_x/\Phi_o\ ,\quad L/L_o < 1\ . \tag{2.7}$$

In Eq. (2.7) Φ_x denotes the externally applied magnetic flux and Φ is the actual flux enclosed in the loop (cf. Eq. (2.11) and Table 1 (FBJ)). The condition $L/L_o < 1$ means that the magetization curve $\Phi(\Phi_x)$ is a single-valued and monotonically increasing function, and the equation of motion can be solved analytically, yielding

$$\Phi(\Phi_x) = \Phi_x + (\hbar/e)\sum_1^{\infty} (J_{-n}(nL/L_o)/n)\sin\left(n2\pi\Phi_x/\Phi_o\right)\ . \tag{2.8}$$

Here J_n denote ordinary Bessel functions of the first kind, and the coefficients $J_{-n}(nL/L_o)/n$ describe the non-linear influence of the ratio L/L_o of the inductance-coefficients on the magnetization of the SQUID as a direct consequence of the stationary sinusoidal supercurrent flowing through the Josephson junction.

If a finite DC-voltage V_o biases the junction the supercurrent $I_c\sin(\varphi(0) + \omega_o t)$ oscillates harmonically with the angular frequency $\omega_o = 2eV_o/\hbar$. This describes the *AC Josephson effect*, and ω_o is denoted as the Josephson oscillator angular frequency. Furthermore, if a microwave signal $\widehat{V}\cos(\omega_{RF}t)$ is superimposed, i.e. $\varphi(t) = \varphi(o) + \omega_o t + (2e\widehat{V}/\hbar\omega_{RF}) \sin(\omega_{RF}t)$, the Josephson oscillator (ω_o) may become 'quantum phase locked' by the microwave signal (ω_{RF}): If the energy $\hbar\omega_o = 2eV_o$ associated with the applied voltage equals an integral multiple of the photon energy $\hbar\omega_{RF}$, i.e.

$$V_o(k) = k\,\hbar\omega_{RF}/2e, \ k \text{ integer}, \tag{2.9}$$

the quantum phase shift becomes a 2π-periodic function of the second kind, which means $\varphi(t + 2\pi/\omega_{RF}) = \varphi(t) + k2\pi$, and, consequently, the supercurrent

$$I_c \sin\varphi = \sum_{-\infty}^{\infty} I_c J_n(2e\widehat{V}/\hbar\omega_{RF}) \sin(\varphi(0) + (n + k)\omega_{RF}t) \tag{2.10}$$

contains the stationary component: $n = -k$, with the amplitude

$I_c J_{-k}(2e\widehat{V}/\hbar\omega_{RF})$,

cf. [10], p. 361. The corresponding $\langle I\rangle - \langle V\rangle$ characteristic ($\langle\ldots\rangle$ are the mean values) shows equidistant constant-voltage steps, cf. Fig. 3. This can be interpreted as a reproduction of the zero-voltage step, I_c, at finite voltages $V_o = k(\hbar\omega_{RF}/2e)$ but modulated in the heights with the Bessel functions J_{-k}. The microwave induced constant-voltage steps are a fundamental manifestation of the AC and the DC Josephson effect, and used for the realization of the quantum-voltage standard.

2.2 Circuit Implications

Basically, under stationary conditions there are only two different ways of connecting the junction to external fields: the first one is realized if the junction is connected to an ideal current source (CBJ: current biased junction); in the second, the junction is included in a superconducting loop and coupled inductively to an external signal (FBJ: flux biased junction), as realized in a single-junction SQUID. - In these cases the current and the flux play the role of macroscopic canonical conjugated variables with respect to the thermodynamical potentials, namely the free energy F_J and the free enthalpy $G = F_J + W_x$ which additionally includes the potential energy of the external signals [11]. Introducing the free enthalpy into the equation of motion, cf. Eq. (2.1), one finds

$$C\ddot{\Phi} + (1/R)\,\dot{\Phi} + \partial G/\partial \Phi = 0 \ . \qquad (2.11)$$

In Table 1 the potentials G, F_J, and W_x are given, and the conditions for the states of equilibrium and stability due to the two possible ways of connecting the junction to external signals.

Table 1
Free enthalpy G and the conditions for dynamical equilibrium and stability for the current (CBJ)- and the flux- biased junction (FBJ); F_J denotes the coupling energy of the junction, Eq. (2.4), and $W_x=-I_x\Phi$ the external potentials - with $LI_x = \Phi_x$ the applied magnetic flux for FBJ. Without screening, the external field would produce Φ_x in the superconducting ring. The stability criterion for FBJ is known to be the condition for the non-hysteretic mode of the single-junction SQUID.

free enthalpy	equilibrium	stability
$G = F_J + W_x$	$\partial G/\partial\Phi = 0$	$\partial^2 G/\partial\Phi^2 > 0$
CBJ: $F_J - I_x\Phi$	$I_c\sin(2\pi\Phi/\Phi_o) - I_x = 0$	$\lvert I_x\rvert < I_c$
FBJ: $F_J + (\Phi^2/2L) - (\Phi\Phi_x/L)$	$I_c\sin(2\pi\Phi/\Phi_o) + (\Phi-\Phi_x)/L = 0$	$2\pi LI_c/\Phi_o < 1$

The duality of current and quantum phase shift equivalent flux is expressed in the following canonical relations of the conjugated variables:

$$\partial G/\partial I_x = -\,\Phi, \ \ \partial F_J/\partial\Phi = I_x \qquad (2.12)$$

which are analogous to position and rate of momentum in classical Hamilton mechanics.

2.3 Damping Equivalent

Under non-stationary conditions, $\dot{\varphi} \neq 0$, the flow of pairs through the junction is superimposed coherently by the transfer of thermally excited quasiparticles. From phenomenological aspects these charge carriers behave like single electrons. The flow of both types of charge carriers is associated with dissipative processes which serve to maintain a finite voltage. Although the dissipation specifically depends upon the particular properties of the junction type under consideration (see Sect. 3), the simple damping equivalent which corresponds to the phase-linearized equation of motion, cf. Eq. (2.1), i.e.

$$\ddot{\varphi} + \omega_{RC}\,\dot{\varphi} + \omega_{po}^2\,\varphi \approx \omega_{po}^2\,(I_x/I_c) \ , \qquad (2.13)$$

reveals the widely varied dynamical behaviour of the junction. Here ω_{po} =

$(2\pi I_c/\Phi_o C)^{1/2}$ and $\omega_{RC} = (1/RC)$ denote the plasma angular frequency and the cut-off angular frequency, respectively. A small signal analysis yields

$$\varphi(t) = \varphi(0) \exp(-\omega_{RC} t/2) \exp i\left((4\,\omega_{po}^2 - \omega_{RC}^2)^{1/2} t/2\right) , \tag{2.14}$$

and the corresponding discriminant is directly related to the *damping equivalent*, the so-called Stewart-Mc Cumber parameter,

$$\beta_C = (\omega_{po}/\omega_{RC})^2 = 2(2\pi R/R_K)^2 (E_P/E_e) \; . \tag{2.15}$$

In the second formula in Eq. (2.15) the influence of the non-dissipative junction parameters, I_c and C, is reflected by means of the ratio of the two energy expressions associated, namely the Josephson coupling energy E_P, Eq. (2.4), which results from the linear combination of the wave functions linked by the junction, and the charging energy $E_e = e^2/2C$ of the relative capacitance with respect to one single electron. The influence of the resistivity on the damping equivalent, Eq. (2.15), is described by the ratio $\alpha = 2\pi R/R_K$ of the effective resistance R and the von Klitzing constant $R_K = h/e^2 = 26$ kΩ being the natural unit of quantum resistance. The damping equivalent determines whether the junction is operated in the overdamped mode: $\beta_C < 0.25$ (aperiodic motion) or in the underdamped mode: $\beta_C > 0.25$ (quasiperiodic motion); $\beta_C = 0.25$ denotes the case of critical damping (aperiodic limit). Two special cases are of formal interest: 1) If any damping is ignored, $R = \infty$, the junction performs harmonic oscillations with the plasma angular frequency. 2) On the assumption that $C = 0$, the full Eq. (2.1) (including the $\sin\varphi$ term) can be solved analytically, yielding the voltage

$$\begin{gathered} v(t) = \langle v \rangle \left(1 + 2 \sum_{1}^{\infty} \left(\left((\Phi_o/2\pi I_c R)\,\omega + (I_{DC}/I_c)\right)^n \cos(n\omega t)\right)\right) \\ \omega = (2\pi I_c R/\Phi_o)\left((I_{DC}/I_c)^2 - 1\right)^{1/2} = 2e\langle v \rangle/\hbar, \quad I_{DC} > I_c \; , \end{gathered} \tag{2.16}$$

which exhibits a rich content of higher harmonic components, $n \geq 2$, as a consequence of the nonlinearity of the supercurrent-to-phase relation ($\sin\varphi$). Here $I_{DC} = I_x$ denotes the external current constant in time. - Under actual physical conditions, however, a finite capacitance and a finite resistance is conceptually involved.

In order to discuss the resistive nature of the coherent charge carriers transfer through the Josephson junction, it is necessary to differentiate between non-shunted junctions and devices involving junctions which are shunted by an external resistor. Naturally, in this context the particular junction type under consideration should be specified, and in this section and the following one, the discussion will be restricted exclusively to tunnel junctions. This particular junction type is distinguished by a non-linear quasistatic $\langle I \rangle - \langle V \rangle$ characteristic as shown in Fig.1.

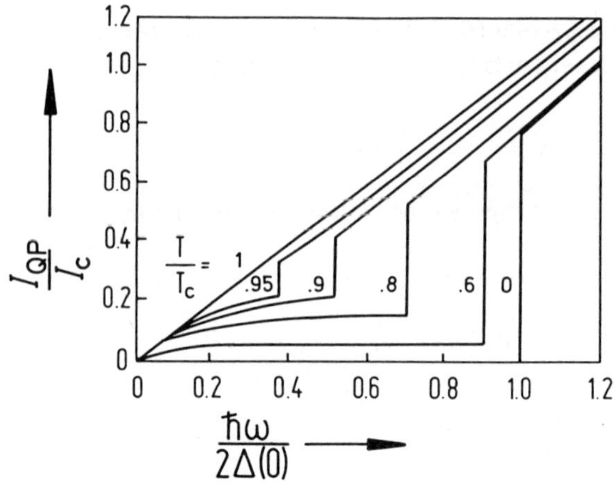

Figure 1

The stationary ⟨ I ⟩ - ⟨ V ⟩ characteristic of the quasi-particle tunnel junction (constant voltage bias) at various values of the reduced temperature.

The mean voltage ⟨ V ⟩ biases the junction either below or above the energy-gap equivalent voltage $V_\Delta = 2\Delta\ (T/T_c)/e$, which corresponds to the sharp rise in the ⟨ I ⟩ - ⟨ V ⟩ characteristic, Fig. 1. In Table 2 the ratio $\alpha = 2\pi R/R_K$ is related to the different positions of the bias point in the ⟨ I ⟩ - ⟨ V ⟩ characteristic, and to the alternative of an externally shunted or non-shunted tunnel junction.

Table 2

resistance ratio α:	$0 < \alpha\ (R_S, R_{QP}) < \alpha_N$ $\alpha_N = 2\pi R_N/R_K$	$\alpha_N < \alpha(R_{QP}) < \infty$
junction:	shunted $R = (R_S R_{QP})/(R_S + R_{QP})$	non-shunted $R \approx R_{QP}$
biased:	$V >> 2\,\Delta\,/e,\ R_{QP} \approx R_N$	$V << 2\,\Delta\,/e$ $R_{QP} >> R_N$ if $T << T_c$ $R_{QP}(0) = \infty,\ R_{QP}(T_c) = R_N$

R_N denotes the normal-state resistance of the tunnel junction, and α_N, which is related to the non-shunted junction biased above V_Δ, reads $\alpha_N = (\pi\ \Delta_{BCS}/4E_P)\tanh(\Delta_{BCS}/2k_BT)$ according to the microscopic theory of AMBEGAOKAR and BARATOFF [12] concerning the stationary tunnel junction.

In usual applications the junctions are operated under moderate damping conditions. In the two limits of low and high damping ($\beta_C \approx 10^3$ and $\beta_C \approx 10^{-2}$, respectively) the appearance of secondary quantum processes has been demonstrated (see Sect. 4).

3. Classical Dynamics of the Quantum Phase Shift in Pair- and Quasiparticle Tunnel Junctions

Superconducting tunnel junctions consist of two superconducting electrodes separated by a thin isolating layer (about 3 nm thickness) which acts as a small energy barrier allowing electrons (with a finite but very small probability) to penetrate it via tunnelling [13]. The BCS-type theory of the Josephson tunnel effect is a single complete theory of non-stationary, non-linear phenomena in (low T_c) superconductors. This makes the tunnel junction a favoured object for non-linear dynamics, particularily if the dissipation is conceptually involved. The dynamical model, also known as the WERTHAMER model [7], significantly generalizes the simple RSJ model, Eq. (2.1), since it takes into account the coherent flow of pairs, $I_P(t)$, and of quasiparticles, $I_{QP}(t)$. It therefore allows the frequency dependent response of the tunnel diode to arbitrary external excitations due to its internal dissipative degrees of freedom to be deduced. The dynamical behaviour of the junction is characterized by the superposition of a finite frequency-dependent conductivity of pairs σ_P and of quasiparticles σ_{QP} if the actual temperature is smaller than the critical temperature of the electrodes, and by the frequency-independent normal-state conductivity $\sigma_N := 1/R_N$ if $T > T_c$.

The frequency-dependent dissipation causes a memory-based friction of the current in the time domain [11]:

$$C\ddot{\Phi} + \sigma_N \dot{\Phi} + \cos(\pi\Phi(t)/\Phi_o) \int_{-\infty}^{t} m_+ (t - t')\, d \sin(\pi\Phi(t')/\Phi_o)$$

$$+ \sin(\pi\Phi(t)/\Phi_o) \int_{-\infty}^{t} m_- (t - t')\, d \cos(\pi\Phi(t'/\Phi_o)$$

$$+ \partial G/ \partial\Phi = 0. \tag{3.1}$$

(It should be noted that the potential G implies the $\sin(2\pi\Phi/\Phi_o)$-term, and the external excitation, Eq. (2.11) and Table 1.) The kernels of the convolution integrals describe the time delay of the junction's response to an external signal. They denote the relaxation functions of the sum and of the diffence of I_P and I_{QP}:

$$m_{+,-} := \begin{cases} \displaystyle\int_{t+0}^{\infty} (I_P(t') \pm I_{QP}(t'))dt' & t \geq 0 \\ 0 & t < 0 \end{cases} , \tag{3.2}$$

cf. article by BRUNK, Eq. (3.19), $m_\pm = \pm N_\pm$. The integral term in Eq. (3.1) is *bilinear* with respect to the circular functions of half the quantum phase shift $\varphi/2 = \pi\Phi/\Phi_o$. In order to clarify the physical meaning, we introduce the two-component flux variable

$$\Phi_+ := (\Phi_o/\pi)\sin(\pi\Phi/\Phi_o)\,,\quad \Phi_- := (\Phi_o/\pi)\cos(\pi\Phi/\Phi_o)\,, \tag{3.3}$$

and, correspondingly, the two-component current variable

$$\begin{aligned} I_+ &:= C\ddot{\Phi}_+ + \sigma_N\dot{\Phi}_+ + (\pi/\Phi_o)\int_{-\infty}^{t} m_+\ (t-t')\,\dot{\Phi}_+\ (t')\,dt'\,,\\ I_- &:= C\ddot{\Phi}_- + \sigma_N\dot{\Phi}_- - (\pi/\Phi_o)\int_{-\infty}^{t} m_-\ (t-t')\,\dot{\Phi}_-\ (t')\,dt'\,. \end{aligned} \tag{3.4}$$

The actual voltage $v(t) = \dot{\Phi}$, and the current balance, Eq. (3.1), now read

$$v(t) = (\pi/\Phi_o)\,(\dot{\Phi}_+\,\Phi_- - \dot{\Phi}_-\Phi_+)\,, \tag{3.5}$$

and

$$I_+\,\Phi_- - I_-\,\Phi_+ + \partial G/\partial(\varphi/2) = 0\,, \tag{3.6}$$

respectively.

The introduction of the two-component flux variable, Eq. (3.3), has led to the decomposition of the current into the two components I_+ and I_-, Eq. (3.4), and each of these is related linearily to Φ_+ and to Φ_-, respectively. This allows a *linear response description* of the junction dynamics [14] to be set up, which is common to all types of superconductors linked. The Fourier transformation of Eq. (3.5) and Eq. (3.4) yields

$$v(\omega) = (1/2\,\Phi_o)\int_{-\infty}^{\infty} d\omega'\left(\dot{\Phi}_+(\omega-\omega')\,\Phi_-(\omega') - \dot{\Phi}_-(\omega-\omega')\,\Phi_+(\omega')\right)\,, \tag{3.7}$$

and

$$\begin{pmatrix} I_+(\omega)\\ I_-(\omega)\end{pmatrix} = \begin{pmatrix} Y_+(\omega) & 0\\ 0 & Y_-(\omega)\end{pmatrix}\begin{pmatrix}\dot{\Phi}_+(\omega)\\ \dot{\Phi}_-(\omega)\end{pmatrix}. \tag{3.8}$$

In Eq. (3.8) $Y_\pm(\omega) = -i\omega C + \sigma_N \pm (\pi/\Phi_o)\,m_\pm(\omega)$ are the components of the admittance matrix, and

$$\begin{aligned} &\mathrm{Re}\,Y_\pm = \pm(\pi/\Phi_o)\,\omega^{-1}\,\mathrm{Im}(I_P(\omega) \pm I_{QP}(\omega)) =: \sigma_\pm = \sigma_{QP} \pm \sigma_P\\ \text{and}\quad & \\ &-\mathrm{Im}\,Y_\pm = \omega C \pm (\pi/\Phi_o)\,\omega^{-1}\,\mathrm{Re}((I_P(\omega) - I_P(0)) \pm (I_{QP}(\omega) - I_{QP}(0))\\ &=: \omega(C + C_\pm(\omega)) \end{aligned} \tag{3.9}$$

denote the *tunnel conductance* and the *tunnel susceptance*, respectively.

Equations (3.7, 3.8) describe the fundamental dynamical relations of the Josephson tunnel effect [14, GB]. (For identical superconductors $I_P(0) = I_{QP}(0) = - I_c$.)

3.1 Quantum Phase Self-Coupling

Analogous to the way in which the relaxation functions $m_{\pm}(t')$, Eq. (3.1), control the development of the quantum phase shift $\varphi(t')$ up to the present state $\varphi(t)$ by sensing the whole system's evolution beginning at the starting point $t' = -\infty$, the Fourier transform of the voltage $v(\omega)$, Eq. (3.7), at any arbitrarily chosen angular frequency ω results from a spectral convolution of $\dot{\Phi}_{\pm}(\omega')$ with the flux variable $\Phi_{\mp}(\omega')$ covering the whole frequency range. This quantum phase self-coupling procedure is energetically neutral since it depends on the junction's dissipation only indirectly by means of the spectral distribution of $\Phi_{\pm}(\omega)$, which determines the special process.

3.2 Tunnel Junction Admittance

The constitutive properties of the tunnel junction enter the dynamical description exclusively by means of the admittance matrix, Eqs. (3.8, 3.9). For a BCS-type tunnel junction the frequency dependence of the conductivity components $\sigma_{\pm}/\sigma_N$ is shown in Fig. 2 at two different reduced temperatures. σ_+ and σ_- represent a low-pass filter and a high-pass filter, respectively, as a

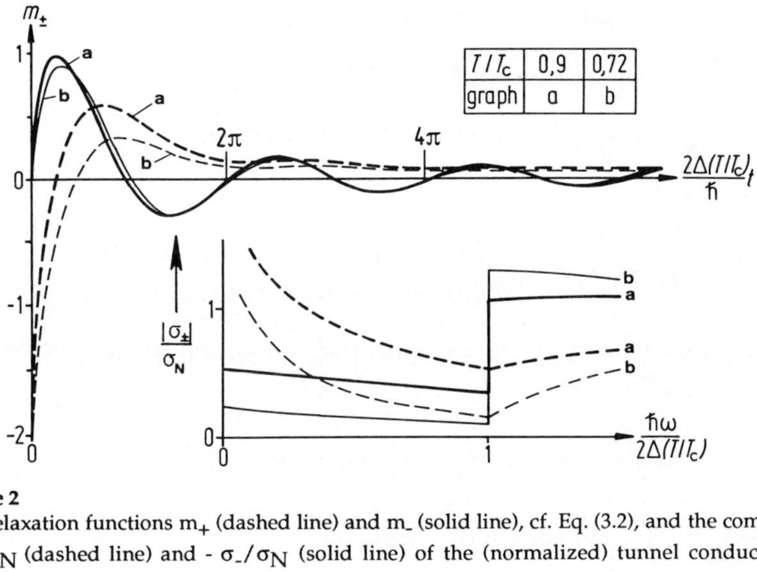

Figure 2
The relaxation functions m_+ (dashed line) and m_- (solid line), cf. Eq. (3.2), and the components σ_+/σ_N (dashed line) and $-\sigma_-/\sigma_N$ (solid line) of the (normalized) tunnel conductivity at different reduced temperatures T/T_c. At $T = 0$ $\sigma_{\pm} = 0$ for $\hbar\omega/2\Delta < 1$; the step of σ_-/σ_N equals $\pi/2$, and $\sigma_{\pm}/\sigma_N = \pm (1 - (1 \pm 2\ln(2\hbar\omega/\Delta))/(\hbar\omega/\Delta)^2)$ as $\omega \approx \infty$.

direct consequence of the constitutive properties of the superconductors coupled.

The fundamental construction principles for equivalent circuits have been discussed by BRUNK et al. [14] (cf. article by G. BRUNK), and the original dispersion relations of the tunnel junction have been used by JABLONSKI, and by JUTZI et al. to set up finite degree-of-freedom circuit models consisting of classical elements [15].

3.3 Special Cases

In order to demonstrate the range of validity of this dynamical concept in a simple way, we will study here the example of the 'small signal analysis' due to the excitation $v(t) = V_o + \widehat{V}\cos\omega_o t$, $\hbar\omega_o = 2eV_o$, $\widehat{V} << Vo$, and will discuss different physical aspects using the expression for the corresponding dissipated energy:

$$D := \int_{-\infty}^{t} i(t')\, v(t')\, dt'$$

$$= (1/2\pi) \int_{-\infty}^{\infty} d\omega \left(\sigma_+(\omega) \left|\dot{\Phi}_+(\omega)\right|^2 + \sigma_-(\omega) \left|\dot{\Phi}_-(\omega)\right|^2\right) . \qquad (3.10)$$

The above assumption leads to the quantum phase shift $\varphi(t) = \varphi(0) + \omega_o t + (\widehat{V}/V_o)\sin(\omega_o t)$, and the flux variables $\Phi_\pm(t)$ contain only odd harmonic components of the fundamental frequency $\omega_o/2$:

$$\Phi_\pm(t) = (\hbar/e) \sum_{-\infty}^{\infty} J_n\left(\widehat{V}/2V_o\right) \begin{cases} \sin\left(\varphi(0)/2 + (2n+1)\,\omega_o t/2\right) \\ \cos\left(\varphi(0)/2 + (2n+1)\,\omega_o t/2\right) \end{cases} . \qquad (3.11)$$

Correspondingly, the voltage components are the sum of all those spectral components:

$$\dot{\Phi}_\pm(\omega) = \sum_{-\infty}^{\infty} J_n(.)\,(\hbar\omega_n/e) \begin{cases} \delta(\omega+\omega_n)\, e^{i\varphi(o)/2} + \delta(\omega-\omega_n)\, e^{-i\varphi(o)/2} \\ i\left[\delta(\omega+\omega_n)\, e^{i\varphi(o)/2} - \delta(\omega-\omega_n)\, e^{-i\varphi(o)/2}\right] \end{cases} , \qquad (3.12)$$

$\omega_n = (2n+1)\omega_o/2$, and for the corresponding dissipated energy per time, cf. Eq. (3.10), this yields:

$$D = 2\pi \sum_{k=\pm 1, \pm 3, \ldots} k^2 V_o^2 \Big[J^2_{\frac{k-1}{2}}(\cdot)\, \sigma_{QP}(keV_o/\hbar)$$

$$+ J_{\frac{k-1}{2}}(\cdot)\, J_{-\frac{k+1}{2}}(\cdot) \cos\varphi(0)\, \sigma_p(keV_o/\hbar)\Big] . \qquad (3.13)$$

The prefactors in Eq. (3.13) express the result of the phase self-coupling effect,

Eq. (3.7), which produces the content of higher harmonic components of the voltage V_o. The appearance of the Bessel functions indicates the non-linearity of this procedure. Each of the harmonic components (k = 1, 3...) is specifically damped by the corresponding conductance of pairs and of quasiparticles, $\sigma_P(keV_o/\hbar)$ and $\sigma_{QP}(keV_o/\hbar)$. In addition, the pair current term is associated with the $\cos\varphi(0)$ term as a consequence of the free energy F_J, Eq. (2.4), necessary to build up the initial state $|\varphi(0)| < \pi/2$. To analyze the range of validity of this dynamical concept it is helpful to study briefly different physical situations:

1) RSJ model: Omit the memory based response of the junction by considering the case $m_\pm(t) = 0$ in Eq. (3.1) to find the RSJ model, Eq. (2.1). It is important to realize that the decomposition of the admittance matrix into one part, i.e. $-i\omega C + \sigma_N$, acting instantaneously and the memory based part, i.e. $m_\pm(\omega)$, can be shown to be unique (at least in the case of BCS-type superconductor tunnel junctions). This means that the following unique interrelationship exists between the type of dynamical flux variable and the type of corresponding admittance:

flux variable	admittance	
$\Phi_+ = (\Phi_o/\pi)\sin(\pi\Phi/\Phi_o)$ $\Phi_- = (\Phi_o/\pi)\cos(\pi\Phi/\Phi_o)$	$\begin{pmatrix} Y_+(\omega) & 0 \\ 0 & Y_-(\omega) \end{pmatrix}$	Werthamer model
$\Phi = (\Phi_o/2\pi)\,\varphi$	$-i\omega C + \sigma_N$	RSJ model

Applying the Werthamer model the current-voltage characteristic of tunnel junctions [16] as well as the theory of thermal fluctuations [14 (GB)] and of quantum noise has been studied extensively [17].

2) Chaos: One of the important features of the non-linear dynamics is the appearance of randomlike solutions of deterministic equations of motion. In dissipation-free systems stochastic motions have been studied intensively in classic mathematics, mainly by H. Poincaré. In recent years, however, the demonstration of chaotic behaviour in systems with dissipation has led to a couple of new basic phenomena. The Josephson junction has been shown to be a good candidate for fundamental studies in this field [18]. Applying the RSJ model numerical simulations of chaotic solutions occurring in microwave-induced steplike current-voltage characteristics of Josephson junctions have been carried out and compared with effects resulting from thermal noise [19]. Fig. 3 illustrates the influence of memory based friction (cf. Eq. (3.1)) on the steplike characteristic, a result obtained by numerical calculations [20]. Curve (a) refers to the RSJ model whereas curve (b) which

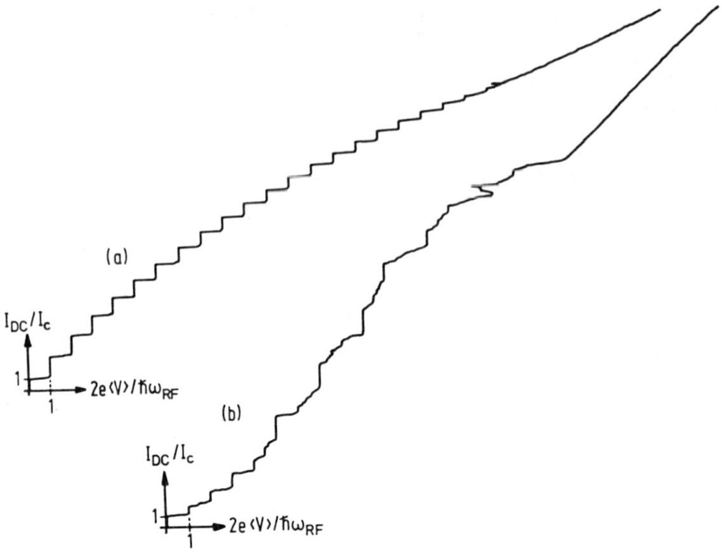

Figure 3

Microwave-induced $I\langle V\rangle$ characteristics of a Josephson tunnel junction at T = 0. ($I_x = I_{DC} + I_{RF}\sin(\omega_{RF}t)$) a) result of the RSJ model, b) chaotic solutions resulting if the retardation in Eq. (3.1) is taken into account additionally. (The origins are shifted.)

partly exhibits chaotic behaviour was calculated with indentical parameters but taking the convolution integrals (Eq. (3.1)) into account. The comparison between both curves demonstrates what may happen if the internal dissipative degrees of freedom of the tunnel junction are taken into account.

3) Quasiparticle-Diode-Mixer: In the following we assume that an even number of magnetic flux quanta, Φ_0, has been included in the tunnel barrier by means of the action of an external magnetic field applied perpendicular to the direction of the current flow. Under these circumstances the distribution of the pair current density is influenced in such a way that the net pair current vanishes. Consequently, the dynamical relations concerning the total current under the conditions considered are reduced to the dispersive tunnel dynamics of quasiparticles ($\sigma_P = 0$), which is the basis of the quasiparticle-diode-mixer theory [21].

4) Approach to Bulk Superconductor: We suppose hypothetically that the junction must not be allowed to localize coupling energy (F_J), a condition which is to be replaced here by the claim to fix $\varphi(0) = \pi/2$ in order to suppress the possibility of stationary states within the range $|\varphi(0)| < \pi/2$. Under this condition one would expect the junction to exhibit bulk-like behaviour. Indeed, in this case the dynamics reported here become equivalent to the Mattis-Bardeen theory which describes the microwave losses in bulk BCS-type superconductors [22].

4. Macroscopic Quantum Phenomena Based on JOSEPHSON Tunnel Dynamics

Despite the quantum nature of its origin the quantum phase shift across the junction behaves like a continuously varying classical variable as long as the mean energy associated with the essential degrees of freedom, k_BT, is larger than the Josephson plasma-oscillation energy, $\hbar\omega_{po}$, which is a characteristic unit for the pair tunnelling dynamics: $k_BT > \hbar\omega_{po}$. In the opposite limit, which is realized in 'small capacitance' tunnel junctions ($C < 1$ pF, $I_c \approx 1\ \mu$A) at very low temperatures ($T < 50$ mK), several macroscopic quantum effects have been predicted and partly experimentally demonstrated. Two groups of macroscopic quantum phenomena have been investigated recently: The first one is concerned with a tunnelling process of a 'quantum particle' out of macroscopically distinct metastable states: *Macroscopic Quantum Tunnelling (MQT)*, and with coherent quantum oscillations between the different metastable states of a double-well potential: *Macroscopic Quantum Coherence (MQC)* [23]. The other group of these effects reflects the properties of *macroscopic energy bands* known as quantum charge oscillations (Bloch oscillations), and single-electron tunnelling (SET) [24]. In both these types of secondary quantum effects, dissipation is involved, accounting for *resistive phenomena on a fully quantum-mechanical level.*

The quantum-mechanical equation of motion is a constructive extension of the classic equation of motion. As a consequence of the canonical conjugation between the current variable I_x and the quantum phase shift equivalent flux variable Φ on the classic level, cf. Eq. (2.12), in the quantum regime, $k_BT < \hbar\omega_{po}$, the charge variable Q and the flux variable Φ

$$Q := (\hbar/i)\,\partial/\partial\Phi\,, \qquad \Phi \tag{4.1}$$

play the role of *non-commutating conjugated variables.* The charging variable and the flux variable are now no longer quantities which can have simultaneously sharp values, but they satisfy the uncertainty relation [17], [23], [24]. Leaving aside the dissipative properties of the tunnel junction for the moment, the Hamiltonian corresponding to an equilibrium state is the sum of the kinetic energy $E_Q = Q^2/2C$ associated with the charging of the capacitor, and the potential energy $F_J(\varphi) = E_P(1 - \cos\varphi) = 2\,E_P \sin^2(\varphi/2)$ which is due to the stationary pair tunnelling:

<u>classical:</u> $\hbar\omega < k_BT$ <u>quantum:</u> $\hbar\omega > k_BT$

$$H_o = Q^2/2C + F_J(\varphi) \qquad \underline{H}_o = -E_e\,\partial^2/\partial(\varphi/2)^2 + 2\,E_P \sin^2(\varphi/2)\,. \tag{4.2}$$

However, the dissipation is involved in these quantum phenomena conceptually, and this is the reason why they are termed macroscopic.

4.1 Macroscopic Quantum Tunnelling

Macroscopic quantum tunnelling has been demonstrated experimentally in the limit of high damping, $\beta_C << 1$, applying an FBJ-configuration [25] and under extremely low-damping conditions, $\beta_C >> 1$, using the CBJ arrangement [26]. In the latter case additionally energy-level quantization of this macroscopic object has also become evident [26].

Let us consider a Brownian particle with mass m moving in a medium with the damping coefficient η around the metastable state q_o of the potential as depicted in the inset of Fig. 4. The motion arises out of the interaction with the heat bath environment of temperature T. The thermal hopping rate out of the potential minimum has been studied in the classic work of H.A. Kramers:

$$\Gamma_{therm.} = (\omega_p/2\pi) \exp - (\Delta G/k_B T), \quad \hbar\omega_p < k_B T . \tag{4.3}$$

$\Gamma_{therm.}$ vanishes for T = 0. The thermally activated fluctuations die out and the transitions to lower-lying minima can only be based on fluctuations by tunnelling through the barrier. Next we identify q with φ, m with $C(\Phi_0/2\pi)$, and η with $(\Phi_0/2\pi)/R$ to find the analog to the equation of motion of the Josephson junction, Eq. (2.1), and consider the quantum regime $\hbar\omega_{po} > k_B T$. For the undamped system the WKB approximation can be applied, yielding

$$\Gamma_{qu.} (WKB) = (\omega_p/2\pi) \exp - (7.2\, \Delta G/\hbar\omega_p), \quad \hbar\omega_p > k_B T . \tag{4.4}$$

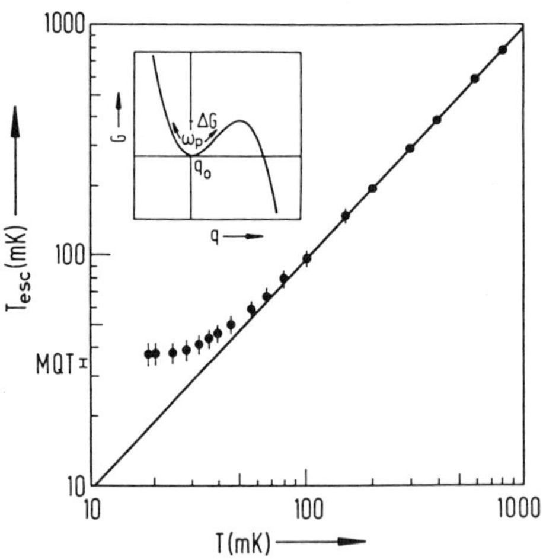

Figure 4
Macroscopic Quantum Tunnelling: T_{esc} vs.T, the dots are experimental results obtained by J. CLARKE and his coworkers [26].The straight line shows the thermal prediction; MQT at the left indicates the result predicted by CALDEIRA and LEGGETT. In the inset is shown the potential well. G denotes the free enthalpy.

However, taking into account the *frictional degrees of freedom* which couple the junction to the heat bath environment leads to a strong suppression of the decay rate $\Gamma_{qu.}$ (WKB) in the very low temperature regime. This is the main result of the Caldeira-Leggett (CL) theory [27]:

$$\Gamma_{qu.}(CL) = \Gamma_{qu.}(WKB)\ \exp - [(6.26\ \Delta G/\hbar\omega_p)\,(1/\tilde{Q})]\,, \tag{4.5}$$

where $\tilde{Q} = \omega_p/\omega_{RC} = (1 - x^2)^{1/4}\beta_C{}^{1/2}$, $x = I_x/I_c < 1$, denotes the quality factor. A corresponding experimental result is shown in Fig. 4, where the escape temperature $T_{esc}: = \Delta G/k_B \ln(\omega_P/2\pi\Gamma)$ is plotted vs. the working temperature T.

The temperature $T = T_{cr}$ ($\approx$ 30 mK in Fig. 4), which determines the cross-over from thermal to quantal behaviour, is directly related to the junction parameters: Following a theory of GRABERT and WEISS [28], the cross-over temperature reads

$$T_{cr} = (\hbar\omega_{po}/2\pi k_B)\left[\left((1/4\,\beta_C) + (1 - x^2)^{1/2}\right)^{1/2} - (1/4\,\beta_C)^{1/2}\right],$$

with the two limits

$$k_B T_{cr} = \begin{cases} \hbar\omega_p/2\pi & \beta_C >> 1 \\ \hbar\omega^2{}_p/2\pi\omega_{RC} & \beta_C << 1 \end{cases} \tag{4.6}$$

for low-damping and high-damping, respectively. This result is in good agreement with the experimental data.

4.2 Quantum Charge Oscillations

The fundamental role of quantum charge oscillations was demonstrated for the first time in 1969 by LAMBE and JAKLEVIC [29] using small capacitance tunnel junctions controlled by a constant-voltage source. More recently random arrays of tunnel junctions have been studied, and very recently the cross-over from the Josephson pair tunnelling regime to the domain dominated by quantum charge oscillations has been demonstrated by varying the ratio E_P/E_e in single junctions in situ [30].

4.2.1 Bloch Oscillations

First, we neglect the quasiparticle tunnelling component and consider a non-shunted junction which is coupled to the external circuit by very weakly conducting leads carrying $I_x = \dot{Q}_x$. The external charge, Q_x, appearing at the capacitor is a continuous variable and the actual charge $Q = Q_x + 2me$ differs from Q_x by an integer of twice the elementary charge as a consequence of the discreteness of the pair-tunnelling transfer. Due to the periodicity of the wave functions $\psi(\varphi + 2\pi) = \psi(\varphi)\exp i\,(2\pi\,Q_x/2e)$, which correspond to the

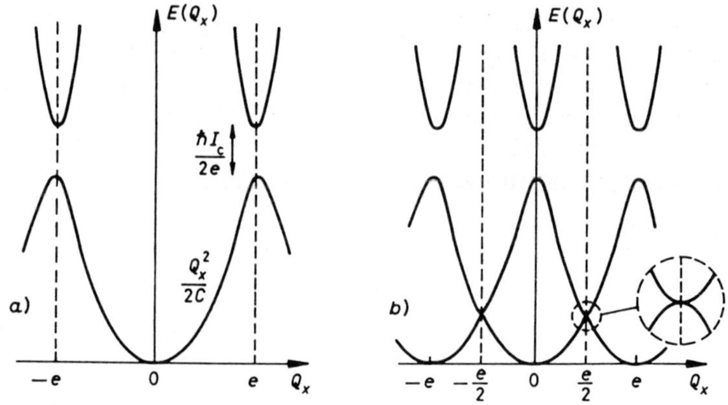

Figure 5
Macroscopic energy bands of a tunnel junction in the limit $E_P << E_e$ without dissipation (a) and with dissipation (b) for weak quasiparticle tunnelling ($\alpha > 1/\pi$); according to GS ref. [24].

Hamiltonian $\underline{H}_0$, Eq. (4.2) , the system can be described as a quantum particle moving in a periodic potential analogous to an electron in a crystal lattice [24]. Consequently, the external charge is the quasimomentum due to this analog. The corresponding energy band spectrum as depicted in Fig. 5 shows 2e-periodicity with respect to Q_x, and band-splitting at the boundaries of the Brillouin zone $Q_x = e$.

The increase of Q_x caused by the external current I_x corresponds to sweeping the quasimomentum through the band, and as a consequence of its periodicity, the voltage appearing across the junction, i.e. $V = \partial E/\partial Q_x$, oscillates with the fundamental frequency $\omega_{Bloch} = \dot{Q}_x/2e$. These oscillations of a charge-controlled tunnel junction are known as Bloch oscillations.

4.2.2 Single-Electron Tunnelling (SET)

The inclusion of damping modifies the picture substantially. As shown in Fig. 5 the energy bands become e-periodic with respect to the quasimomentum Q_x and consequently, the fundamental frequency is now $\omega_{SET} = \dot{Q}_x/e$. Energy band-doubling occurs and the band-splitting at the new boundaries of the Brillouin zone is inhibited. It depends on the detailed structure of the dissipation whether the bands cross at $Q_x = e/2$ (weak tunnelling, $\alpha > 1/\pi$) or wether they become flat there (strong tunnelling, $\alpha < 1/\pi$), GS [24]. Increasing the external charge means sweeping the quasimomentum through the bands and - depending on the rate $\dot{Q}_x/e$ - the system may then either be transformed into excited states and perform BLOCH oscillations, or relax incoherently to the ground state. One important feature in the dynamics of charge-controlled small capacitance tunnel junctions is concerned with the appearance of equilibrium states of constant-current voltage drops: at temperatures low enough the junction remains in the minimum state of the charge even if $Q_x > e/2$. The reason for this phenomenon, known as *Coulomb blockade*, is the fact that the energy associated with charging the capacitor blocks the external current to a sequence of equilibrium states [24], [31], [32]. The current-voltage characteristic under these conditons exhibits constant-current steps in a sequence of mean voltage states. This effect reverses the AC Josephson effect:

charge oscillations:	pair current oscillations:
$E_P << E_e$	$E_P >> E_e$
$I_{m,n} = (m/n)\, 2ef$	$V_{m,n} = (m/n)\, (hf/2e)$, m, n integers.
constant-current steps	constant-voltage steps

A detailed understanding of these phenomena requires the external circuit to be specified (whether it is a constant-current source or a constant-voltage source) [33] and specification of the internal dissipative degrees of freedom of the tunnel junction.

The possibility of coherent oscillations based upon SET was noticed for the first time by BEN-JACOB, GEFEN, MULLEN, and SCHUSS, and by AVERIN and LIKHAREV [31] and a considerable number of recent investigations are concerned with experimental and theoretical aspects cf. [32], [33].

I want to conclude with a historical retrospect. Quantum-phase coherent transport properties of junctions weakly coupling two bulk superconductors have been a favoured subject of fundamental and applied physics for a long period of time: From the records of the Kuratorium of the Physikalisch-Technische Reichsanstalt (March 1926) it follows that the very first mention of this subject was by Einstein: "The question whether the point of contact between two superconductors is also superconducting should be of particular interest". In 1932 R. HOLM and W. MEISSNER [34] observed macroscopic phase coherence across superconducting point contacts. But only the breakthrough of the BCS theory in understanding the nature of the superconducting state in 1956 provided the fundamental basis for Josephson's prediction in 1962. In the following two decades the fundamentals of the 'classical' Josephson effect have been worked out in detail and numerous applications have been suggested or made. The recognition of the duality between coherent pair current oscillations on the one hand and charge oscillations on the other hand is attributable to P.W. ANDERSON, to A. WIDOM, and to T.D. CLARK and his coworkers [35]. Nearly twenty years after Josephson's prediction, insight into the quantum nature of the phenomenon has been gained, a progress made possible mainly by modern junction fabrication techniques and by the fundamental theoretical work on macroscopic quantum dynamics. The experimental demonstration of MQT and of macroscopic energy level quantization [25], [26] basically have verified this concept of the 'macroscopic quantum dynamics based Josephson effect'. The tunnel junction still remains a good candidate in this field.

References

[1] M. Tinkham, Introduction to Superconductivity, McGraw-Hill, New York, 1975; W. Buckel, Supraleitung, Physik-Verlag, Weinheim, 1984

[2] S. Tanaka, Proc. **LT-18** (1987) pp. 2005-2010, and references therein

[3] A. Barone, G. Paterno, Physics and Applications of the Josephson Effect, John Wiley and Sons, New York, 1982

[4] K.K. Likharev, Dynamics of Josephson junctions and circuits, Gordon and Breach, New York, 1986

[5] D. Rainer, J.A Sauls, Proc. **LT-18** (1987) pp. 1804-1808

[6] B.D. Josephson, Phys. Lett. **1** (1962), pp. 251-253

[7] N.R. Werthamer, Phys. Rev. **147** (1966) pp. 255-263; A.I Larkin, Yu.N. Ovchinnikov, Sov. Phys. JETP **24** (1967) pp. 1035-1040

[8] A. Leggett, Prog. Theor. Phys. **Suppl. 69** (1980) pp. 80-100

[9] G. Brunk, H. Lübbig, J. Low Temp. Phys. **47** (1982) pp. 13-26

[10] M. Abramowitz, I.A. Stegun, (eds.), Hdb. Math. Functs., Dover Publ., New York, 1972

[11] H. Lübbig, Proc. IC SQUID'80, (1980) pp. 1-28

[12] V. Ambegaokar, A. Baratoff, Phys. Rev. Lett. **10** (1963) pp. 486-489; **E**:11, 104

[13] E. Burstein, S. Lundquist, (eds.) Tunneling Phenomena in Solids, Plenum Press, New York, 1969

[14] G. Brunk, J. Non-Equilib. Thermodyn. **5** (1980) pp. 339-360; G. Brunk, H. Lübbig, J. Low Temp. Phys. **42** (1981) pp. 179-185; G. Brunk, Ch. Zurbrügg, H. Lübbig, IEEE Trans. Magn. **MAG-25** (1989) pp. 1420-1423

[15] D. Jablonski, J. Appl. Phys. **53** (1982) pp. 7458-7463; G, Brunk, Proc. IC SQUID'85 (1985) pp. 77-82; H. Kratz, W. Jutzi, IEEE Trans. Magn. **MAG-23** (1987) pp. 731-734; W. Jutzi, E. Crocoll, J. Marz, Jap. J. Appl. Phys. **26, Suppl. 26-3** (1987) pp. 1589-1590; W. Jutzi, E. Crocoll, D. Drung, G. Krämer, IEEE Trans. Magn. **MAG-25** (1989) pp. 1046-1049

[16] W.A. Schlup, J. Phys. C: Solid State Phys. **7** (1974) pp. 736-748; W.A. Schlup, J. Phys. (Paris) **39** (Suppl.) (1979) pp. 567-568; G. Brunk, Ch. Zurbrügg, H. Lübbig, Proc. IC SQUID'85 (1985) pp. 83-88

[17] D. Rogovin, D.J. Scalapino, Annals of Phys. **86** (1974) pp. 1-90; G. Schön, Proc. IC SQUID'85 (1985) pp. 251-287

[18] J. Kurkijärvi, Proc. IC SQUID'85 (1985) pp. 377-407

[19] R.L. Kautz, J. Appl. Phys. **58** (1985) pp. 424-440

[20] S.N Erné, H. Lübbig, unpublished

[21] J.R. Tucker, IEEE **QE-15** (1979) pp. 1234-1258

[22] R.E. Harris, Phys. Rev. **11** (1975) pp. 3329-3333

[23] H. Grabert, Proc. IC SQUID'85 (1985) pp. 289-315; U. Eckern, G. Schön, V. Ambegaokar, Phys. Rev. **B30** (1984) pp. 6419-6431; V. Ambegaokar, U. Eckern, Z. Phys. **B69** (1987) pp. 399-407

[24] F. Guinea, G. Schön, J. Low Temp. Phys. **69** (1987) pp. 219-243; K.K. Likharev, A.B. Zorin, J. Low Temp. Phys. **59** (1984) pp. 347-382

[25] R. De Bruyn Ouboter, Proc. Int. Symp. Foundations of Quantum Mechanics, Tokyo (1983) pp. 83-93; D.B. Schwartz, B. Sen, C.N. Archie, J.E. Lukens, Phys. Rev. Lett. **55** (1985) pp. 1547-1550

[26] R.F. Voss.R.A. Webb, Phys. Rev. Lett. **47** (1981) pp. 265-268; J.M. Martinis, M.H. Devoret, J. Clarke, Phys. Rev. **B35** (1987) pp. 4682-4698

[27] A.O. Caldeira, A. Leggett, Phys. Rev. Lett. **46** (1981) 211-214

[28] H. Grabert, U. Weiss, Phys. Rev. Lett. **53** (1984) pp. 1787-1790

[29] J. Lambe, R.C. Jaklivic, Phys. Rev. Lett. **22** (1969) pp. 1371-1375

[30] M. Iansiti, A.T. Johnson, C.J. Lobb, M. Tinkham, Phys. Rev. Lett. **60** (1988) 2414-2417

[31] D.V. Averin, K.K. Likharev, Proc. IC SQUID'85 Proc. (1985) pp.197-202; E. Ben-Jacob, Y. Gefen, K. Mullen, Z. Schuss, Proc. IC SQUID'85 (1985) pp. 203-218.

[32] L.S. Kuz'min, K.K. Likharev, JETP Lett. **45** (1987) pp. 495-497;
K.K. Likharev, IBM J. Res. Develop. **32** (1988) pp. 144-158
D.V. Averin, K.K. Likharev, in: Quantum Effects in Small Disordered Systems, B.L. Altshuler, P.A. Lee, R.A. Webb (eds.) Elsevier, Amsterdam, to be published

[33] M. Büttiker, Phys. Rev. **B36** (1987) pp. 3548-3555

[34] R. Holm, W. Meißner, Z. Phys. **74** (1932) pp. 715-735

[35] P.W. Anderson, in: Lectures on the Many Body Problem, E.R. Canianiello (ed.) Academic Press **2** (1964) p. 113;
A. Widom, J. Low Temp. Phys. **37** (1979) pp. 449-460;
A.P. Long, R.J. Prance, T.D. Clark, A. Widom, J.E. Mutton, J. Sacco, M.W. Potts, G. Megaloudis, Proc. IC SQUID'80 (1980) pp. 599-602

Special Abbreviations:
Proc. LT-18: Jpn. J. Appl. Phys **26, Suppl. 3**;
Proc. IC SQUID: Superconducting Quantum Interference Devices and their Applications, De Gruyter, Berlin, New York, H.-D. Hahlbohm, and H. Lübbig (eds.).

Modelling of Resistive Networks for Dispersive Tunnel Processes

G. Brunk
Technische Universität Berlin, 1. Institut für Mechanik, D-1000 Berlin

1. Introduction

The phenomenological description of the dynamics of the Josephson tunnel junction as given by the theory of WERTHAMER [1] is central to our considerations, the reason for which pre-eminence is well substantiated in the introduction to H. Lübbig's article in this book. We start with a formulation of the constitutive relation in the time domain, cf. [2], [3], and use notations similiar to those in Lübbig's article:

$$i(t) = 2\,\frac{\mathrm{d}\mathbf{u}}{\mathrm{d}\varphi} \cdot \mathbf{s}\,[\psi] \tag{1.1}$$

Here i denotes the applied current and

$$\mathbf{u} := \begin{bmatrix} \sin\frac{\varphi}{2} \\ \cos\frac{\varphi}{2} \end{bmatrix} \tag{1.2}$$

the state vector of the junction, φ is the quantum phase shift and $\mathbf{s}\,[\psi]$ is a two component current vector linearily dependent on the transformed state vector

$$\psi \equiv \begin{bmatrix} \varphi_+ \\ \varphi_- \end{bmatrix} := \frac{\Phi_0}{\pi}\,\mathbf{u}, \qquad \Phi_0 = \frac{h}{2e} \tag{1.3}$$

which is scaled by magnetic flux units. Eq. (1.1) presents the essential multiplicative decomposition of the constitutive relation into the nonlinear term $\mathrm{d}\mathbf{u}/\mathrm{d}\varphi$ related to Eq. (1.2) and the linear functional $\mathbf{s}\,[\psi]$. The resistive part of $\mathbf{s}\,[\psi]$ is characterized by certain relaxation functions in the time domain and assigned conductivities in the frequency domain as shown in Fig. 2 of Lübbig's article. Later (Sect. 2), we shall not restrict ourselves a priori to this Werthamer model valid for tunnel junctions, but design a frame possibly also suited for other types of weak links by using macroscopic non-equilibrium thermodynamics of relaxation systems.

2. Classification of Different Essential Processes

By means of some special excitations $i(t)$ we can generate distinguished responses of the Josephson tunnel diode, thus we are able to study the interaction of the quantumphase selfcoupling caused by the nonlinearity $\mathbf{u}(\varphi)$ and of the dispersion involved in the linear relation $\mathbf{s}\,[\psi]$.

First, let be $i(t)$ = const. =: I_0 and I_0 exceed a certain limiting current I_h (McCumber limit). The junction may then respond with a running state or - in mathematical terms - a periodical solution of the second kind to the assigned integro-differential equation. The existence of such a solution or the related physical behaviour is not a matter of course, but has been proved only recently [4]. When there is indeed stationarity then we have the balance of the energy $I_0\ \Phi_0$ supplied in each turn and of the energy loss during a period caused by the dissipation in the tunnel diode, cf. Eq. (4.29) in Sect. 3. If the rotatory frequency $\langle\dot{\varphi}\rangle$ falls short of the double gap frequency, the dissipation of the fundamental harmonic vanishes with the frequency-dependent conductivity exactly at $T = 0$ and approximately at $T/T_c < 0{,}4$. Therefore the balance of the supplied and the dissipated energy is governed by the higher harmonics with frequencies above the double gap frequency due to the nonlinearity, at least at these low temperatures. Numerical simulations completely confirm this conclusion (Fig. 1). The amplitude spectrum of the harmonics of the running state of a tunnel junction is therefore quite different in comparison with the RSJ model at low rotatory frequencies.

Secondly, let $i(t)$ = const. =: $I_0 < I_c$ (critical current). Then equilibrium states φ = const. =: $\alpha + 2\ n\pi$ with $|\alpha| < \pi/2$ are possible. The junction will now respond to a sufficiently small impulsive excitation with damped generally nonlinear plasma oscillations or with an aperiodic evolution. This kind of motion is likewise controlled by the interaction of the nonlinearity $\mathbf{u}(\varphi)$ and the impedance assigned to the linear current operator $\mathbf{s}\,[\psi]$ including the effect of the geometrical capacitance C_0 . If we study the case of oscillations with the ("fundamental") plasma frequency below the gap frequency, we observe that the relative energy loss per period decreases in time. For evidence we look at a relatively small damped oscillation with a large amplitude. This motion behaves almost like an undamped, periodic motion and for this reason can be discussed in terms of "fundamental" and "higher harmonics". Due to the dispersion first the higher harmonics with supra-gap frequencies are damped away and then the resulting amplitude decreases.The remaining sub-gap harmonics are much less damped because of the small sub-gap conductivity and in the limit $T = 0$ they are not damped at all.

Thirdly, we study arbitrary transients between stationary states. This type of evolution can be characterized by its Fourier transform. The associated dissipation can also be described by integration over the frequency axis with the dispersive conductivities as weighting factors, cf. Eq. (4.31) in

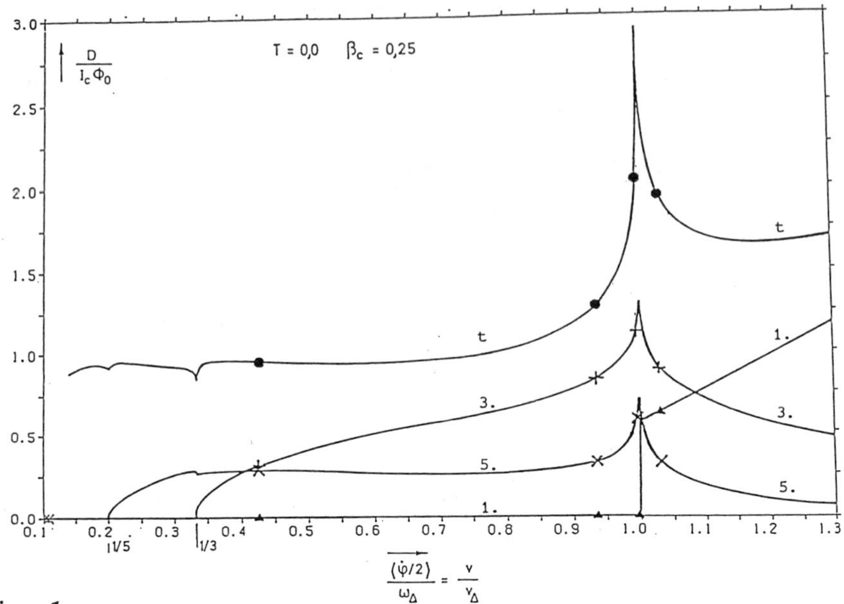

Figure 1
Energy D dissipated in a Josephson tunnel diode during one turn 2π of the quantum phase φ at temperature $T = 0$ versus rotatory frequency or mean voltage for $\beta_C = 0.25$. t: Total Energy loss and simultaneously I-V-characteristic. 1., 3. and 5.: Contributions of the first, third and fifth harmonic to the energy loss. $\omega_\Delta := 2\pi \cdot 2\Delta/h$, $v_\Delta := 2\Delta/e$.

Sect. 4. For this reason the Fourier transforms make the main contribution to the dissipation in the supra-gap domain where the conductivities have the order of magnitude of the normal conductivity $G_N := 1/R_N$. The Fourier transforms thus prevail in a regime of higher frequencies in comparison with the RSJ model, which means that the transition times are shorter and the transient motion runs faster, cf. [5].

3. The Macroscopic Dynamical Structure of Superconductive Tunnel Diodes

The phenomenological description of the dynamics of the Josephson tunnel junction given by Eqs. (1.1 - 1.3) is completed by the famous general relation for the voltage

$$v = \frac{\Phi_0}{2\pi} \dot{\varphi} \tag{3.1}$$

due to gauge invariance of the quantum state with respect to electromagnetic fields. For the differential of work Eqs. (1.1) - (1.3) and (3.1) yield

$$dA = iv\,dt = \frac{\Phi_o}{2\pi}\,i\,d\varphi = \frac{\Phi_o}{\pi}\,\mathbf{du}\cdot\mathbf{s}[\psi] = \mathbf{s}[\psi]\cdot d\psi\ . \tag{3.2}$$

This means that the work differential is completely described in terms of the flux-like state vector ψ. This structure suggests embedding the phenomenological description of the Josephson tunnel junction in the general frame of the thermodynamics of nonlinear relaxation systems. Let $x \equiv (x^1, x^2 \dots x^n)$ be the set of work variables of such a system. We then have the work differential

$$dA = \mathbf{y}[x]\cdot \mathbf{dx} \equiv y_k\,dx^k \text{ (sum over k)} \tag{3.3}$$

with $\mathbf{dx} \in V_n$ (n dimensional vector space) and the generalized force $\mathbf{y} \in \widehat{V}_n$ (dual space of V_n). $\mathbf{y}$ is related to x by the constitutive relation

$$\mathbf{y}[x] = \mathbf{y}_e(x) + \mathbf{y}_v[x] \tag{3.4}$$

with the equilibrium contribution

$$\mathbf{y}_e(x) = \partial_x F(x, T) \tag{3.5}$$

determined by the free energy F and the resistive or viscous portion $\mathbf{y}_v\,[x]$ obeying the condition

$$\mathbf{y}_v[x_o] \equiv 0 \tag{3.6}$$

for all constant processes x = const. =: x_o. To each process $x(t-\tau)$, t fixed, we can assign a class of "motions" delayed or accelerated in the past ($\tau > 0$) by the transformation

$$x(t-\tau) \mapsto x_\alpha(t-\tau) := x(t-\alpha\tau) \tag{3.7}$$

with $\alpha < 1$ for delay and $\alpha > 1$ for acceleration. For the fixed functional form $x(t-\tau)$ of the process, the resistive response

$$\mathbf{y}_v[x_\alpha] =: y_{v,x}(\alpha) \tag{3.8}$$

then becomes a function of α with the property $y_{v,x}\,(0) = \mathbf{0}$. For a suitably restricted set of motions it may be admissible to approximate the function $y_{v,x}\,(.)$ by a linearization

$$y_{v,x}(\alpha) \mapsto y_{\ell,x}(\alpha) = \alpha\, y_{\ell,x}(1)\ . \tag{3.9}$$

If we assume that the work variables or non-thermal state coordinates x enter the constitutive description with respect to the work expression by means of a state vector $g(x) \in V_m$, we have the general representation by a convolution integral

$$y_{\ell,x}(1) \equiv \mathbf{y}_\ell[x]$$
$$:= \partial_x \mathbf{g} \cdot \int_{-\infty}^{t} \mathbf{Q}(\gamma(t), t-\tau) \cdot \dot{\gamma}(\tau)\, d\tau \tag{3.10}$$

with the evolution of the state vector $\gamma(t) := \mathbf{g}(x(t))$ and the tensor-valued relaxation kernel $\mathbf{Q}(\mathbf{g},t) \in \widehat{V}_m \otimes \widehat{V}_m$ ($\widehat{V}_m$ means the dual space of V_m) [6].

If we presuppose that in a step-like excitation energy is stored instantaneously and subsequently dissipated in finite times, it follows that

$$\exists\, \mathbf{m}(\mathbf{g}, t) \mapsto \mathbf{Q}(\mathbf{g}, t) = \partial_{\mathbf{g}} \mathbf{m}(\mathbf{g}, t)\ , \tag{3.11}$$

i.e. the tensor-valued kernel $\mathbf{Q}$ is derived from a vector-valued function $\mathbf{m}$. Generally, Eq. (3.10) provides a strong nonlinear mapping of the motion $x(\tau)$, $\tau < t$, in the past on the present value of generalized force $\mathbf{y}_v(t)$. The irreversible portion of work reads

$$dA_v = \mathbf{y}_v[x] \cdot dx = \boldsymbol{h}_v[\mathbf{g}] \cdot d\boldsymbol{g} \tag{3.12}$$

with the internal force vector

$$\boldsymbol{h}_v[\mathbf{g}] := \int_{-\infty}^{t} \mathbf{Q}(\gamma(t), t-\tau) \cdot \dot{\gamma}(\tau)\, d\tau \equiv \mathbf{Q} * \dot{\gamma}\ . \tag{3.13}$$

Equation (3.12) has the same structure as Eq. (3.2). Therefore we can identify

$$x \equiv \frac{\Phi_0}{2\pi}\varphi, \qquad y \equiv i \qquad (n = 1) \tag{3.14.1}$$

and

$$g \equiv \psi, \qquad h \equiv \mathbf{s} \qquad (m = 2)\ . \tag{3.14.2}$$

If the kernel $\mathbf{Q}$ does not depend on $\gamma(t)$ the convolution integral in Eq. (3.13) represents a linear operator applied on $\gamma(\tau) = g(x(\tau))$. The thermodynamic theory of linear passive systems is then valid as stated by MEIXNER and coworkers, cf. e. g. [7], and rearranged and generalized by BÖCKER [8] using the distribution theory. This applies precisely in the case of the Josephson tunnel junction. Here we have the current vector

$$\mathbf{s}[\psi] = C_0 \cdot \ddot{\psi} + Y(t) * \dot{\psi} + L_0^{-1} \cdot \psi\ . \tag{3.15}$$

$\mathbf{C}_0 := C_0\mathbf{1}$ is the geometrical capacitance matrix, the inverse inductance matrix

$$L_o^{-1} := L_o^{-1}\begin{bmatrix} 1 & 0 \\ 0 & 0 \end{bmatrix} \qquad \left(L_o^{-1} := 2\pi \frac{I_c}{\Phi_o}\right) \tag{3.16}$$

enters the (equilibrium) free energy by the relation

$$F = \frac{1}{2}\psi \cdot L_o^{-1} \cdot \psi = \frac{\Phi_o I_c}{2\pi}(1 - \cos\varphi) \ , \tag{3.17}$$

and the relaxation matrix $\mathbf{Q}(t) \equiv \mathbf{Y}(t)$ has the form

$$\mathbf{Y}(t) = G_N \mathbf{1}\, \delta(t) + \frac{\pi}{\Phi_o}\begin{bmatrix} N_+(t) & 0 \\ 0 & N_-(t) \end{bmatrix} \ . \tag{3.18}$$

Here, $\mathbf{1}$ denotes the unity matrix, $\delta(.)$ the DIRAC distribution and $G_N \equiv 1/R_N$ the normal state conductance. The functions $N_+(.)$ and $N_-(.)$ are related to the kernels $I_{qp}(.)$ and $I_J(.)$ representing the flow of quasiparticles and Cooper pairs [2], [3] by the definition

$$\begin{aligned} M_\pm(t) &\equiv \frac{\Phi_o}{\pi} G_N\, \delta(t) + N_\pm(t) := \int_{t-0}^{\infty} \left(I_{qp}(\tau) \pm I_J(\tau)\right) d\tau, \qquad && t > 0 \\ M_\pm(t) &= 0 && t < 0 \, . \end{aligned} \tag{3.19}$$

Inserting Eqs. (3.16) and (3.18) into Eq. (3.15) yields the relations

$$s_+ = C_o\, \ddot{\varphi}_+ + G_N\, \dot{\varphi}_+ + \frac{\pi}{\Phi_o} N_+ * \dot{\varphi}_+ + L_o^{-1}\, \varphi_+ \tag{3.20.1}$$

and

$$s_- = C_o\, \ddot{\varphi}_- + G_N\, \dot{\varphi}_- + \frac{\pi}{\Phi_o} N_- * \dot{\varphi}_- \tag{3.20.2}$$

for the components of the current vector (* denotes the convolution product). For $T \neq 0$ the relaxation functions $N_\pm$ or their Fourier transforms are not given analytically but must be placed ready in tables as it has been done by SCHLUP [9], GRIGORIEFF and ZURBRÜGG [10].

The linearization of the phenomenological description in the sense of Eqs. (3.9) and (3.10) is generally the counterpart of a corresponding approximation procedure at the microscopic level, such as a first order perturbation technique for example. Since the macroscopic approach requires no detailed knowledge of the microscopic model, as does the form of the Hamiltonian for example, it may be expected that this method is also convenient to describe the relaxation dynamics of other types of weak links provided that we know the corresponding form of the state vector. Naturally phenomenological relaxation thermodynamics gives only a qualitative frame for constitutive relations and no quantitative information, i. e. no numerical values of material parameters.

4. The Mapping of the Dynamical Structure on Technical Equivalent Systems

4.1 Mechanical Analogies

If we insert Eqs. (3.15) or (3.20) into Eq. (1.1) using the definitions Eqs. (1.2), (1.3), (3.16) and (3.18), we find the integro-differential equation with respect to the phase shift φ

$$C_o \frac{\Phi_o}{2\pi} \ddot{\varphi} + G_N \frac{\Phi_o}{2\pi} \dot{\varphi} + I_c \sin\varphi$$

$$+ \cos\frac{\varphi}{2} \int_{-\infty}^{t} N_+(t-\tau)\left[\sin\frac{\varphi(t)}{2}\right]^{\cdot} d\tau$$

$$- \sin\frac{\varphi}{2} \int_{-\infty}^{t} N_-(t-\tau)\left[\cos\frac{\varphi(\tau)}{2}\right]^{\cdot} d\tau = i(t)\,. \tag{4.1}$$

With a few different notations this is Lübbig's Eq. (3.1). Setting $N_\pm(t) \equiv 0$ we obtain the familiar differential equation of the RSJ model, cf. Eq. (1.1) of Lübbig's article. Equation (4.1) can be interpreted as the equation of motion of a *pendulum* with nonlinear retarded damping driven by the "turning moment" i(t) or as the equation of motion of a *classical particle* moving on a *sinusoidal potential* hill chain. Since Eq. (4.1) is strongly nonlinear and the kernel $N_-(t)$ decreases slowly as $(\sin \omega_\Delta t)/t$ $(\omega_\Delta/2\pi = 2\Delta/h)$, the direct numerical evaluation of Eq. (4.1) for calculating the processes discussed in Sect. 1 is rather difficult and cumbersome, cf. [4].

We obtain another mechanical situation together with a new starting point for calculations by studying the properties of the state vector **u**, Eq. (1.2), or ψ Eq. (1.3). For the norm of **u**

$$|\mathbf{u}|^2 = \sin^2\frac{\varphi}{2} + \cos^2\frac{\varphi}{2} = 1 \tag{4.2}$$

is valid and we find likewise

$$(2\,d\mathbf{u}/d\varphi)^2 = \cos^2\frac{\varphi}{2} + \sin^2\frac{\varphi}{2} = 1\,, \tag{4.3}$$

whereas from Eq. (4.2)

$$(2\,d\mathbf{u}/d\varphi)\cdot\mathbf{u} = 0 \tag{4.4}$$

follows.

This means the vectors

$$\mathbf{e}_1 := \mathbf{u}, \qquad \mathbf{e}_2 := 2\,\mathbf{du}/d\varphi \tag{4.5}$$

are orthogonal unit vectors. In a **u**-plane the equation $|\mathbf{u}| = 1$ determines the unit circle, $\mathbf{e}_1$ is the radial vector and $\mathbf{e}_2$ is the corresponding tangent vector. Equation (1.1) defines the component of the internal current vector s with respect to $\mathbf{e}_2$, namely

$$\mathbf{e}_2 \cdot \mathbf{s} = i(t) \ . \tag{4.6}$$

The other component

$$\mathbf{e}_1 \cdot \mathbf{s} =: \lambda \frac{\Phi_o}{\pi} \tag{4.7}$$

is unknown. We therefore have the vector equation

$$\mathbf{s}[\psi] = \lambda \frac{\Phi_o}{\pi} \mathbf{e}_1 + i\mathbf{e}_2 = \lambda\psi + 2\,i\,\mathbf{du}/d\varphi \tag{4.8}$$

together with the constraint, Eq. (4.2), to determine the state vector and the quantity λ. Inserting $\mathbf{s}[\psi]$, Eq. (3.13), we obtain

$$\boldsymbol{C}_o \cdot \ddot{\psi} + \boldsymbol{Y}(t) * \dot{\psi} + \boldsymbol{L}_o^{-1} \cdot \psi - \lambda\psi = \frac{2\pi}{\Phi_o}\, i\, d\psi/d\varphi \ . \tag{4.9}$$

Equation (4.9) can be interpreted as a equation of motion of a *classical particle restricted* by the constraint

$$|\psi| = \frac{\Phi_o}{\pi} \tag{4.10}$$

on a *circle in the ψ-plane*. It is in the form of a Lagrange equation of the first kind with constraint parameter λ.

When the relation

$$2\frac{d\psi}{d\varphi} = \frac{\Phi_o}{\pi} \begin{bmatrix} \cos\frac{\varphi}{2} \\ -\sin\frac{\varphi}{2} \end{bmatrix} = \begin{bmatrix} \varphi_- \\ -\varphi_+ \end{bmatrix} \tag{4.11}$$

and Eq. (3.20) are taken the component representation of Eq. (4.9) reads

$$C_o\,\ddot{\varphi}_+ + G_N\,\dot{\varphi}_+ + \frac{\pi}{\Phi_o} N_+ * \dot{\varphi}_+ + L_o^{-1}\,\varphi_+ - \lambda\varphi_+ - \frac{\pi}{\Phi_o} i\varphi_- = 0 \tag{4.12.1}$$

and

$$C_o\,\ddot{\varphi}_- + G_N\,\dot{\varphi}_- + \frac{\pi}{\Phi_o} N_- * \dot{\varphi}_- - \lambda\varphi_- + \frac{\pi}{\Phi_o} i\varphi_+ = 0 \ . \tag{4.12.2}$$

The nonlinearity of the Josephson tunnel junction enters these equations of motion only by the terms $\lambda\varphi_+$ and $\lambda\varphi_-$, but it is additionally involved in the constraint, Eq. (4.10) or

$$\varphi_+^2 + \varphi_-^2 = \left(\frac{\Phi_o}{\pi}\right)^2 . \tag{4.13}$$

By the use of simple and double differentiation of Eq. (4.13), we obtain

$$\lambda = \left(\frac{\pi}{\Phi_o}\right)^2 (\varphi_+ s_+ + \varphi_- s_-)$$

$$= \left(\frac{\pi}{\Phi_o}\right)^2 \left[-C_o\left(\dot\varphi_+^2 + \dot\varphi_-^2\right) + L_o^{-1}\varphi_+^2 + \frac{\pi}{\Phi_o}\left(N_+ * \dot\varphi_+\right)\varphi_+ + \frac{\pi}{\Phi_o}\left(N_- * \dot\varphi_-\right)\varphi_-\right] \tag{4.14}$$

for the constraint "force".

The three Eqs. (4.12.1), (4.12.2) and (4.14) form a good basis for numerical calculations. To avoid instability with respect to the constraint, Eq. (4.2) or (4.13), it is advisable to correct the results after each computation step according to the prescription

$$\varphi_\pm \to \varphi_\pm \left[\varphi_+^2 + \varphi_-^2\right]^{-1/2} \frac{\Phi_o}{\pi} \tag{4.15}$$

resulting from Eq. (4.13).

4.2 Electrical Equivalent Circuits

First let us return to discuss the expressions for the work, Eq. (3.2), in terms of electrical quantities. By definition of an internal voltage vector

$$\mathbf{v} \equiv \begin{bmatrix} v_+ \\ v_- \end{bmatrix} := \dot\psi \equiv \begin{bmatrix} \dot\varphi_+ \\ \dot\varphi_- \end{bmatrix} \tag{4.16}$$

Eq. (3.2) yields

$$P = iv = \mathbf{s} \cdot \mathbf{v} \equiv s_+ v_+ + s_- v_- \tag{4.17}$$

for the power supplied.

From the definitions of ψ, Eq. (1.3), and $\mathbf{u}$, Eq. (1.2), it follows that by using Eq. (3.1)

$$\mathbf{v} = \frac{\Phi_o}{\pi}\frac{d\mathbf{u}}{d\varphi}\dot\varphi = 2\frac{d\mathbf{u}}{d\varphi}v , \tag{4.18}$$

or written by components

$$v_+ = (2\, d_\varphi\, u_+)\, v\,, \tag{4.19.1}$$

$$v_- = (2\, d_\varphi\, u_-)\, v\,. \tag{4.19.2}$$

Equations (4.18) or (4.19) can be interpreted as transmission rules from the outer voltage v to the internal ones v_+ and v_-. These transformations together with the power invariance statement, Eq. (4.17), suggest the representation by two ideal phase controlled transducers with [11] transmission ratios $2\, du_+/d\varphi$ and $2\, du_-/d\varphi$, Fig. 2. The dual relation of Eq. (4.18) is Eq. (1.1) defining the current transmission

$$i = (2\, d_\varphi\, u_+)\, s_+ + (2\, d_\varphi\, u_-)\, s_-\,, \tag{4.20}$$

thus securing the validity of Eq. (4.17). On the other hand, it is also possible to interpret Eq. (4.20) as a definition of a current source in the external circuit controlled by the phase and the internal currents and to read Eqs. (4.19.1) and (4.19.2) for prescriptions of voltage sources in the two internal subcircuits, as KRATZ and JUTZI have done [12]. Since both pictures give equivalent representations it is not possible to decide definitively between them. In this context we remember the general thermodynamical background given by Eqs. (3.10), (3.12) and (3.13).

Whereas the transmission Eqs. (4.19) and (4.20) represent the nonlinear properties of the Josephson diode mathematically and as equivalent network model the corresponding nonlinear transducers represent them completely at this stage of our considerations we lack a detailed discussion of the characteristics of the linear current operator $\mathbf{s}[\psi]$, Eq. (3.15) or (3.20). For this purpose it is convenient to study the linear current operator $\mathbf{s}[\psi]$ in the frequency domain of the Fourier transform. We use the notation

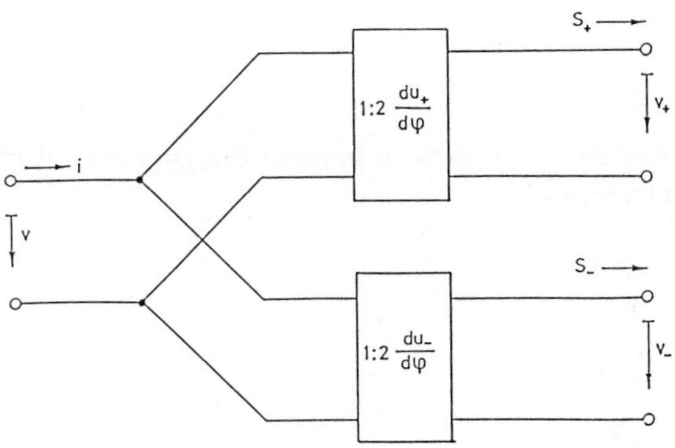

Figure 2
Representation of the nonlinear selfcoupling of the Josephson tunnel junction as an equivalent circuit: Ideal transducers with transmission controlled by the quantum phase according to Eqs. (1.1) and (4.19).

$$f(t) \mapsto \tilde{f}(\omega) := \int_{-\infty}^{+\infty} dt\, f(t)\, e^{-j\omega t}, \qquad j^2 \equiv -1\,, \tag{4.21.1}$$

$$\tilde{f}(\omega) \mapsto f(t) := \frac{1}{2\pi} \int_{-\infty}^{+\infty} dt\, \tilde{f}(\omega)\, e^{j\omega t}. \tag{4.21.2}$$

With the correspondence

$$\mathbf{v} = \dot{\psi} \leftrightarrow \tilde{\mathbf{v}} = j\omega\, \tilde{\psi} \tag{4.22}$$

it follows from Eq. (4.18) that

$$\tilde{\mathbf{s}} = \mathbf{Y}_t(\omega) \cdot \tilde{\mathbf{v}} \tag{4.23}$$

with the total admittance matrix

$$\mathbf{Y}_t(\omega) = j\omega\, \boldsymbol{C}_o + \tilde{\boldsymbol{Y}}(\omega) + \frac{1}{j\omega} \boldsymbol{L}_o^{-1}\,. \tag{4.24}$$

Written in components using Eq. (3.20) this reads

$$Y_{t+} = j\omega\, C_o + G_N + \frac{\pi}{\Phi_o} \tilde{N}_+ + 1/j\omega L_o \tag{4.25.1}$$

and

$$Y_{t-} = j\omega\, C_o + G_N + \frac{\pi}{\Phi_o} \tilde{N}_-\,. \tag{4.25.2}$$

These equations describe two linear subcircuits of the system as shown in Fig. 3. The first one, marked by +, is a connection in parallel of an $L_o\, C_o$ parallel resonance circuit representing the plasma oscillations and a frequency-dependent admittance

$$\tilde{Y}_+ = (\omega) := G_N + \frac{\pi}{\Phi_o} \tilde{N}_+(\omega)\,. \tag{4.26.1}$$

The second one marked by -, connects in parallel the geometrical capacity C_o and the dispersive admittance

$$\tilde{Y}_- = (\omega) := G_N + \frac{\pi}{\Phi_o} \tilde{N}_-(\omega)\,. \tag{4.26.2}$$

The conductances assigned are

$$G_+(\omega) := \mathrm{Re}\,\{Y_{t+}\} = \mathrm{Re}\,\{\tilde{Y}_+(\omega)\}, \tag{4.27.1}$$

$$G_-(\omega) := \mathrm{Re}\,\{Y_{t-}\} = \mathrm{Re}\,\{\tilde{Y}_-(\omega)\}. \tag{4.27.2}$$

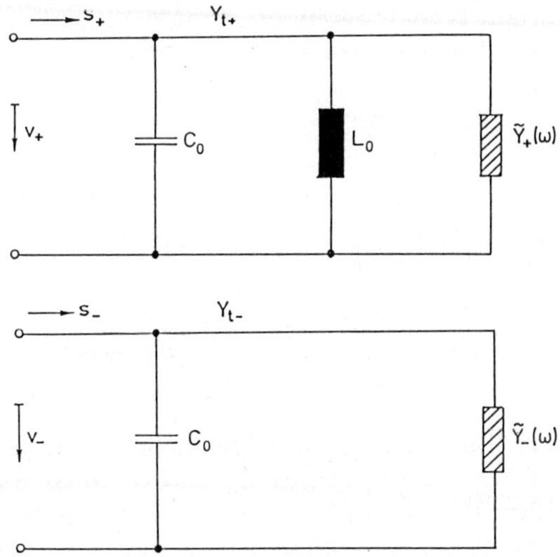

Figure 3
The structure of the two linear equivalent subcircuits of a Josephson tunnel junction.

Their immediate physical significance is the control of dissipation in an arbitrary situation: First let us look at the periodical processes

$$v_{\pm}(t) = v_{\pm}\left(t + \frac{2\pi}{\Omega}\right) = \sum_{n=-\infty}^{\infty} \widehat{V}_{\pm n}\, e^{nj\Omega t} \tag{4.28}$$

with complex amplitudes $\widehat{V}_{\pm n}$. In this case the mean power loss is given by

$$\overline{P} = \sum_{n=-\infty}^{\infty} \left[G_{+}(n\Omega)\left|\widehat{V}_{+n}\right|^{2} + G_{-}(n\Omega)\left|\widehat{V}_{-n}\right|^{2}\right] \tag{4.29}$$

whereas for transition processes

$$v_{\pm}(t) \neq 0 \qquad \text{for} \qquad t > t_o \tag{4.30}$$

we find the energy dissipated up to time t to be

$$D(t_o, t) := \int_{t_o}^{t} P(\hat{t})\, d\hat{t}$$

$$= \frac{1}{2\pi}\int_{-\infty}^{+\infty} d\omega\left[G_{+}(\omega)\left|\tilde{v}_{+t}(\omega)\right|^{2} + G_{-}(\omega)\left|\tilde{v}_{-t}(\omega)\right|^{2}\right] \tag{4.31}$$

with

$$\tilde{v}_t(\omega) := \int_{t_o}^{t} v(\hat{t})\, e^{-j\omega\hat{t}}\, d\hat{t}\ . \tag{4.32}$$

The third important type of evolutions consists of stationary noise processes v(t) with correlation functions or assigned spectra

$$S_{\pm vv}(\tau) := \langle v_{\pm}(t)\, v_{\pm}(t+\tau)\rangle \leftrightarrow \widetilde{S}_{\pm vv}(\omega) \; . \tag{4.33}$$

Here we obtain the expectation value of the dissipated power of fluctuations

$$\langle P\rangle := \langle s_{+}(t)\, v_{+}(t) + s_{-}(t)\, v_{-}(t)\rangle$$

$$= \frac{1}{2\pi}\int_{-\infty}^{+\infty} d\omega \left[G_{+}(\omega)\, \widetilde{S}_{+vv}(\omega) + G_{-}(\omega)\, \widetilde{S}_{-vv}(\omega)\right] \; . \tag{4.34}$$

According to the fluctuation dissipation theorem applied to the Josephson tunnel diode [3], the conductances G_{+} and G_{-} must for simple statistical reasons be positive. The dissipation expressions of Eqs. (4.29), (4.31) and (4.34) therefore become positive, which means that the two subcircuits form passive systems. Passivity always involves causality, cf. [7] or [8], p. 24. With this property the susceptances

$$H_{\pm}(\omega) := \operatorname{Im} \widetilde{Y}_{\pm}(\omega)$$

are determined by the Kramers-Kronig relations as Hilbert transforms of the conductances $G_{\pm}(\omega)$. In this context we also wish to emphasize that the structure of the total admittance, Eq. (4.24) or (4.25), corresponds to the general representation of passive systems, cf. [8], p. 44. Just these properties make it possible to map the two mathematically defined subcircuits on equivalent networks consisting of passive continuous or lumped elements.

4.2.1 Circuit Model with Infinite Degree of Freedom

The mapping of the current operator **s**[ψ] on equivalent circuits requires a thorough study of the frequency characteristics $G_{\pm}(\omega)$, curves W in Fig. 4 corresponding to a medium temperature $T/T_c = 0.72$. The conductance $G_{+}(\omega)$ shows a logarithmic singularity at $\omega = 0$, which means low pass-filter behaviour. After passing through a minimum at the gap frequency ω_Δ it tends asymptotically from below to the normal state conductance G_N (weak high-pass behaviour). The frequency characteristics $G_{-}(\omega)$ is jumping by a step at the gap frequency ω_Δ from small values (zero at T = 0) below to greater ones of the order of magnitude of G_N above and tends asymptotically to G_N at infinity, but from above (strong high-pass behaviour). The step at ω_Δ corresponds to the logarithmic Riedel singularity of the susceptance $H_{-}(\omega)$ and to the gaposcillations of the relaxation function $N_{-}(t)$ which behave assymptotically as $\sin \omega_\Delta t/t$. Thus the - subcircuit represents the gap resonance whereas the + subcircuit determines the plasma resonance as we

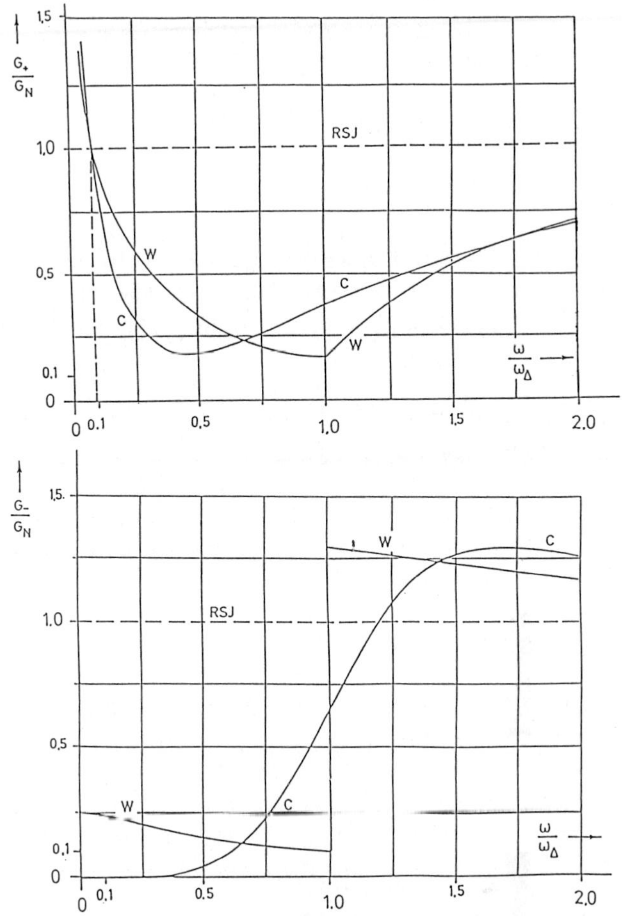

Figure 4

Frequency characteristics of the conductances $G_\pm(\omega) := \mathrm{Re}\left\{\widetilde{Y}_\pm(\omega)\right\}$ of the Werthamer model of a tunnel diode (W) in comparison to the circuit model (C) of Fig. 6 and the RSJ model.

have found above. The admittances $\widetilde{Y}_+(\omega)$ and $\widetilde{Y}_-(\omega)$, both contain logarithmic singularities. These transcendent functions correspond to systems with infinite numbers of degree of freedom and for this reason they cannot be exactly represented by equivalent networks consisting of finite numbers of lumped elements. The intensity of the singularities depends on the temperature. If we wish to preserve the main feature of the gap singularity, but drop out its temperature dependence, we can approximate the admittance $\widetilde{Y}_-(\omega)$ by

$$\widetilde{Y}_-(\omega) \approx G_N \frac{j}{\pi} \ln \frac{\omega_\Delta + \omega}{\omega_\Delta - \omega} \tag{4.35}$$

or the conductance by

$$G_-(\omega) = \begin{cases} 0 & |\omega| < \omega_\Delta \\ G_N & |\omega| > \omega_\Delta \end{cases} . \tag{4.36}$$

By means of the continued fraction representation

$$\ln \frac{1+z}{1-z} = \frac{2z}{1-}\,\frac{z^2}{3-}\,\frac{4z^2}{5-}\,\frac{9z^2}{7-} \cdots \tag{4.37}$$

we can reproduce the admittance $\widetilde{Y}_-$, Eq. (4.35), by an infinite inhomogeneous reactive chain like network, Fig. 5, consisting of lengthwise connected capacitors C_k and inductors L_k shunting crossways. The normalized values

$$\widehat{C}_k := \omega_\Delta R_N C_k \text{ and } \widehat{L}_k := \omega_\Delta G_N L_k$$

of these capacitances and inductances are

$$\widehat{C}_1 = \frac{2}{\pi} \cdot 1 \qquad \widehat{L}_1 = \frac{\pi}{2} \cdot \frac{1}{3}$$

$$\widehat{C}_2 = \frac{2}{\pi} \cdot \frac{1}{5} \cdot 4 \qquad \widehat{L}_2 = \frac{\pi}{2} \cdot \frac{1}{7} \cdot \frac{9}{4}$$

$$\widehat{C}_3 = \frac{2}{\pi} \cdot \frac{1}{9} \cdot \frac{4}{9} \cdot 16 \qquad \widehat{L}_3 = \frac{\pi}{2} \cdot \frac{1}{11} \cdot \frac{9}{4} \cdot \frac{25}{16}$$

$$\begin{aligned} \widehat{C}_n &= \frac{2}{\pi} \cdot \frac{1}{4n-3} \cdot \frac{4}{9} \cdot \frac{16}{25} \cdots \frac{(2n-4)^2}{(2n-3)^2} (2n-2)^2 \to \frac{1}{2} \\ \widehat{L}_n &= \frac{\pi}{2} \cdot \frac{1}{4n-1} \cdot \frac{9}{4} \cdot \frac{25}{16} \cdots \frac{(2n-3)^2}{(2n-4)^2} \frac{(2n-1)^2}{(2n-2)^2} \to \frac{1}{2} . \end{aligned} \tag{4.38}$$

The behaviour $\widehat{C}_n \to 1/2$ and $\widehat{L}_n \to 1/2$ in the limit $n \to \infty$ shows the circuit corresponds practically (the first elements excepted) to a homogeneous chain or recurrence of the cutoff frequency $\omega_\Delta = 1/(2\sqrt{LC})$. As the exact frequency characteristics $\widetilde{Y}_\pm(\omega)$ are given only numerically the problem to map this dispersive response property on equivalent circuits must generally be solved by computation, too.

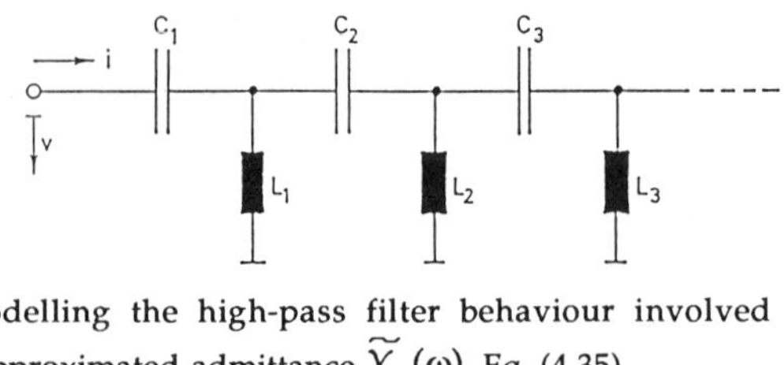

Figure 5
Reactive chain modelling the high-pass filter behaviour involved by the logarithmic singularity of the approximated admittance $\widetilde{Y}_-(\omega)$, Eq. (4.35).

4.2.2 Circuit Model with Finite Degree of Freedom

If we dispense with a true reproduction of the singularities we can design finite degree-of-freedom networks consisting of finite numbers of elements. The accuracy required with respect to the class of processes studied determines the degree of freedom and accordingly, the number of two-terminal elements. A criterion for mapping may be the minimal number of elements necessary to produce the low-pass filter behaviour and the weak high-pass filter feature of $\tilde{Y}_+(\omega)$ as well as the gap resonance connected with the strong high-pass filter feature of $\tilde{Y}_-(\omega)$. The resulting equivalent subcircuits [11] are shown in Fig. 6. The corresponding parameters depend on temperature and have been plotted in a diagram, Fig. 7. The normalizing frequency used is the temperature dependent gap frequency $\omega_\Delta = 2\pi \cdot 2\Delta/h$ with the exception of the curve of the normalized inductivity $\omega_1 L_1/R_N$. Here, ω_1 is the limiting frequency of the low-pass domaine defined by $G_+(\omega_1) := G_N$. It is plotted in the additional diagram, Fig. 8. The frequency characteristics of the conductances $G_\pm(\omega)$ approximated in such a way are plotted by curves C in Fig. 3 for the temperature $T/T_c = 0.72$. Higher accuracy is achieved but therefore more elements are used by KRATZ and JUTZI [12], but unfortunately there is no information on the temperature dependence of the parameters.

The admittance of a circuit built up by a finite number of capacitors, inductors and resistors is described mathematically by a so-called "positive"

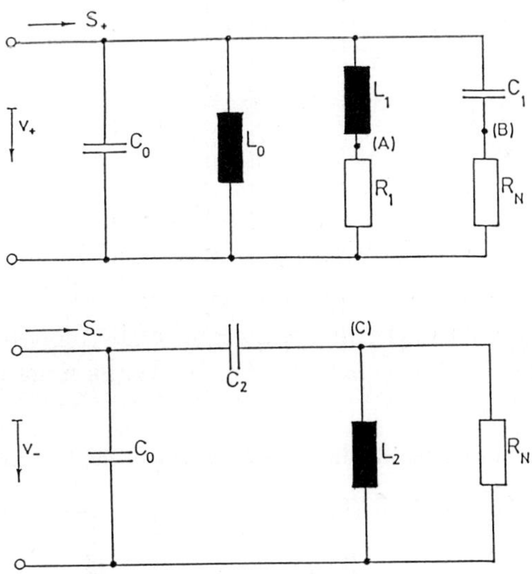

Figure 6
Circuits modelling the total admittances $Y_{t\pm}(\omega)$, Eq. (4.25), by a minimal number of passive linear two-terminal elements. The upper one represents the plasma resonance by the L_0C_0 connection in parallel, the lower one involves the gap resonance by C_2 and L_2 with damping by R_N.

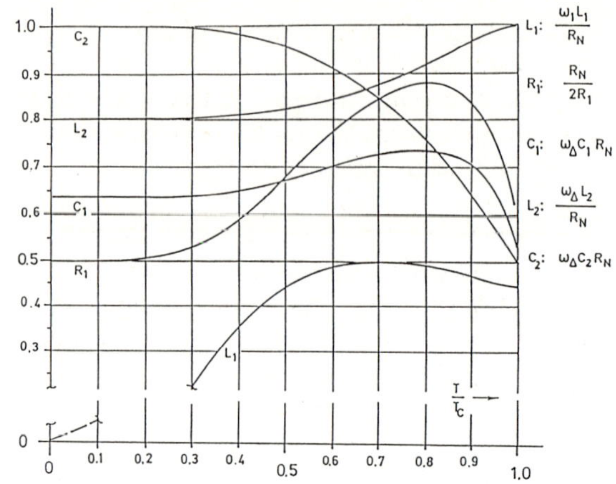

Figure 7
Temperature dependence of the circuit parameters reffering to the networks of Fig. 6.

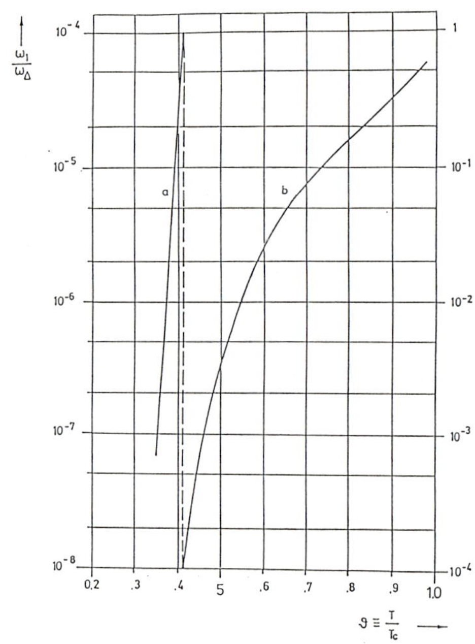

Figure 8
Temperature dependence of the limiting frequency ω_1 of the low-pass filter regime of the admittance $\widetilde{Y}_+$ defined by $G_+(\omega_1) = G_N$.

rational function, cf. [8], p. 59. Physical realisations of such a function other than these networks are certainly possible, e. g. they may be constructed by means of operational amplifiers as in analog computers or in particular, in JABLONSKI's electronic simulator [13] of the Josephson tunnel junction.

Besides this, the modelling of the dispersive admittance also can be executed in the time domain, e. g. by use of an expansion of the relaxation matrix **Y**(t) by exponential functions. This method has been chosen by SEMENOV and coworkers [14]. The truncation of the resulting infinite series in the time domain corresponds to the approximation of the transcendental admittance in the frequency domain by a rational function. However, not any subtotal of the truncated expansion formed by a linear combination of exponentialfunctions can be mapped on a subcircuit consisting of linear passive elements. The a priori modelling by a network is therefore preferrable to the a posteriori interpretation of formal approximations connected with necessary corrections referring to causality or passivity. Ensuring passivity is a necessary condition of modelling. The same is true, of course, if we do not restrict ourselves to rational function models in the frequency domain. For example, we may state a singularity equivalent description [4], [15] by the equations

$$\tilde{Y}_+(\omega) := G_N\left[A \ln\left(1 + \frac{\alpha}{j\omega}\right) + \frac{j\omega\tau_s}{1 + j\omega\tau_s}\right] \tag{4.39.1}$$

$$\tilde{Y}_+(\omega) := G_N\left[1 - j\frac{g_o}{\pi}\ln\frac{\omega - \omega_\Delta}{\omega + \omega_\Delta} + g_1\frac{1}{1 + j\omega\tau_L}\right] \tag{4.39.2}$$

containing only few temperature-dependent parameters A, α, τ_s, g_o and $g_1 = (1-2/\pi)\, g_o$, τ_L. In the first equation, Eq. (4.39.1), both terms have positive real components, thus we can try to assign to each of them a subcircuit consisting of passive elements. But on the right-hand side of the second equation, Eq. (4.39.2), the logarithmic term has a negative real component and therefore lacks correspondence to a passive subcircuit.

5. Conclusion and Outlook

After giving various methods for approximating the constitutive relations of the Josephson tunnel junction and emphasizing the suitability of circuit models with respect to dispersion and passivity we recommend the further use of this approach. One advantage is that it makes it possible to describe the macroscopic junction dynamics by means of a Hamiltonian and generalized non conservative friction forces, also with respect to the internal degrees of freedom [11]. This description may serve as a basis for various theoretical considerations such as the stability analysis of the phenomenological dynamics and the extended studies of secondary quantum effects. A second advantage results from the mapping of the original

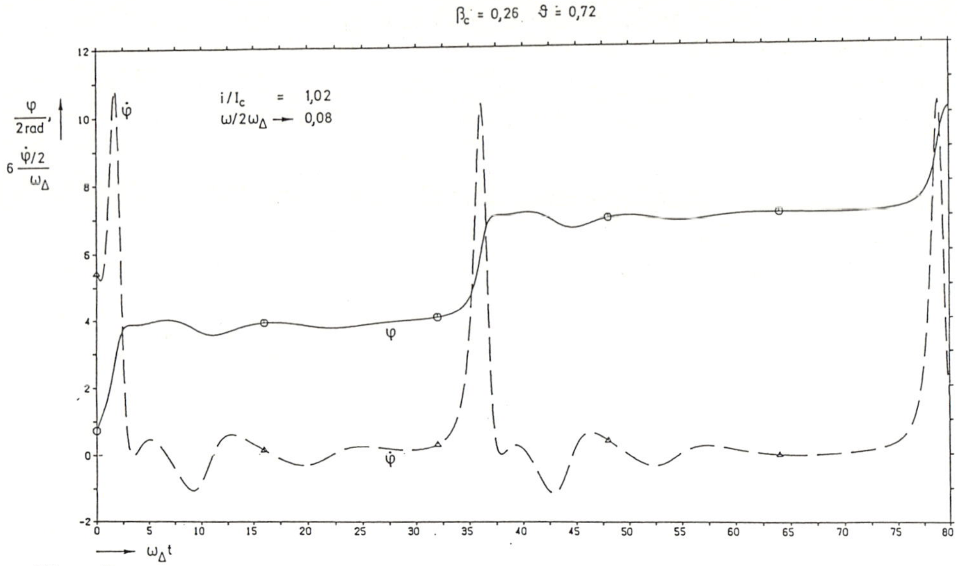

Figure 9
Interaction of energy storage by internal degress of freedom, nonlinearity and dissipation controlling the transition in the running state at low rotatory frequency $\langle\dot{\varphi}/2\rangle = 0.08\ \omega_\Delta$ driven by current $i = 1.02\ I_c$ ($T/T_c = 0.72$, $\beta_c = 0.26$, circuit model Fig. 7).

integro-differential equation, Eq. (4.1), on a system of ordinary differential equations. This makes the computation of transient processes and the calculation of periodical solutions in the time domain much easier. As an example,we demonstrate the transient of the circuit-modelled system [11] from starting conditions $\dot{\varphi}(0) / 2\omega_\Delta = 0.9$ and $\sin\varphi(0) = 0.96$ to the running state at $i = 1.02\ I_c$ and $\langle\dot{\varphi}/2\rangle = 0.08\ \omega_\Delta$ ($\beta_c = 0.26$ and $T/T_c = 0.72$), Fig. 9. Here the dissipation is governed by the low-pass filter behaviour of $G_+(\omega)$ and the internal degrees of freedom effect a superimposition of the rotatory motion by damped oscillations.

References

[1] N.R. Werthamer, Phys. Rev. **147** (1966) p. 255.

[2] H. Lübbig, in: Superconducting Quantum Interference Devices and their Applications. W. de Gruyter & Co., Berlin - New York 1977.

[3] G. Brunk, J. Non-Equilib. Thermodyn., Vol. **5** (1980) p. 339.

[4] Chr. Zurbrügg, G. Brunk, Bericht über das DFG-Forschungsvorhaben Br 693/3 - 1, TU Berlin 1988.

[5] S.N. Erné, H. Lübbig, J. Appl. Phys. **51** (1980 a) p. 4927.

[6] G. Brunk, in: Recent development in nonequilibrium thermodynamics, 446. Lecture Notes in Physics 199, Springer, Berlin - Heidelberg - New York - Tokyo 1984.

[7] J. Meixner, Arch. Rat. Mech. Anal., Vol. **17** (1964) p. 278.

[8] J. Böcker, Reihe 8: Meß-, Steuerungs- und Regelungstechnik, Nr. 159, VDI Verlag, Düsseldorf 1988.
[9] W. Schlup, Tables of functions entering the current-phase relation of a Josephson junction in the weak coupling limit. IBM Thomas J. Watson Research Center, 1972.
[10] R.D. Grigorieff, Chr. Zurbrügg, Bericht zu einem DFG-Forschungsvorhaben, TU Berlin 1985.
[11] G. Brunk, in: Superconducting Quantum Interferene Devices and their Applications, SQUID '85, p. 77. W. de Gruyter & Co., Berlin - New York 1985.
[12] H. Kratz, W. Jutzi, IEEE, Vol. **MAG-23** (1987) p. 731.
[13] D.G. Jablonski, J. Appl. Phys. **53** (1985) p. 7458.
[14] Y.K. Semenov, A.A. Odintsov, A.B. Zorin, in: Superconducting Quantum Interference Devices and their Applicaions, SQUID '85. p. 71. W. de Gruyter & Co., Berlin - New York 1985.
[15] G. Brunk, Chr. Zurbrügg, H. Lübbig, IEEE Trans. Magn. **MAG-25** (1989) pp. 1420-1423

Electromagnetic Properties of Superconductors

Exact Solution of the Mattis-Bardeen Equations for Bulk Material and Thin Films

R. Pöpel
Physikalisch-Technische Bundesanstalt, D-3300 Braunschweig

Commit thy work unto the Lord, and thy thoughts shall be established. (Proverbs 16,3)

1. Introduction

The main objective of this work is to show, that the theory for the anomalous skin effect in normal and superconducting metals by Mattis and Bardeen is capable of describing the electromagnetic behavior of superconductors. It is necessary to do justice to it in this way, because in many applications of this theory the equations are solved only approximately, leading to apparent discrepancies between the theory and measurement results.

In the following many statements from the literature are summarized and sometimes extended or generalized, in order to achieve a complete solution. Thus the present state in this area is represented providing a foundation for future work.

Section 2.1 describes the historical development in the field of electromagnetic properties of superconductors before the Mattis-Bardeen theory appeared and introduces most of the parameters used in this field.

The complete solution of the Mattis-Bardeen equations will be derived mathematically for bulk superconductors in Sect. 2.2 and simplified to the extreme anomalous limit in Sect. 2.3 .

With the help of the surface impedance as a fundamental quantity, in Sect. 2.4 we then are able to make some important applications. In Sect. 3.1 computed results will be compared to other published results. Section 3.2 presents the application to measurements on resonant cavities in the microwave region and Sect. 3.3 the application to measurements on bulk superconductors in the far infrared region.

The exact solution of the Mattis-Bardeen equations for thin films is treated in Sect. 4.1. It provides a knowledge of the spatial variation of the vector potential, from which the complex conductivity is calculated in Sect. 4.2.

A comparison to bulk conductors is carried out in Sect. 5.1, and a comparison to measured transmission spectra in the far infrared region in Sect. 5.2 .

2. Bulk Superconductors

2.1 Theories of the Normal and Anomalous Skin Effect

Due to the skin effect, an electromagnetic field penetrates into the normal, plane and bulk conductor in Fig.1 with a finite skin penetration depth δ, inducing a current density J_n. At room temperature and not too high frequencies we get a local relationship between J_n and the electrical field E. The current density at a given point depends only on the field in this point:

$$J_n = \frac{\sigma_{dc}}{1 + j\omega\tau} E \tag{2.1}$$

σ_{dc} is the DC-conductivity and τ is the relaxation time of the electrons, connected by $v_o = l/\tau$ with the Fermi velocity v_o and the mean free path l.

Equation (2.1) is valid as long as the spatial field distribution varies little over the region of the mean free path ($\delta > l$). δ decreases by going to higher frequencies and l increases with decreasing temperatures. Hence at high enough frequencies or low enough temperatures a nonlocal relationship between J_n and E arises. According to CHAMBERS [1] it is given by

$$J_n(r) = \frac{3\sigma_{dc}}{4\pi l} \int_V \frac{R\, R\, E(r')\, e^{-R/l}}{R^4} \, dV' , \tag{2.2}$$

where $R = r - r'$ is the vector from the point for which the current density is to be calculated to the volume element dV'. (2.2) is the fundamental equation for the anomalous skin effect in normal conductors and a nice illustration by a measurement by CHAMBERS [2] can be seen in Fig. 2. Here R is the surface resistance and T is the absolute temperature, the dashed line corresponds to the classical skin effect (local relationship) and the circles are measurement points. They deviate from the classical calculation where the mean free path l becomes of the order of the classical skin penetration depth δ_c.

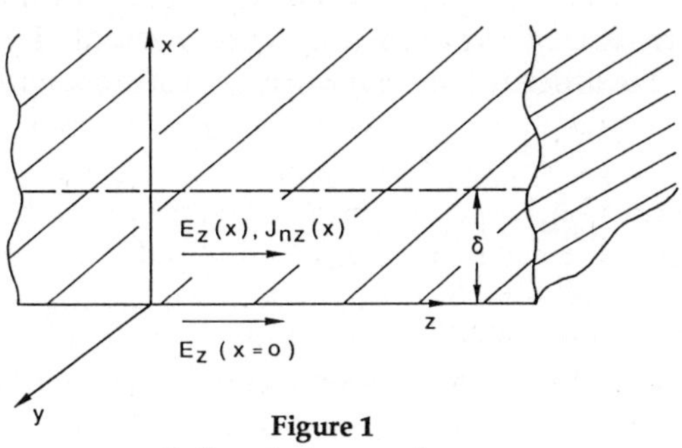

Figure 1
Bulk and plane conductor.

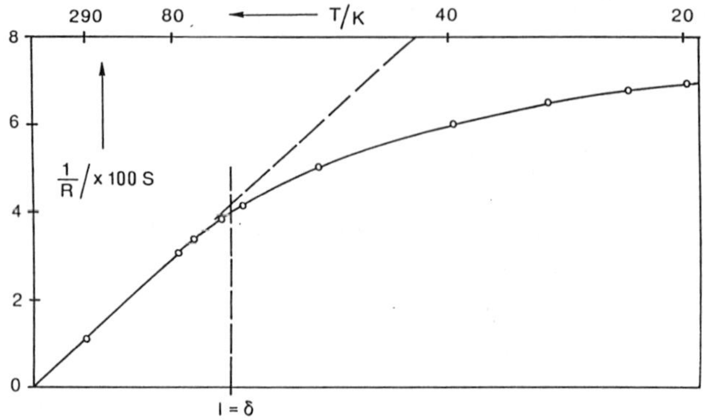

Figure 2
Temperature dependence of the surface resistance R of Cu at 1.2 GHz (CHAMBERS [2]).

Due to the Meißner-Ochsenfeld effect, an electromagnetic field penetrates into a superconductor with the finite superconducting penetration depth λ inducing a current density J. Assuming the Two-Fluid model with $J = J_s + J_n$, in the local limit the superconducting current density J_s can be described by the two London equations

$$J_s = \frac{1}{j\omega\mu_o\lambda_L^2} E \qquad \text{and} \tag{2.3}$$

$$\text{rot}\, J_s = -\frac{1}{\lambda_L^2} H\,. \tag{2.4}$$

where μ_o is the magnetic field constant, H the magnetic field and λ_L the London penetration depth at a given temperature T. With a special choice of the vector potential A [3] Eq. (2.4) can be rewritten as

$$J_s = -\frac{1}{\lambda_L^2} A\,. \tag{2.5}$$

As shown in Fig. 3, PIPPARD [4] observed an increase in the superconducting penetration depth with increasing impurity content, i.e. with decreasing mean free path. To describe his measurement he expanded Eq. (2.5) by a factor ξ/ξ_o

$$J_s = -\frac{\xi}{\xi_o}\frac{1}{\lambda_L^2} A \quad \text{with} \quad \frac{1}{\xi} = \frac{1}{\xi_o} + \frac{1}{\alpha_p l}\,, \tag{2.6}$$

where ξ_o is the coherence length of the pure superconductor, ξ is the coherence length of the impure superconductor and α_p is an empirical constant. In analogy to Eq. (2.1) and Eq. (2.2) he gave a nonlocal relationship between the superconducting current and the vector potential:

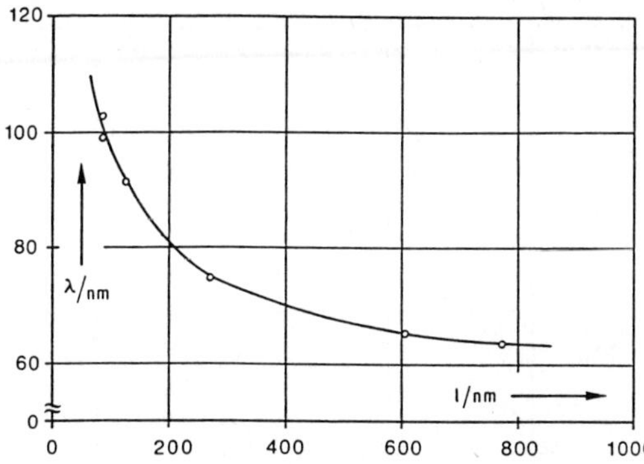

Figure 3

PIPPARD's measurement on diluted tin specimens [4].

$$J_s(r) = -\frac{3}{4\pi\xi_o\lambda_L^2}\int_V \frac{R\,R\,A(r')\,e^{-R/\xi}}{R^4}\,dV', \tag{2.7}$$

Replacing l by ξ and δ by λ in the condition for the validity of Eq. (2.1), the condition for the validity of the London theory becomes $\lambda > \xi$.

With Eqs. (2.1), (2.2), (2.6) and (2.7) a complete theory describing the normal and anomalous skin effect in normal conductors and superconductors is available.

In the well known BCS-theory, BARDEEN, COOPER and SCHRIEFFER [5] introduce the energy gap 2Δ at a given temperature as a new material parameter. As shown in Fig. 4, 2Δ goes to $2\Delta_o$ at $T = 0$ K and to zero at the critical temperature T_c. This temperature dependence was calculated by MÜHLSCHLEGEL [6]. It can also be obtained to within 2 % from the simple equation

$$\frac{2\Delta}{2\Delta_o} = \left[\cos\left(\frac{\pi}{2}t^2\right)\right]^{1/2} \tag{2.8}$$

with $t = T/T_c$.

To take account of the BCS results, the Pippard kernel $\exp(-R/\xi_o)$ is replaced with a BCS-kernel $J(R,T)$, which may be written as

$$J(R,T) = \frac{\lambda_L^2}{\lambda_{L0}^2}\frac{\Delta}{\Delta_o}\left\{\tanh\left(\frac{\Delta}{2k_BT}\right) - \frac{2}{\pi}\Delta\int_\Delta^\infty [1-2f(E)]\frac{1}{E^2-\Delta^2}\sin\left[\frac{2R}{\hbar v_o}\left(E^2-\Delta^2\right)^{1/2}\right]dE\right\}, \tag{2.9}$$

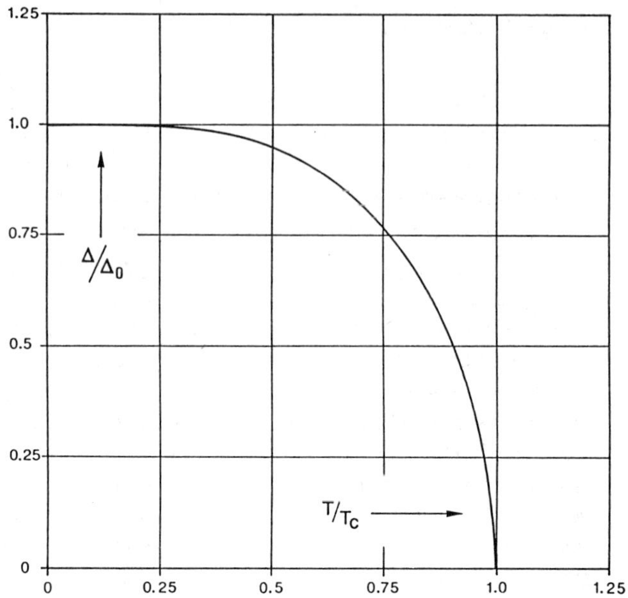

Figure 4
BCS-Temperature dependence of the energy gap 2Δ (MÜHLSCHLEGEL [4]).

where λ_{Lo} is the London penetration depth at $T = 0$ K, k_B is the Boltzmann constant, $\hbar$ is the Plank constant divided by 2π, E is the energy and $f(E) = 1/(1 + \exp(E/k_BT))$ is the Fermi function. In addition, α_p is set to be 1.

Since $J(R,T)$ is defined such that it has the same integral as the Pippard kernel, the simple relation

$$\xi_o = \frac{\hbar v_o}{\pi \Delta_o} \tag{2.10}$$

results. Measured values of $2\Delta_o$ listed in Table 1 differ somewhat from the theoretically predicted energy gap $2\Delta_o = 3.52\ k_BT_c$.

Table 1
Material parameters

	$2\Delta_0/k_BT_c$	T_c/K	λ_{L0}/nm	$v_0/10^6$ m/s	ξ_0/nm	l/nm	Reference
Sn	3.53	3.75	35.5	0.65	239	350	[11]
Pb	4.10	7.19	30.8	0.6	99	710	[11]
Pb	4.10	7.22	28.0	0.68	111	1000	[16]
Nb	3.80	9.20	33.3	0.28	39	20	[27]
Al	3.20	1.178	15.4	1.34	1729	10000	[33]

As an expansion of the BCS-theory the theory of MATTIS and BARDEEN [7] includes a frequency dependence. It provides a relationship between the over all current density and the vector potential:

$$J(r) = \frac{3}{4\pi^2 v_0 \hbar \lambda_{L0}^2} \int_V \frac{R\, R\, A(r')\, I(\omega,R,T)\, e^{-R/l}}{R^4}\, dV' , \tag{2.11}$$

with

$$\begin{aligned} I(\omega,R,T) = &- j\pi \int_{\Delta-\hbar\omega}^{\Delta} [1 - 2f(E+\hbar\omega)]\left[g(E)\cos\alpha\Delta_2 - j\sin\alpha\Delta_2\right] e^{j\alpha\Delta_1}\, dE \\ &- j\pi \int_{\Delta}^{\infty} [1 - 2f(E+\hbar\omega)]\left[g(E)\cos\alpha\Delta_2 - j\sin\alpha\Delta_2\right] e^{j\alpha\Delta_1}\, dE \\ &+ j\pi \int_{\Delta}^{\infty} [1 - 2f(E)]\left[g(E)\cos\alpha\Delta_1 + j\sin\alpha\Delta_1\right] e^{-j\alpha\Delta_2}\, dE , \end{aligned}$$

where $\Delta_1 = (E^2-\Delta^2)^{1/2}$, $\Delta_2 = [(E+\hbar\omega)^2-\Delta^2]^{1/2}$, $g(E) = (E^2+\Delta^2+\hbar\omega E)/(\Delta_1\Delta_2)$, and $\alpha = R/(\hbar v_0)$.

It can be shown [3] that at $\omega = 0$ the relation between $I(0,R,T)$ and $J(R,T)$ is

$$I(0,R,T) = -\frac{\lambda_{L0}^2}{\lambda_L^2} \frac{\hbar v_0 \pi}{\xi_0} J(R,T) . \tag{2.12}$$

2.2 Solution of the Mattis-Bardeen Kernel $K(q)$

Considering a plane, bulk superconductor as in Fig. 1 with an incident plane wave, and introducing with

$$J_z(x) = \int_{-\infty}^{+\infty} J_z(q)\, e^{jqx}\, dq$$

$$A_z(x) = \int_{-\infty}^{+\infty} A_z(q)\, e^{jqx}\, dq \tag{2.13}$$

a onedimensional Fourier transformation for the current density and the vector potential, the Fourier component $A_z(q)$ can be taken out of the integral in Eq. (2.11), leading to

$$J_z(q) = -K(q)\, A_z(q) , \tag{2.14}$$

which is a local relationship in Fourier space.

The above procedure is common to the other nonlocal theories as well [3]. For the Mattis-Bardeen kernel $K(q)$ one finds

$$K(q) = -\frac{3}{4\pi v_o \hbar \lambda_{L0}^2} \int_o^{\infty} \int_{-1}^{+1} (1-u^2)\, I(\omega,R,T)\, e^{jqRu}\, e^{-R/l}\, du\, dR\,, \tag{2.15}$$

an equation mentioned by TINKHAM [8].

The integration with respect to u can easily be done. With

$$\int_{-1}^{+1} (1-u^2)\, e^{jqRu}\, du = \frac{4}{(qR)^2}\left[\frac{\sin(qR)}{qR} - \cos(qR)\right] \tag{2.16}$$

and the substitution $qR = x$ the kernel $K(q)$ becomes

$$K(q) = -\frac{3}{\pi \hbar v_o \lambda_{L0}^2 q} \int_o^{\infty} \left[\frac{\sin x}{x^3} - \frac{\cos x}{x^2}\right] I(\omega,x/q,T)\, e^{-x/ql}\, dx\,. \tag{2.17}$$

Considering $I(\omega,x/q,T)$ in Eq. (2.11) we have to distinguish the two cases $\hbar\omega < 2\Delta$ and $\hbar\omega > 2\Delta$. At photon energies below the energy gap $g(E)$ goes to infinity at $E = \Delta - \hbar\omega$ and at $E = \Delta$. In the integration intervall from $\Delta - \hbar\omega$ to Δ, Δ_1 becomes imaginary and can then be replaced by $\pm j|\Delta_1|$. At photon energies above the energy gap $g(E)$ additionally goes to infinity at $E = -\Delta$. Therefore the first integral of $I(\omega,x/q,T)$ in Eq. (2.11) has to be split into two parts and the radicand of Δ_1 is negative in the integration region from $-\Delta$ to Δ.

Noting this and splitting $I(\omega,x/q,T)$ into real and imaginary parts we get

$$\mathrm{Re}\left\{I(\omega,x/q,T)\right\} = \tag{2.18}$$

$$-\frac{\pi}{2}\int_{\Delta-\hbar\omega}^{-\Delta} [1-2f(E+\hbar\omega)]\left\{[g(E)+1]\sin(a^- x) - [g(E)-1]\sin(a^+ x)\right\} dE$$

$$-\pi \int_{\Delta-\hbar\omega,\,-\Delta}^{\Delta} [1-2f(E+\hbar\omega)]\left\{\frac{E^2+\Delta^2+\hbar\omega E}{(\Delta^2-E^2)^{1/2}[E+\hbar\omega]^{1/2}}\cos(a_2 x) + \sin(a_2 x)\right\} e^{-a_1 x}\, dE$$

$$+\pi \int_{\Delta}^{\infty} [1-f(E)-f(E+\hbar\omega)][g(E)-1]\sin(a^+ x)\, dE$$

$$-\pi \int_{\Delta}^{\infty} [f(E)-f(E+\hbar\omega)][g(E)+1]\sin(a^- x)\, dE \qquad \text{and}$$

$$\mathrm{Im}\left\{I\left(\omega,x/q,T\right)\right\} = \tag{2.19}$$

$$-\frac{\pi}{2}\int_{\Delta-\hbar\omega}^{-\Delta}\left[1-2f(E+\hbar\omega)\right]\left\{\left[g(E)+1\right]\sin(a^{-}x)-\left[g(E)-1\right]\sin(a^{+}x)\right\}dE$$

$$-\pi\int_{\Delta}^{\infty}\left[f(E)-f(E+\hbar\omega)\right]\left\{\left[g(E)+1\right]\cos(a^{-}x)+\left[g(E)-1\right]\cos(a^{+}x)\right\}dE$$

with $a^{+} = a_1 + a_2$, $a^{-} = a_2 - a_1$, $a_1 = \Delta_1/(\hbar v_o q)$ and $a_2 = \Delta_2/(\hbar v_o q)$.

Below the gap frequency ($\hbar\omega < 2\Delta$), the first integrals in Eq. (2.18) and in Eq. (2.19) have to be set to zero and the lower integration limit of the second integral in Eq. (2.18) is $\Delta - \hbar\omega$. Above the gap frequency all integrals have to be considered and the lower limit of the second integral in Eq. (2.18) becomes $-\Delta$.

After inserting $I(\omega,x/q,T)$ into the Mattis-Bardeen kernel $K(q)$, the following integrals appear:

$$\int_0^{\infty} e^{-bx}\left[\frac{\sin x}{x^3}-\frac{\cos x}{x^2}\right]\cos(ax)\,dx = R(a,b) \quad \text{and} \tag{2.20}$$

$$\int_0^{\infty} e^{-bx}\left[\frac{\sin x}{x^3}-\frac{\cos x}{x^2}\right]\sin(ax)\,dx = S(a,b) \tag{2.21}$$

with $b = \frac{1}{ql}$.

With the help of partial integration they can be transformed into tabulated integrals ([3], GRÖBNER and HOFREITER [9]). The solutions are

$$R(a,b) = -\frac{b}{2}+\frac{ab}{4}\ln\left[\frac{b^2+(1+a)^2}{b^2+(1-a)^2}\right] + \frac{1}{4}\left(1+b^2-a^2\right)\left\{\arctan\left[\frac{2b}{b^2+a^2-1}\right]+n_x\pi\right\}, \tag{2.22}$$

$$S(a,b) = \frac{a}{2}-\frac{ab}{2}\left\{\arctan\left[\frac{2b}{b^2+a^2-1}\right]+n_x\pi\right\} + \frac{1}{8}\left(1+b^2-a^2\right)\ln\left[\frac{b^2+(1+a)^2}{b^2+(1-a)^2}\right], \tag{2.23}$$

$$n_x = 0 \quad \text{for} \quad b^2+a^2-1 \geq 0$$

$$n_x = 1 \quad \text{for} \quad b^2+a^2-1 < 0 \;.$$

With $b = 0$ (infinitely large mean free path l), Eq. (2.22) and Eq. (2.23) are equivalent to the result given by MILLER [10].

Finally, the solution of the kernel $K(q)$ becomes

$$\mathrm{Re}\left\{K(q)\right\} = \frac{3}{\hbar v_o \lambda_{L0}^2 q} \tag{2.24}$$

$$\left\{\int_{\Delta-\hbar\omega,\,-\Delta}^{\Delta} [1-2f(E+\hbar\omega)]\left\{\frac{E^2+\Delta^2+\hbar\omega E}{\left(\Delta^2-E^2\right)^{1/2}\left[(E+\hbar\omega)^2-\Delta^2\right]^{1/2}} R(a_2,a_1+b) + S(a_2,a_1+b)\right\} dE\right.$$

$$+\frac{1}{2}\int_{\Delta-\hbar\omega}^{-\Delta} [1-2f(E+\hbar\omega)]\left\{[g(E)+1]\,S(a^-,b)-[g(E)-1]\,S(a^+,b)\right\} dE$$

$$-\int_{\Delta}^{\infty} [1-f(E)-f(E+\hbar\omega)]\,[g(E)-1]\cdot S(a^+,b)\,dE$$

$$\left.+\int_{\Delta}^{\infty} [f(E)-f(E+\hbar\omega)]\,[g(E)+1]\,S(a^-,b)\,dE\right\} \quad \text{and}$$

$$\mathrm{Im}\left\{K(q)\right\} = \frac{3}{\hbar v_o \lambda_{L0}^2 q} \tag{2.25}$$

$$\left\{-\frac{1}{2}\int_{\Delta-\hbar\omega}^{-\Delta} [1-2f(E+\hbar\omega)]\left\{[g(E)+1]\,R(a^-,b)+[g(E)-1]\,R(a^+,b)\right\} dE\right.$$

$$\left.+\int_{\Delta}^{\infty} [f(E)-f(E+\hbar\omega)]\left\{[g(E)+1]\,R(a^-,b)+[g(E)-1]\,R(a^+,b)\right\} dE\right\},$$

which, for $\hbar\omega < 2\Delta$ is identical to the result of TURNEAURE [11].

One can show that in the normal conducting state at $T = T_c$ where $2\Delta = 0$ Eq. (2.24) and Eq. (2.25) lead to the kernel $I(\omega,x/q,T_c)$ as given by MATTIS and BARDEEN [7]:

$$I(\omega,x/q,T_c) = -\,\mathrm{j}\pi\,\hbar\omega\,\exp\left(-\,\mathrm{j}\frac{x\omega}{q v_o}\right), \tag{2.26}$$

corresponding to a kernel $K_n(q)$ in the normal conducting state:

$$K_n(q) = \frac{3}{v_o \lambda_{L0}^2 q}\left[S(a_n,b) + \mathrm{j}R(a_n,b)\right] \quad \text{with} \tag{2.27}$$

$$a_n = \frac{\omega}{v_o q} \text{ and } b = \frac{1}{ql}\,.$$

2.3 Extreme Anomalous Skin Effect

According to the conditions for a local relationship between current and field in Sect. 2.1, for the extreme anomalous skin effect the mean free path has to be much greater than the skin penetration depth ($l >> \delta$), and the Pippard coherence length has to be much greater than the superconducting penetration depth ($\xi >> \lambda$). Mattis and Bardeen, in this case, set $\alpha = R = 0$. Therefore the kernel $I(\omega,0,T)$ follows as

$$\mathrm{Re}\left\{I(\omega,0,T)\right\} = -\pi \int_{\Delta-\hbar\omega,\,-\Delta}^{\Delta} \left[1-2f(E+\hbar\omega)\right] \frac{E^2+\Delta^2+\hbar\omega E}{\left(\Delta^2-E^2\right)^{1/2}\left[(E+\hbar\omega)^2-\Delta^2\right]^{1/2}}\,dE\,,$$

$$\mathrm{Im}\left\{I(\omega,0,T)\right\} = -\pi \int_{\Delta-\hbar\omega}^{\Delta} \left[1-2f(E+\hbar\omega)\right] g(E)\,dE \qquad (2.28)$$

$$-2\pi \int_{\Delta}^{\infty} \left[f(E)-f(E+\hbar\omega)\right] g(E)\,dE$$

in the superconducting state and as

$$I(\omega,0,T_c) = -j\pi\hbar\omega \qquad (2.29)$$

in the normal conducting state.

Calculating $J_\infty/J_{n\infty}$ in the extreme anomalous limit using the Mattis-Bardeen integral in Eq. (2.11), the corresponding kernels can be taken out of the integral, and according to GLOVER and TINKHAM [12] a complex conductivity is introduced:

$$\frac{J_\infty}{J_{n\infty}} = \frac{I(\omega,0,T)}{I(\omega,0,T_c)} = \frac{\sigma_1 - j\sigma_2}{\sigma_n}\,, \qquad (2.30)$$

where σ_n is the conductivity in the normal conducting state at a given frequency. The kernels $K_\infty(q)$ and $K_{n\infty}(q)$ than can be written as

$$K_\infty(q) = -\frac{3}{4}\frac{I(\omega,0,T)}{\hbar v_o \lambda_{L0}^2 q} = \frac{3}{4}\frac{\pi\omega}{v_o\,\lambda_{L0}^2 q}\left(\frac{\sigma_2}{\sigma_n} + j\,\frac{\sigma_1}{\sigma_n}\right) \text{ and} \qquad (2.31)$$

$$K_{n\infty}(q) = -j\,\frac{3}{4}\frac{\pi\omega}{v_o\lambda_{L0}^2 q}\,. \qquad (2.32)$$

2.4 Surface Impedance

Having solved the Mattis-Bardeen kernel $K(q)$ in Sect. 2.2 we are now able to calculate the surface impedance using an equation of REUTER and SONDHEIMER [13] for diffuse scattering of the electrons at the surface of the conductor:

$$Z = j\omega\mu_o\pi \frac{1}{\int_0^\infty \ln\left(1 + \frac{K(q)}{q^2}\right) dq} = R + jX = R + j\omega\mu_o\lambda \ . \tag{2.33}$$

This quantity contains all information at the surface available from the Mattis-Bardeen theory. The surface resistance R is a measure of conductor losses, and the surface reactance X is directely connected to the superconducting penetration depth λ.

3. Applications to Bulk Superconductors

To calculate the surface impedance Z, a numerical double integration has to be carried out, and five material parameters are needed: The energy gap $2\Delta_o$ at $T = 0$ K, the London penetration depth λ_{L0} at $T = 0$ K, the mean free path l, the coherence length ξ_o or the Fermi velocity v_o (connected by Eq. (2.10)), and the critical temperature T_c.

With the help of a FORTRAN computer program the surface impedances can be calculated for any circular frequency ω, for any mean free path l as long as only non-magnetic impurities are involved and for any temperature T.

There are some important restrictions to the validity of the present solution. The superconductors must be isotropic, bulk and plane, and the theory of Mattis and Bardeen is only valid in the weak coupling limit.

3.1 Other Calculations

As mentioned in Sect. 2.2, below the gap frequency the present solution is identical with that of TURNEAURE [11]. Table 2 shows results for tin at 11.2 GHz. While our values for the surface resistance R are greater by about 0.1 % to 0.7 %, the results for the surface reactance X agree within ± 0.1 %.

Similarly good agreement is found in the case of lead, and in the case of niobium (TURNEAURE and WEISSMAN [14]), where our resistances are from 1.4 % to 2.7 %, and the reactances from 0.5 % to 0.9 % greater.

For frequencies below the gap frequency HALBRITTER [15] has obtained a solution, using Greens functions. The lines in Fig. 5a and Fig. 5b show the temperature dependence of the surface resistance of lead at 1 GHz and 100 GHz, while the crosses were calculated by WILSON [16] using Halbritters program.

Table 2
Comparison with calculations of Turneaure [11] for Sn at 11.2 GHz

$t = T/T_c$	R_T/Ω	R/Ω	X_T/Ω	X/Ω
0.96	$3.117\ 10^{-3}$	$3.121\ 10^{-3}$	$1.269\ 10^{-2}$	$1.268\ 10^{-2}$
0.90	$1.192\ 10^{-3}$	$1.195\ 10^{-3}$	$9.787\ 10^{-3}$	$9.707\ 10^{-3}$
0.86	$7.777\ 10^{-4}$	$7.799\ 10^{-4}$	$8.673\ 10^{-3}$	$8.674\ 10^{-3}$
0.80	$4.674\ 10^{-4}$	$4.691\ 10^{-4}$	$7.689\ 10^{-3}$	$7.691\ 10^{-3}$
0.76	$3.491\ 10^{-4}$	$3.504\ 10^{-4}$	$7.241\ 10^{-3}$	$7.243\ 10^{-3}$
0.70	$2.335\ 10^{-4}$	$2.344\ 10^{-4}$	$6.750\ 10^{-3}$	$6.752\ 10^{-3}$
0.60	$1.223\ 10^{-4}$	$1.228\ 10^{-4}$	$6.222\ 10^{-3}$	$6.224\ 10^{-3}$
0.50	$5.994\ 10^{-5}$	$6.032\ 10^{-5}$	$5.907\ 10^{-3}$	$5.913\ 10^{-3}$
0.46	$4.304\ 10^{-5}$	$4.333\ 10^{-5}$	$5.824\ 10^{-3}$	$5.830\ 10^{-3}$
0.40	$2.400\ 10^{-5}$	$2.413\ 10^{-5}$	$5.736\ 10^{-3}$	$5.740\ 10^{-3}$
0.36	$1.484\ 10^{-5}$	$1.492\ 10^{-5}$	$5.700\ 10^{-3}$	$5.700\ 10^{-3}$

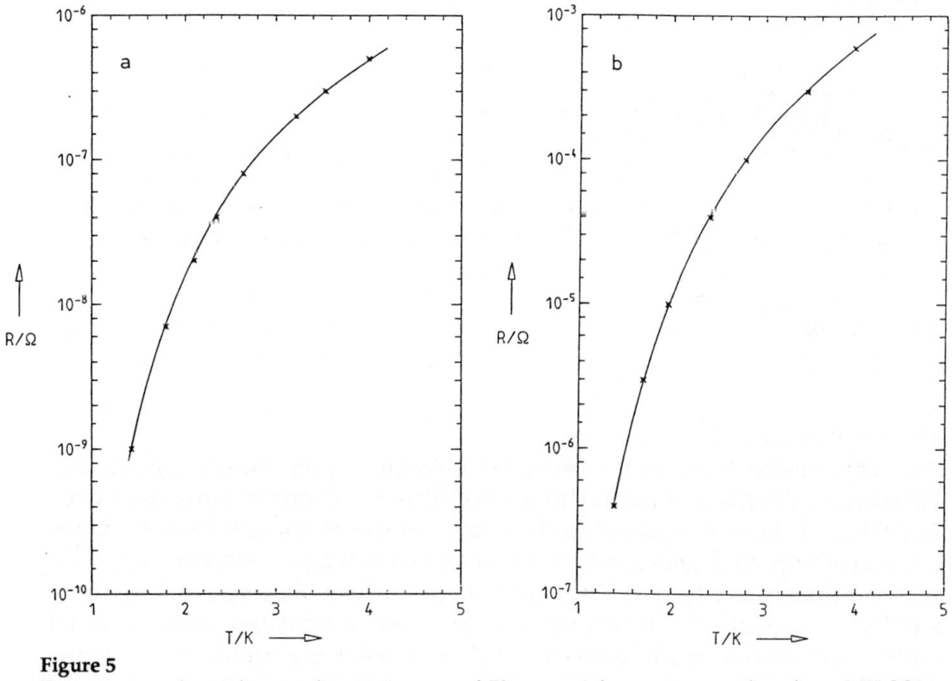

Figure 5
Temperature dependent surface resistance of Pb, material parameters taken from WILSON [16], x calculated by Wilson, a. 1 GHz, b. 100 GHz.

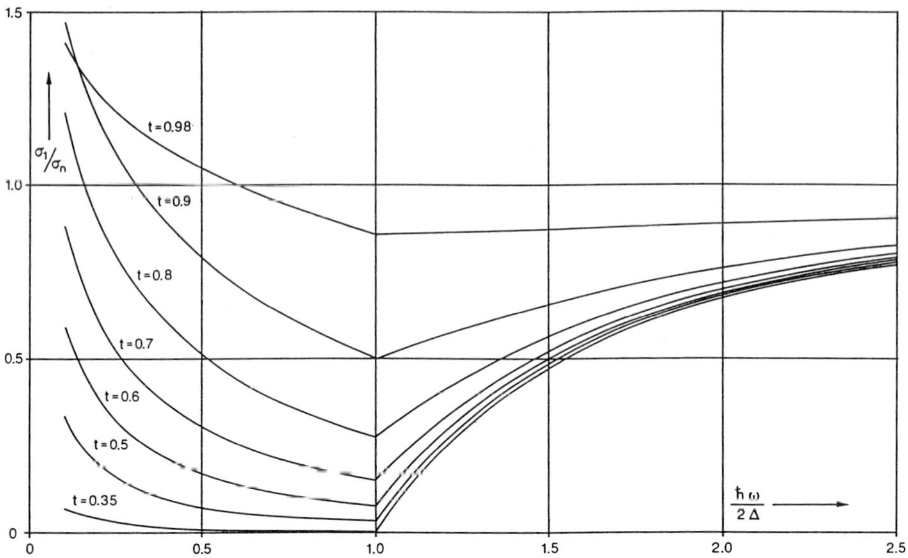

Figure 6
Real part σ_1/σ_n of the complex conductivity in the extreme anomalous limit ($2\,\Delta_0 = 3.52\,k_B T_c$).

MILLER [10] calculated the complex conductivity $(\sigma_1 - j\sigma_2)/\sigma_n$ in the extreme anomalous limit for a wide frequency and temperature range. Our results agree to within ± 1 % with his, except of some points, where deviations up to ± 8 % occur.

The frequency dependence of the complex conductivity at $T = 0$ K is shown in a figure by TINKHAM [17]. The complex conductivity at several different temperatures can be seen in Fig. 6 and Fig. 7.

3.2 Microwave Region

In connection with a suitable solution of Maxwells equations, the surface impedance can be used to describe measurements on resonant cavities or striplines in the microwave region.

The measurement of the unloaded Q-factor Q_0 for a lead cavity at 2.85 GHz by BRUYNSERAEDE et. al. [18] yielded $Q_0 R = 780$, and is indicated by the symbols in Fig. 8. Going to low temperatures, a residual Q-factor corresponding to a temperature independent surface resistance appears, which cannot be described by the Mattis-Bardeen theory. Adding those residual losses to the theoretical dashed line, calculated with material parameters taken from WILSON [16], the resulting solid line fits the measurement very well.

Many measurements of the temperature dependence of the Q-factor on resonant cavities have been done, for example with In and PbIn at 2.85 GHz (BRUYNSERAEDE et. al. [18]), with Nb at 12.6 GHz, 14.4 GHz and 17.4 GHz

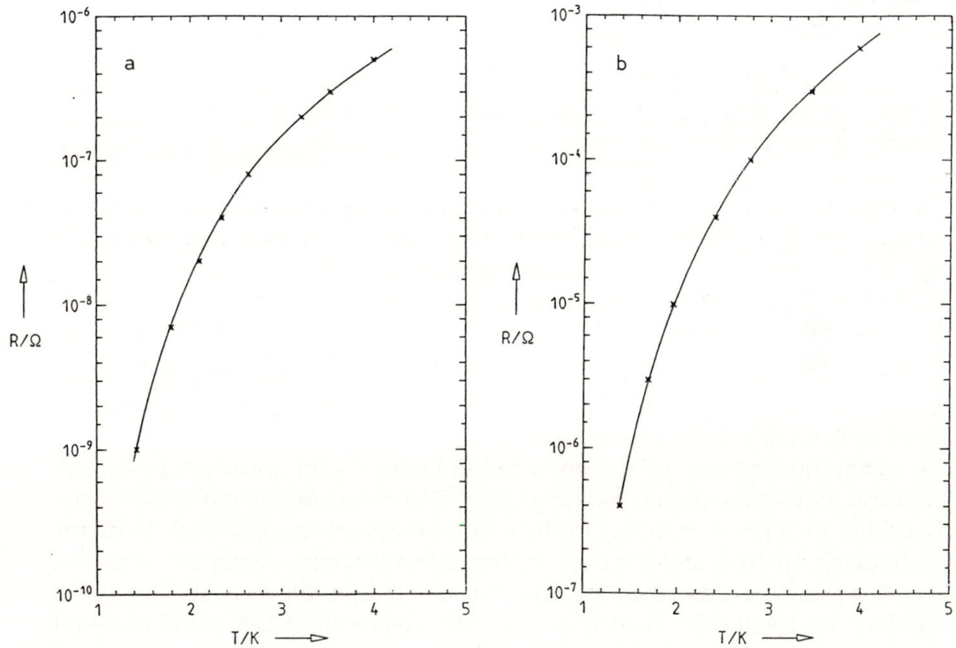

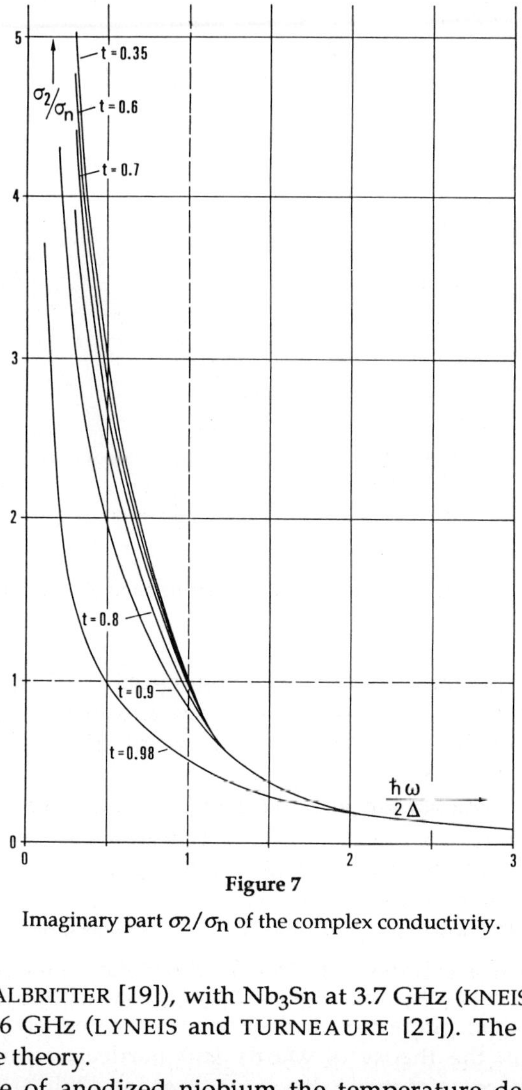

Figure 7

Imaginary part σ_2/σ_n of the complex conductivity.

(PHILIPP and HALBRITTER [19]), with Nb_3Sn at 3.7 GHz (KNEISEL et. al. [20]) or $Nb_{30}Ta_{70}$ at 8.6 GHz (LYNEIS and TURNEAURE [21]). The results are well described by the theory.

In the case of anodized niobium the temperature dependent surface resistance was fitted by assuming that 2 % of the RF penetration region consists of material with a lower energy gap and a lower critical temperature [19].

The frequency dependence of the surface resistance of tin at $t = T/T_c =$ 0.82 in Fig. 9 was calculated by taking the material parameters of Turneaure, listed in Table 1. It agrees well with the experimental data of STURGE [22] obtained on single crystals.

In the case of lead, represented in Fig. 10, the calculated curve (assuming the parameters of Turneaure in Table 1) fits the measured values of SZECSI [23] and FLECHER [24] well, but deviates at higher frequencies from

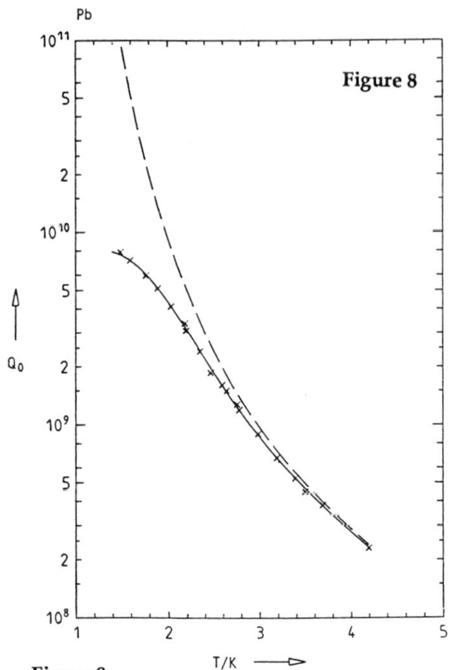

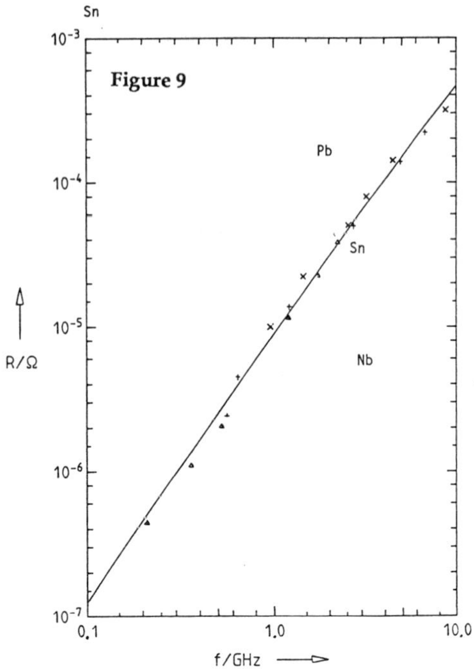

Figure 8
Unloaded Q-factor versus temperature, Pb, $f = 2.85$ GHz, x measured values, - - - calculated with material parameters from WILSON [16], — taking into account the residual losses at 1.4 K ($R = 9.4 \times 10^{-8}\ \Omega$).

Figure 9
Frequency dependence of the surface resistance of Sn at $t = 0.82$, symbols indicate measurement points of STURGE [22] from single crystals with different crystal orientation.

the data of PIERCE [25] and TURNEAURE [11]. Here the calculations of BLASCHKE and BLOCKSDORF [26] give a better agreement, but on the other side they deviate at low frequencies. Blaschke and Blocksdorf took into account one strong coupling effect, namely the finite lifetime of the electrons due to their interaction with thermally excited phonons.

For niobium, the theory of Mattis and Bardeen cannot describe the frequency behavior measured by BAUER et. al. [27], TURNEAURE and WEISSMAN [14] and PHILIPP and HALBRITTER [19] as shown in Fig. 11. The material parameters published by Turneaure and Weissman gave to large results, so we took the values of Bauer et. al. given in Table 1. Also at low frequencies, other experimental data with lower values than the values of PIOSCZYK [28] exist. The theory of BLASCHKE and BLOCKSDORF [26] fits the data a little better at low frequencies and a little worse at high frequencies.

With increasing impurity content the electromagnetic field can penetrate deeper into the conductor, as already mentioned in Sect. 2.1. This behavior is demonstrated by the solid line in Fig. 12 for lead at 4.2 K and 1 GHz. Going to 500 GHz, still below the gap frequency, the superconducting

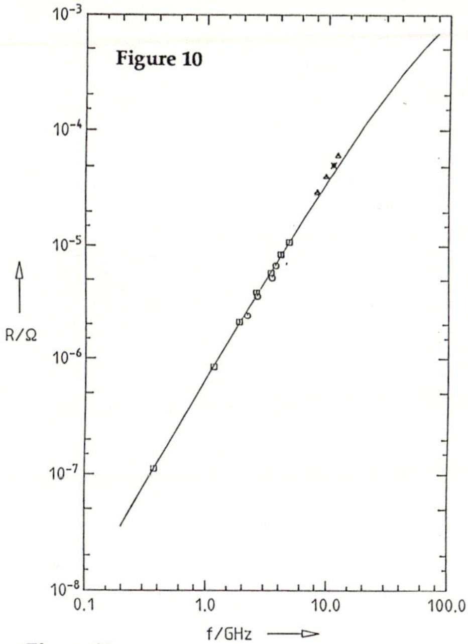

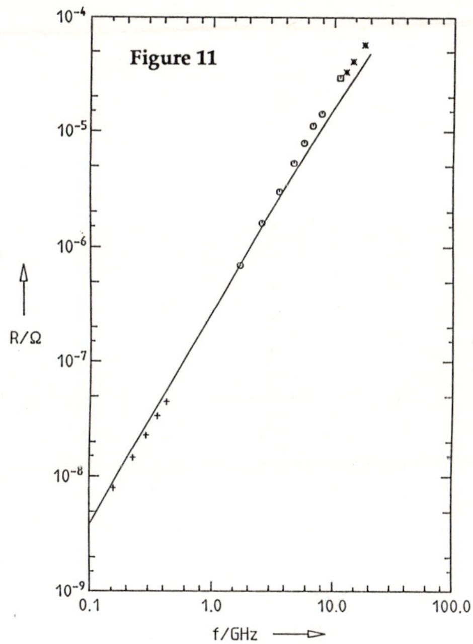

Figure 10
Frequency dependence of the surface resistance of Pb at T = 4.2 K, □ Szecsi [23], o Flecher [24], Δ Pierce [25], * Turneaure [11].

Figure 11
Frequency dependence of the surface resistance of Nb at T = 4.2 K, + PIOSCZYK [28], o BAUER et. al. [27], □ TURNEAURE and WEISSMAN [14], * PHILIP and HALBRITTER [19].

penetration depth does not change very much. Additionally drawn are measured points from HASSE and LACHMANN [29], who used a lead cavity with up to 5 at % Bi at 9.58 GHz and a result of HENKELS and KIRCHER [30] obtained using PbBi thin-film striplines below 30 MHz.

Annother effect, which appears at sufficiently low frequencies is, that the mean free path dependence of the surface resistance can have a minimum as has been noted by HALBRITTER [31]. In Fig. 13 this minimum appears at 1 GHz but not at 500 GHz, where R increases to 8.0 x 10^{-7} Ω for l going to infinity.

BIONDI and GARFUNKEL [32] measured the ratio R/R_n (R_n-surface resistance in the normal conducting state at $T = T_c$) of aluminium over a wide frequency and temperature range, also above the gap frequency (symbols in Fig. 14). From those values they calculated the frequency dependence of the superconducting penetration depth at different temperatures using Kronig-Kramers integral transforms [33] (symbols in Fig. 15).

The solid lines in Fig. 14 and Fig. 15 were obtained by taking the material parameters given by Biondi and Garfunkel listed in Table 1 apart

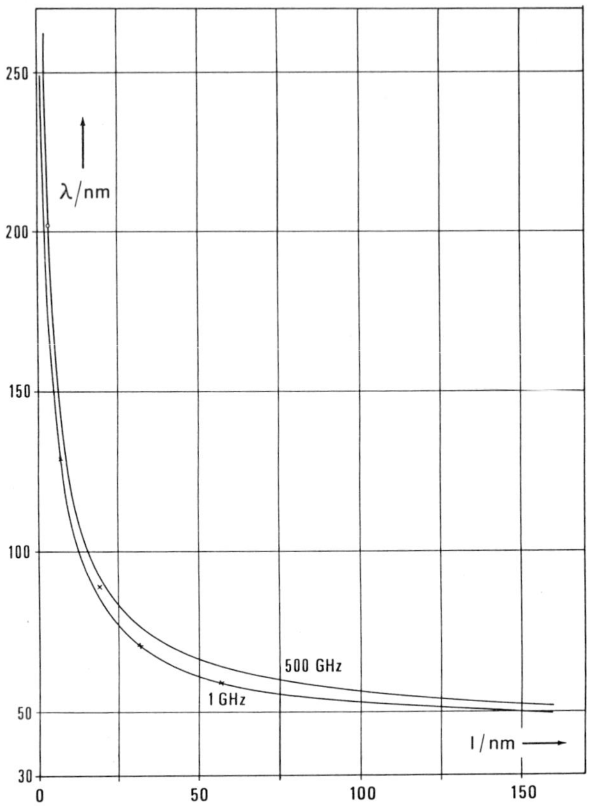

Figure 12
Superconducting penetration depth versus mean free path, Pb, T = 4.2 K, x HASSE and LACHMANN [29], o HENKELS and KIRCHER [30].

from the energy gap. This parameter was changed to 3.40 $k_B T_c$ to fit the values R/R_n at low temperatures and high frequencies, and has only a small influence on the rest of the curves.

3.3 Far Infrared Region

In many cases, it has been tried to fit measurements in the far infrared region on bulk material and thin films with the Mattis-Bardeen theory in the extreme anomalous limit and considerable discrepancies appeared.

The question, to what extent this limit can be used for lead, is answered in Fig. 16. r corresponds to R/R_n and $r_\infty = R_\infty / R_{n\infty}$ is the appropriate quotient in the extreme anomalous limit. If the curves reach the value one, the extreme anomalous limit can be applied without error, but this is only the case at very high frequencies.

Even for aluminium, which fulfills the conditions for the extreme anomalous limit quite well (Sect. 2.3, BIONDI and GARFUNKEL [33]), a satis-

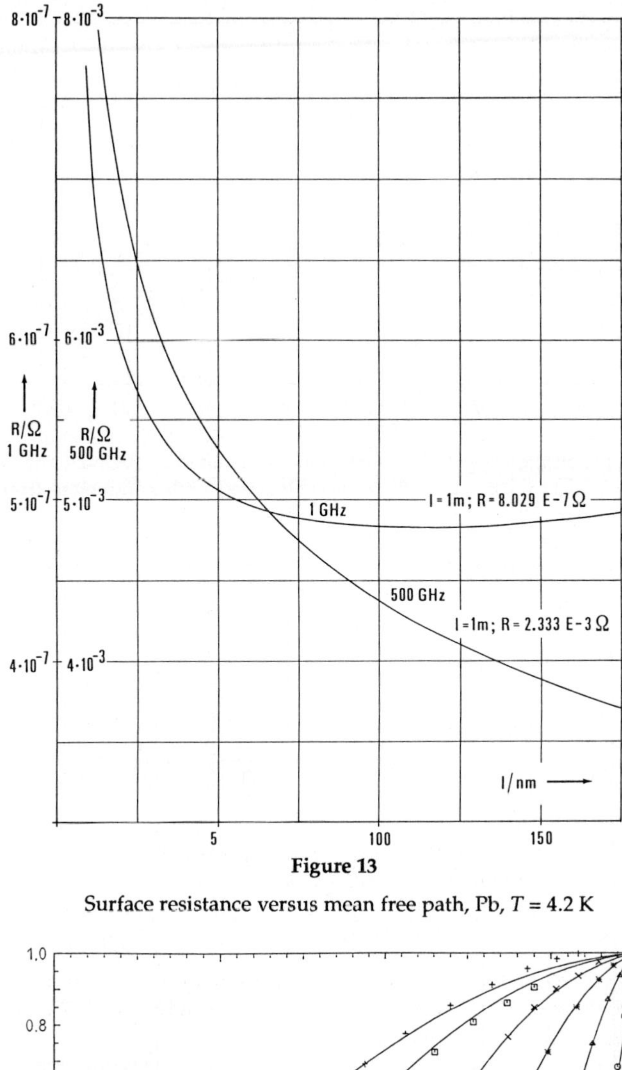

Figure 13

Surface resistance versus mean free path, Pb, $T = 4.2$ K

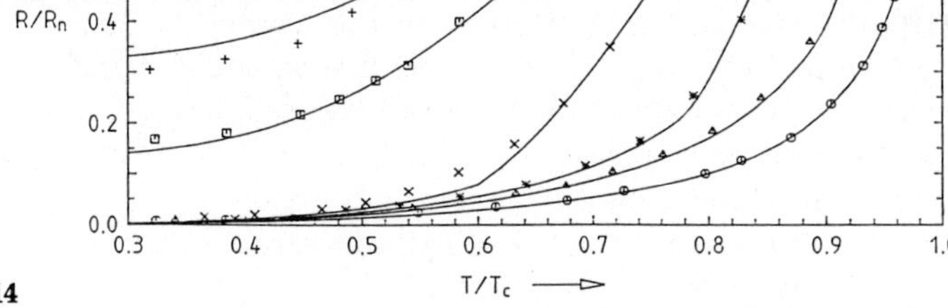

Figure 14

R/R_n versus T/T_c measured by BIONDI and GARFUNKEL [32], o 0.64 k_BT_c, Δ 1.66 k_BT_c, ∗ 2.46 k_BT_c, × 3.08 k_BT_c, □ 3.63 k_BT_c, + 3.91 k_BT_c, curves calculated with parameters from [33] except $2\,\Delta_0 = 3.40\,k_BT_c$.

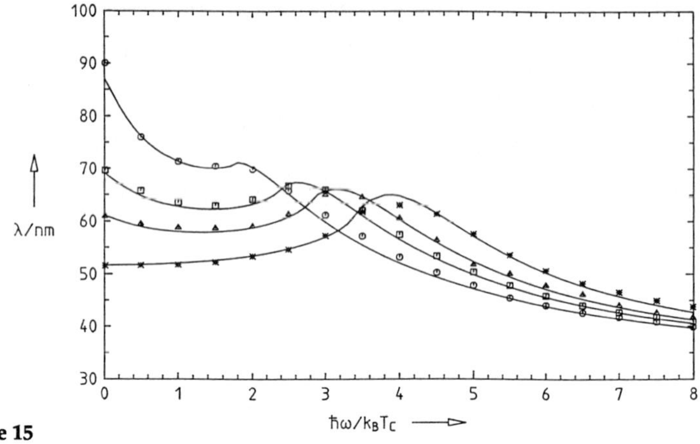

Figure 15
Superconducting penetration depth versus frequency calculated by BIONDI and GARFUNKEL [33], * $t = 0$, Δ $t = 0.7$, □ $t = 0.8$, ○ $t = 0.9$, curves calculated with parameters from [33] except $2\,\Delta_0 = 3.40\,k_B T_c$.

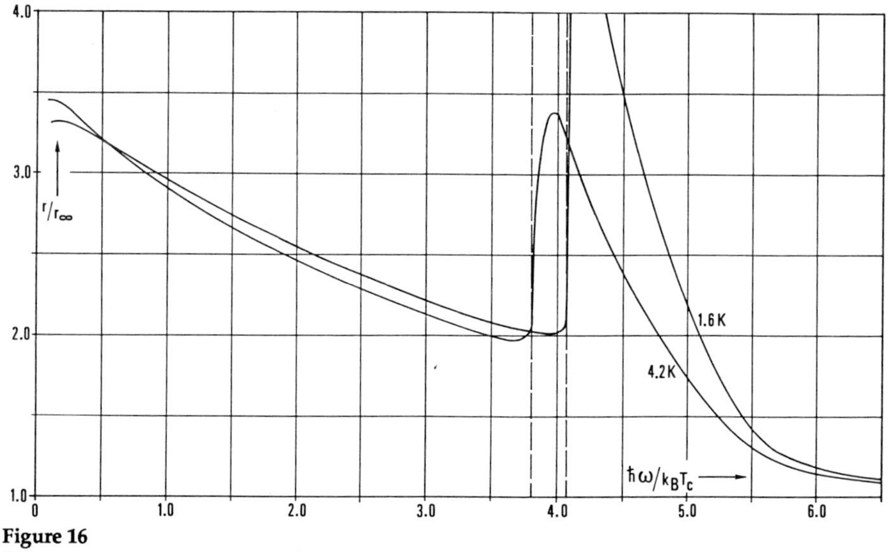

Figure 16
Frequency dependence of r/r_∞, $r = R/R_n$, $r_\infty = R_\infty/R_{n\infty}$, Pb, parameters taken from TURNEAURE [11]

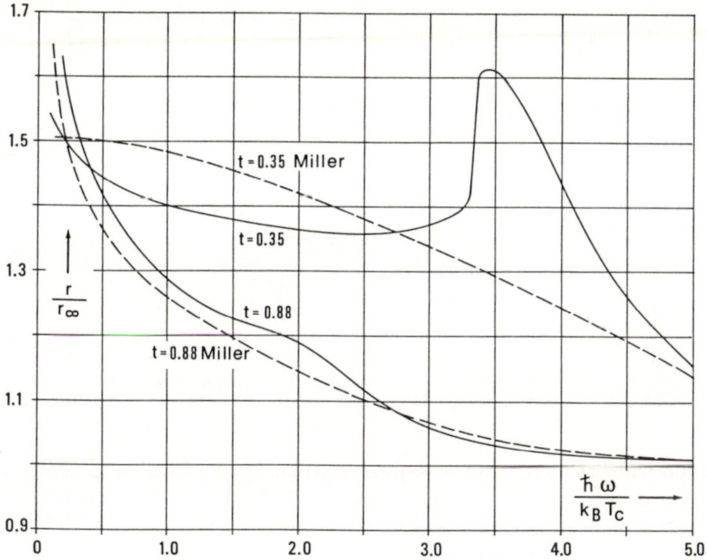

Figure 17
Frequency dependence of r/r_∞ for Al, solid lines calculated with parameters from BIONDI and GARFUNKEL [33], dashed lines are results from MILLER [10].

factory application is possible only at high frequencies and high temperatures below T_c (Fig. 17). Nevertheless the deviations are much smaller than in the case of lead.

LESLIE and GINSBERG [34] measured the far infrared absorption in bulk lead alloys. The points in Fig. 18 show the result on PbTl (1 %). The worst fit is obtained in the extreme anomalous limit (long dashed line) and the local limit (short dashed line) does not give an appreciably better result.

GINSBERG [35] later tried to fit his experiment with the help of LEPLAEs theory [36] (long/short dashed line in Fig. 18). The principle of this evaluation is to calculate Im$\{K(q)\}$ in Eq. (2.25) at $T = 0$ K by neglecting the second integral ($\sigma_{1n}(|\varepsilon'| - |\varepsilon| q)$ in Eq. (4) in [35] corresponds to $R(a^-, b)$), to obtain Re$\{K(q)\}$ by a Kramers-Kronig transform and to calculate the surface impedance using Eq. (2.33). So the only important difference to our calculations seems to be that the second integral in Eq. (2.25) is not considered, while taking $T = 0$ K is quite well justified.

The solid line in Fig. 18 was obtained using Ginsberg's formula for the absorption spectrum

$$P = \frac{R_n - R}{R_n(\omega_g) + R/K} , \tag{3.1}$$

where $R_n(\omega_g)$ is the surface resistance in the normal conducting state at the gap frequency and K is chosen such that the height of the absorption curve at

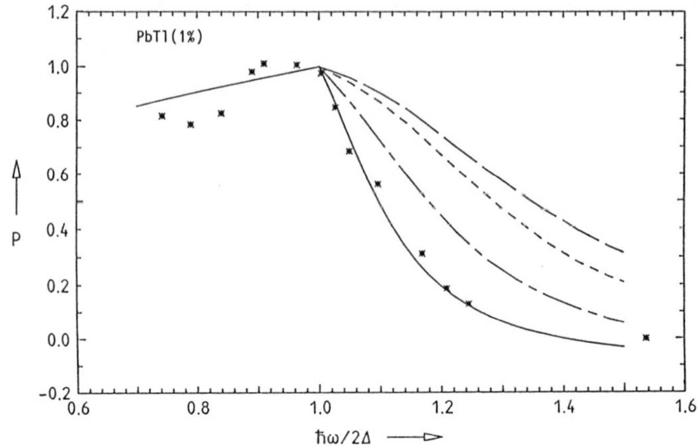

Figure 18
Absorption spectrum of PbTl (1 %), * measured points from LESLIE and GINSBERG [34], — — extreme anomalous limit, - - - - local limit, — - — calculated by GINSBERG [35] using LEPLAEs theory [36], solid line calculated with material parameters from TURNEAURE [11], $2\,\Delta_0 = 4.332\,k_B T_c$, $l = 170$ nm, $K = 4.673$, $T = 1.3$ K.

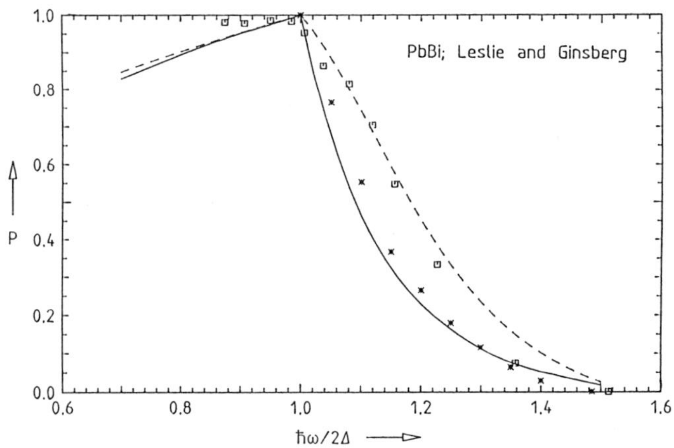

Figure 19
Absorption spectra of two diluted Pb specimens, * PbBi (0.01 %), □ PbBi (6 %), measured by Leslie and Ginsberg, —— $2\,\Delta_0 = 4.362\,k_B T_c$, $l = 9400$ nm, $K = 6.803$, - - - - - $2\,\Delta_0 = 4.40\,k_B T_c$, $l = 16$ nm, $K = 2.915$, other parameters taken from TURNEAURE [11], $T = 1.3$ K.

ω_g is unity. It should be mentioned that for this purpose we used no fitting parameter. Apart from the energy gap, only Turneaures material parameters and the mean free path measured by LESLIE and GINSBERG [34] were used.

In addition, Fig. 19 shows two measurement results by Leslie and Ginsberg on a weakly and a strongly diluted specimen, taking again the measured values of $2\Delta_0$ and l.

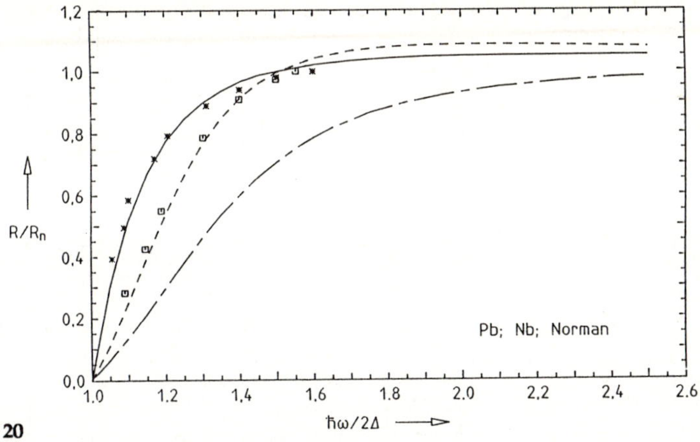

Figure 20

Absorption spectra measured by NORMAN [37], * Pb, □ Nb, —— $2\,\Delta_0 = 4.31\ k_B T_c$, parameters taken from TURNEAURE [11], - - - - - $2\,\Delta_0 = 3.60\ k_B T_c$, l = 9nm, parameters taken from BAUER et. al. [27], — - — extreme anomalous limit.

Also the measured values of NORMAN [37] on lead and niobium in Fig. 20 are well described by the Mattis-Bardeen theory. In the case of niobium we changed l = 20 nm from BAUER et. al. [27] to l = 9 nm to get a better fit.

The two main results in the far infrared region are, that first the extreme anomalous skin effect is not applicable and second the exact solution of the Mattis-Bardeen equations seems to fit well also measurement results on strong coupling superconductors such as lead. Closing this Section, Fig. 21 shows the frequency dependence of the skin penetration depths, formally defined as

$$\lambda = \frac{X}{\omega\,\mu_0} \quad \text{and} \tag{3.2}$$

$$\delta_r = \frac{R}{\omega\,\mu_0} \tag{3.3}$$

for lead at 1.6 K and 4.2 K. The material parameters used were taken from WILSON [16], and are listed in Table 1.

The curves for λ reach maxima at frequencies somewhat higher than the respective gap frequencies and go to the London penetration depths λ_L at very high frequencies. Below the gap frequency, the δ_r values remain small, increase strongly above the gap frequency and finally reach the temperature independent behavior predicted by the theory of the anomalous skin effect in normal conductors by CHAMBERS [1]. I.e. the Mattis-Bardeen theory includes Chambers theory.

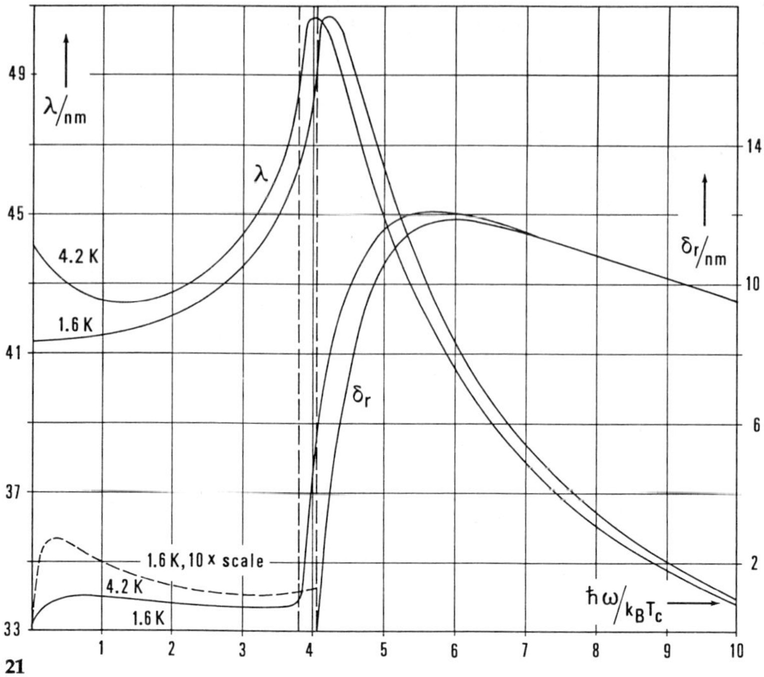

Figure 21
Frequency dependence of the skin penetration depths λ and δ_r, material parameters taken from WILSON [16].

4. Thin Films

4.1 Theoretical Treatment

Our solution of the Mattis-Bardeen equations for thin films is similar to that of HOOK [38], but extended to frequencies higher than the gap frequency and no term is neglected.

Considering a plane conductor as in Fig. 1, but with finite thickness d, the integral (2.11) can be written as

$$J_z(x) = \frac{3}{4\pi \hbar v_o \lambda_{L0}^2} \int_o^d K^*(\eta)\, A_z(x')\, dx' \tag{4.1}$$

with $\eta = |x - x'|$ and $K^*(\eta)$ corresponding to

$$K^*(\eta) = \int_1^\infty \left(\frac{1}{u} - \frac{1}{u^3}\right) I(\omega, \eta u, T)\, e^{-\eta u / l}\, du \ . \tag{4.2}$$

Inserting the kernel $I(\omega, \eta u, T)$ according to Eq. (2.18) and Eq. (2.19) into Eq. (4.2), the then appearing integrals

$$\int_1^\infty \left(\frac{1}{u}-\frac{1}{u^3}\right) e^{-b_1 u} \sin(b_2 u)\, du = CS(b_1, b_2) \quad \text{and}$$

$$\int_1^\infty \left(\frac{1}{u}-\frac{1}{u^3}\right) e^{-b_1 u} \cos(b_2 u)\, du = CC(b_1, b_2) \tag{4.3}$$

can be calculated with the help of repeated partial integrations and the introduction of exponential integrals. This leads to

$$CS(b_1,b_2) = \frac{1}{2} e^{-b_1}\left[(b_1-1)\sin b_2 - b_2 \cos b_2\right] - \left(1-\frac{b_1^2}{2}+\frac{b_2^2}{2}\right) \mathrm{Im}\left\{E_1(b_1+jb_2)\right\} + b_1 b_2 \,\mathrm{Re}\left\{E_1(b_1+jb_2)\right\} \tag{4.4}$$

and

$$CC(b_1,b_2) = \frac{1}{2} e^{-b_1}\left[(b_1-1)\cos b_2 + b_2 \sin b_2\right] + \left(1-\frac{b_1^2}{2}+\frac{b_2^2}{2}\right) \mathrm{Re}\left\{E_1(b_1+jb_2)\right\} + b_1 b_2 \,\mathrm{Im}\left\{E_1(b_1+jb_2)\right\}. \tag{4.5}$$

Therefore the result corresponding to Eq. (4.2) is

$$\begin{aligned}
\mathrm{Re}\left\{K^*(\eta)\right\} = {} & -\frac{\pi}{2}\int_{\Delta-\hbar\omega}^{-\Delta} \left[1-2f(E+\hbar\omega)\right]\left\{\left[g(E)+1\right] CS(b,a^-) - \left[g(E)-1\right] CS(b,a^+)\right\} dE \\
& -\pi \int_{\Delta-\hbar\omega,\,-\Delta}^{\Delta} \left[1-2f(E+\hbar\omega)\right]\left\{\frac{E^2+\Delta^2+\hbar\omega E}{\left(\Delta^2-E^2\right)^{1/2}\left[(E+\hbar\omega)^2-\Delta^2\right]^{1/2}} CC(b+a_1, a_2) + CS(b+a_1, a_2)\right\} dE \\
& +\pi \int_{\Delta}^{\infty} \left[1-f(E)-f(E+\hbar\omega)\right]\left[g(E)-1\right] CS(b,a^+)\, dE \\
& -\pi \int_{\Delta}^{\infty} \left[f(E)-f(E+\hbar\omega)\right]\left[g(E)+1\right] CS(b,a^-)\, dE \qquad \text{and}
\end{aligned} \tag{4.6}$$

$$\begin{aligned}
\mathrm{Im}\left\{K^*(\eta)\right\} = {} & -\frac{\pi}{2}\int_{\Delta-\hbar\omega}^{-\Delta} \left[1-2f(E+\hbar\omega)\right]\left\{\left[g(E)+1\right] CC(b,a^-) + \left[g(E)-1\right] CC(b,a^+)\right\} dE \\
& -\pi \int_{\Delta}^{\infty} \left[f(E)-f(E+\hbar\omega)\right]\left\{\left[g(E)+1\right] CC(b,a^-) + \left[g(E)-1\right] CC(b,a^+)\right\} dE \quad \text{with}
\end{aligned} \tag{4.7}$$

$$b = \frac{\eta}{l},\; a_1 = \frac{\Delta_1}{\hbar v_o}\eta,\; a_2 = \frac{\Delta_2}{\hbar v_o}\eta,\; a^+ = a_1 + a_2 \text{ and } a^- = a_2 - a_1 .$$

If in Maxwells equation rot $\boldsymbol{H} = j\omega\varepsilon_0 \boldsymbol{E} + \boldsymbol{J}$ with $\boldsymbol{H}$ = rot $\boldsymbol{A}$ one neglects the displacement current (which has been done by Mattis and Bardeen), it follows that

$$J_z(x) = -\frac{d^2A_z(x)}{dx^2} \tag{4.8}$$

Thus Eq. (4.1) provides an analytical, not solvable integro-differential equation for the vector potential in the film and we have two boundary conditions:

$$Z = -j\omega\mu_o \left. \frac{A_z(x)}{\frac{dA_z(x)}{dx^2}} \right|_{x=0} \tag{4.9}$$

is the surface impedance we are looking for and

$$Z_o = -j\omega\mu_o \left. \frac{A_z(x)}{\frac{dA_z(x)}{dx^2}} \right|_{x=d} \tag{4.10}$$

the field wave impedance that the wave propagating through the film sees at the boundary $x = d$ (i.e. 120 π Ω in free space).

As shown in Fig. 22, HOOK [38] suggested to divide the film into N equal steps and to introduce three-point formulas taken from ABRAMOWITZ and STEGUN [39]. Then the equations (4.1), (4.9) and (4.10) can be written as

$$\frac{A_{n+1} - 2A_n + A_{n-1}}{H^2} = -\frac{3H}{4\pi\hbar v_o \lambda_{L0}^2} \sum_{n'=0}^{N} c_{nn'} A_{n'} \tag{4.11}$$

with n = 1, 2, 3 ... N-1 and

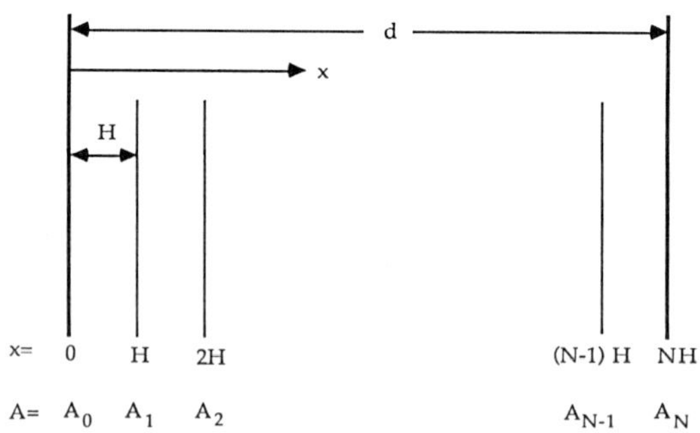

Figure 22

Superconducting film devided into N equal steps.

$$
c_{nn'} = \begin{cases} \frac{1}{2} K^* (|n - n'| H) & \text{if } n' = 0 \text{ or } N \\ \frac{1}{H} \int_{(n-1/2)H}^{(n+1/2)H} K^* (|nH - x'|)\, dx' & \text{if } n' = n \\ K^* (|n - n'| H) & \text{otherwise}, \end{cases}
$$

$$
Z = \frac{j\omega\mu_o H A_o}{\frac{3}{2} A_o - 2A_1 + \frac{1}{2} A_2} \quad \text{and} \tag{4.12}
$$

$$
A_N = - \frac{Z_o}{j\omega\mu_o H} \left(\frac{1}{2} A_{N-2} - 2A_{N-1} + \frac{3}{2} A_N\right) . \tag{4.13}
$$

Equation (4.11) and Eq. (4.13) provide N complex equations for $N+1$ unknown complex vector potentials $A_o, A_1 \dots A_N$. A further equation is obtained by setting

$$
A_o = 1 . \tag{4.14}
$$

The integral in the explanation to Eq. (4.11), which is necessary because of the divergence of $K^*(\eta)$ as η goes to zero, can be written as

$$
c_{nn} = \frac{2}{H} \int_0^{H/2} K^* (\eta)\, d\eta \tag{4.15}
$$

The only terms in Eq. (4.6) and Eq. (4.7) depending on η are $CS(b_1, b_2)$ and $CC(b_1, b_2)$. Therefore we can consider the integration of those parts only, and in the special case, where $n = n'$ replace $CS(b_1, b_2)$ by $CSS(d_1, d_2)$ and $CC(b_1, b_2)$ by $CCS(d_1, d_1)$:

$$
\begin{aligned} CSS(d_1, d_2) = {} & \frac{2}{3} \frac{d_2}{d_1^2 + d_2^2} - \frac{1}{6} \left[\frac{4d_1}{d_1^2 + d_2^2} + 1 - d_1\right] e^{-d_1} \sin(d_2) \\ & - \frac{1}{6} \left[\frac{4d_2}{d_1^2 + d_2^2} + d_2\right] e^{-d_1} \cos(d_2) + \frac{1}{3} d_1\, d_2 \operatorname{Re} \{E_1(d_1 + jd_2)\} \\ & + \frac{1}{6} \left[d_1^2 - d_2^2 - 6\right] \operatorname{Im} \{E_1(d_1 + jd_2)\} \quad \text{and} \end{aligned} \tag{4.16}
$$

$$
\begin{aligned} CCS(d_1, d_2) = {} & \frac{2}{3} \frac{d_1}{d_1^2 + d_2^2} - \frac{1}{6} \left[\frac{4d_1}{d_1^2 + d_2^2} + 1 - d_1\right] e^{-d_1} \cos(d_2) \\ & - \frac{1}{6} \left[\frac{4d_2}{d_1^2 + d_2^2} + d_2\right] e^{-d_1} \sin(d_2) + \frac{1}{3} d_1\, d_2 \operatorname{Im} \{E_1(d_1 + jd_2)\} \end{aligned} \tag{4.17}
$$

$$+\frac{1}{6}\left[d_1^2 - d_2^2 - 6\right] \mathrm{Re}\left\{\mathrm{E}_1(d_1 + jd_2)\right\} \quad \text{with}$$

$$d_1 = \frac{H}{2\eta} b_1 \quad \text{and} \quad d_2 = \frac{H}{2\eta} b_2 \, .$$

The first order exponential integral for small complex arguments is treated by ABRAMOWITZ and STEGUN [39]:

$$\mathrm{E}_1(z) = -\gamma - \ln z - \sum_{n=1}^{\infty} \frac{(-1)^n z^n}{n \; n!} \, , \tag{4.18}$$

$$|\arg z| < \pi \, , \; \gamma \text{- Euler constant} \, .$$

For large arguments a continued fraction is used:

$$\mathrm{E}_1(z) = e^{-z}\left[\frac{1}{z+}\,\frac{1}{1+}\,\frac{1}{z+}\,\frac{2}{1+}\,\frac{2}{z+}\,\cdots\right] , \tag{4.19}$$

the numerical evaluation of which is described by Barnett et. al. [40]. We found that, even for for extemely small arguments, Eq. (4.19) yielded the same result as Eq. (4.18).

4.2 Complex Conductivity

Knowing the dependence of the vector potential $A_z(x)$ in the anomalous limit from the solution of Eq. (4.11), the conductivity in the superconducting state may be obtained with the help of the local relationship

$$J_z(x) = \left[\sigma_1(x) - j\sigma_2(x)\right] E_z(x) = \sigma_s(x)\, E_z(x) \, . \tag{4.20}$$

Here Eq. (4.8) and $E_z(x) = -j\omega\mu_o A_z(x)$, following from Maxwells equations, can be inserted together with $\sigma_n = \mathrm{Re}\{\sigma_{dc}/(1+j\omega\tau)\}$ from Eq. (2.1), the Drude formula for the conductivity in the normal conducting state at a given frequency. From a comparison of the Mattis-Bardeen kernel $K(q)$ with the Chambers kernel [3] the equation

$$\sigma_{dc} = \frac{1}{\lambda_{L0}^2}\,\frac{1}{\mu_o v_o} \tag{4.21}$$

results. Integrating finally over the film thickness, the following equation for the ratio of the superconducting to the normal conducting conductivity appears:

$$\frac{\sigma_s}{\sigma_n} = \frac{1 + (\omega\tau)^2}{j\omega\tau}\,\frac{\lambda_{L0}^2 v_o}{d} \int_o^d \frac{\dfrac{d^2 A_z(x)}{dx^2}}{A_z(x)}\, dx \, . \tag{4.22}$$

It can be calculated by replacing the integral by a sum and $d^2A_z(x)/dx^2$ by a three-point formula.

5. Applications to Thin Films

In the following sections a second FORTRAN computer program is tested, having little similarity to the one mentioned in Sect. 3. The step number N was taken to be 48, and three different interpolation routines for calculating the 48 complex equations were compared to each other and also with the correct solution of the equations. The integration over energy was achieved with the help of Simpsons rule, the integration limits being approached to within 10^{-9}.

5.1 Transition to Bulk Superconductors

Fig. 23 and Fig. 24 show the thickness depending surface resistance and the superconducting penetration depth for lead at 1 GHz and 4.2 K. The dashed lines were calculated, using the first program for bulk superconductors. Investigating the frequency range up to 1500 GHz we found an agreement between bulk and thin film calculations within ± 0.5 % in the bulk limit.

Taking the bulk value $\lambda = 45.72$ nm, the film in Fig. 24 may be treated as a bulk conductor when its thickness is more than three times the superconducting penetration depth. This is in good agreement with the coth-dependence of SWIHART and KAUTZ [41, 42], also measured by HENKELS and KIRCHER [30]:

Figure 23

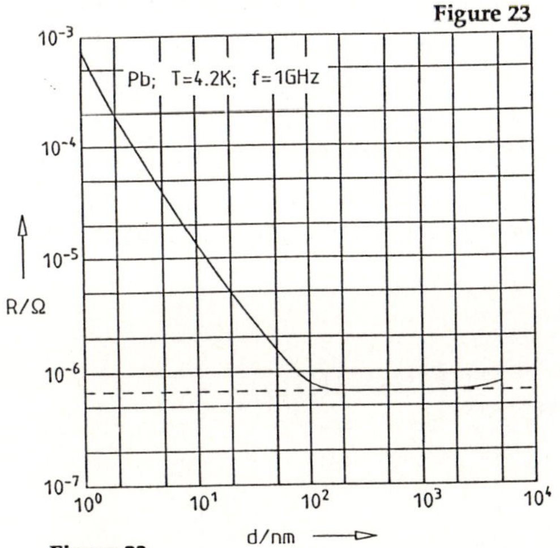

Figure 24

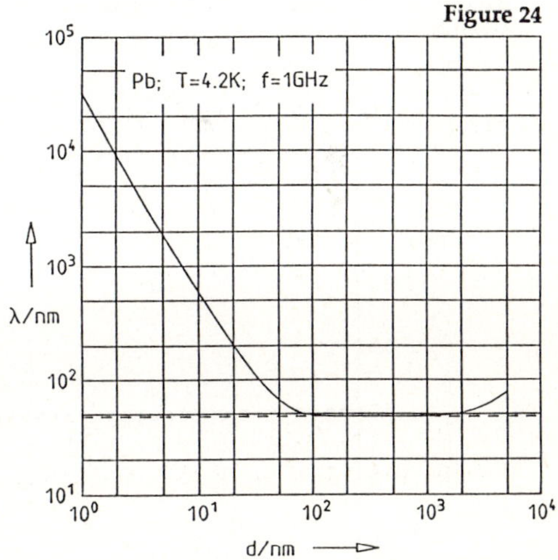

Figure 23
Surface resistance versus film thickness, Pb, 10 GHz, 4.2 K, parameters taken from TURNEAURE [11], —— thin film calculation, - - - - bulk calculation.

Figure 24
Superconducting penetration depth versus film thickness, according to Fig. 23.

$$\lambda_{eff} = \lambda \coth\left(\frac{d}{\lambda}\right) \tag{5.1}$$

In the case of the surface resistance in Fig. 23, the film can be treated as a bulk conductor if it is at least five times thicker than the superconducting penetration depth.

The behaviour of the electric field, the magnetic field and the current density inside a 300 nm thick film can be seen in Fig. 25 and Fig. 26, normalized to the corresponding values at the boundary $x = 0$. In Fig. 25 the

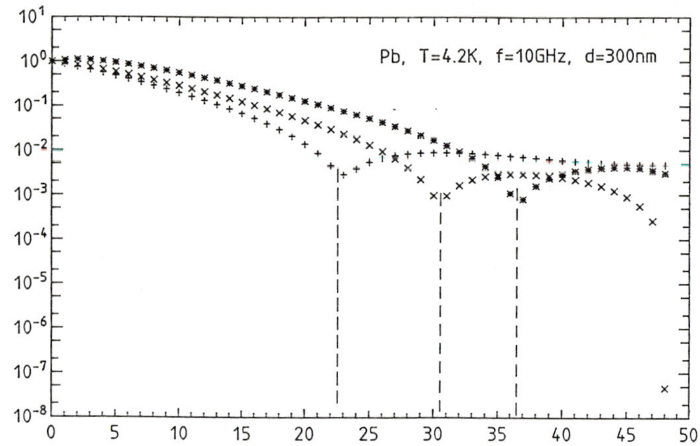

Figure 25
Amount of the fields and current density in a 300 nm thick Pb film at 4.2 K and 10 GHz, + $|En/E0|$, x $|Hn/H0|$, * $|Jn/J0|$.

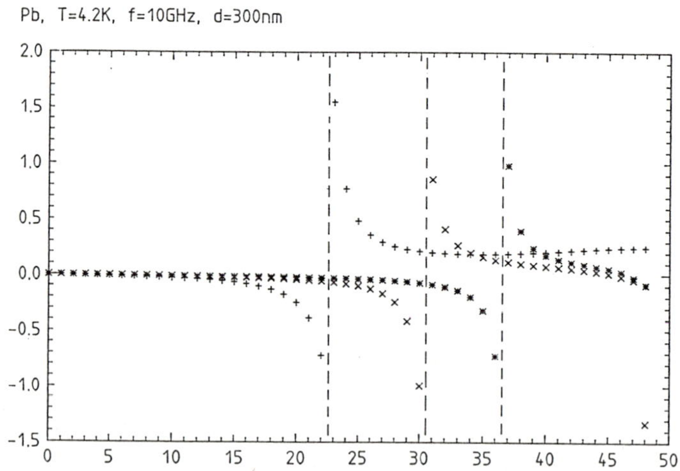

Figure 26

Phases of the fields and current density according to Fig. 25.

electrical field decreases more strongly than the magnetic field and the crossing point between the superconducting penetration depth and 1/e lies between the two curves. The current density has a maximum inside the film in the range of x = 12.5 nm, produced by the nonlocal relationship and all three curves pass through minima. At those minima the corresponding phases shown in Fig. 26 are turned to 180°, and the values of E_n/E_o, H_n/H_o and J_n/J_o become negative.

A change in the sign of the magnetic field was also calculated by SOMMERHALLER and THOMAS [43, 44] using the theory of Pippard.

5.2 Transmission Spectra

As in the case of absorption, also in the case of transmission through thin films, the theory of Mattis and Bardeen had been applied previously only in the extreme anomalous limit.

Instead of using a correct solution of the Mattis-Bardeen equations, NAM [45,46] developed a theory for strong coupling introducing a complex and frequency dependent energy gap. Calculations, based on this complicated theory are also available only in the extreme anomalous limit, but the results agree nicely with measurements of PALMER and TINKHAM [47], and of HARRIS and GINSBERG [48]. We will see in this section, that the correct solution of the Mattis-Bardeen equations does just as well, also for strong coupling superconductors.

The conductivity necessary to calculate the transmission and up to now only determined in the local and extreme anomalous limit, may be obtained in the anomalous limit from the vector potential calculated in Sect. 4.2.

We started to calculate the complex conductivity and the transmission T_s/T_n in the bulk limit (d = 200 nm) as Nam did. The results are indicated by the circles in Fig. 27 to Fig. 29. We used a formula derived from formulae given in [47]:

$$\frac{T_s}{T_n} = \frac{R_1\left(\frac{Z_o}{R_2} + 1\right) + \frac{1-R_1}{4n}\left[\frac{Z_o}{R_2} + n + 1\right]^2}{R_1\left(\frac{Z_o}{R_2}\frac{\sigma_1}{\sigma_n} + 1\right) + \frac{1-R_1}{4n}\left[\left(\frac{Z_o}{R_2}\frac{\sigma_2}{\sigma_n}\right)^2 + \left(\frac{Z_o}{R_2}\frac{\sigma_1}{\sigma_n} + n + 1\right)^2\right]} \tag{5.2}$$

with $R_1 = \left(\frac{1-n}{1+n}\right)^2,$

n = 2.106 - refractive index of quartz,

R_2 = 252 Ω/square - film resistance per square and

Z_o = 120 πΩ - wave impedance of free space,

and took the material parameters of TURNEAURE [11] for lead but with $2\,\Delta_o = 4.48\,k_BT_c$ and l = 20 nm.

In Fig. 27 the real part of the conductivity measured by PALMER and TINKHAM [47] is fitted a little better by the circles than by the extreme

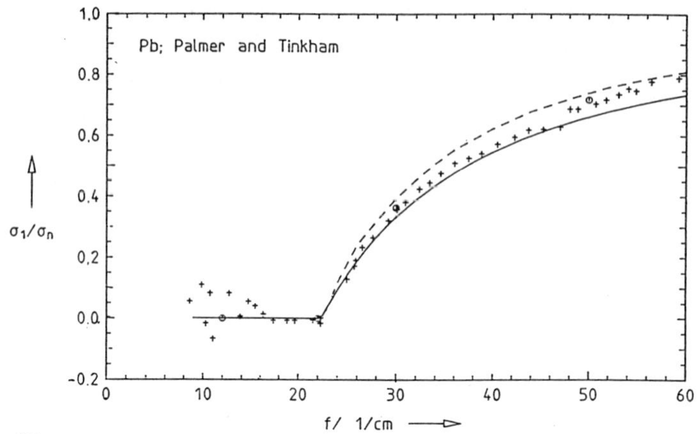

Figure 27
Real part of the complex conductivity for Pb, T = 2 K, + measured points of PALMER and TINKHAM [47], o correct evaluation d = 200 nm, —— correct evaluation d = 60 nm, - - - - extreme anomalous limit.

anomalous limit of the Mattis-Bardeen theory (dashed line). Nam has published no results for σ_1/σ_n, but they must be very similar to the dashed line and slightly higher at higher frequencies.

For the imaginary part of the conductivity in Fig. 28 we get good agreement to the theory of Nam (small dashed line) below the gap frequency. It is remarkable to note that, allthough the big discrepancy between the extreme anomalous limit of the Mattis-Bardeen theory (long dashed line) and the theory of Nam was generally considered to be a strong coupling effect, the correct evaluation of the Mattis-Bardeen theory gives the same low values of σ_2/σ_n.

Above the gap frequency the circles fit the experimental data better and deviate from Nams theory. The question which dependence is right cannot be answered, for in this region the transmission spectrum depends only on the real part of the conductivity. The values σ_2/σ_n calculated according to Nams theory by SHAW and SWIHART [49] here lie between the two curves.

The circles in Fig. 29 are nearly identical with the results of Nams theory (not shown) apart from a slightly lower peak height at 22.5 cm^{-1}. The solid lines in Fig. 27 to Fig. 29 represent changes in the film thickness from 200 nm to 60 nm.The theoretical curve in Fig. 28 is fitted in a optimal way to the measured values for σ_2/σ_n and because of this, the curve for σ_1/σ_n is a little too low. The transmission data in Fig. 29 are described better below the gap frequency, than above the gap frequency and the measured peak height is nearly reached.

Fig. 30 shows the transmission spectrum of the amorphous bismuth sample Bi-2 measured by HARRIS and GINSBERG [48]. This measurement was described using the formula

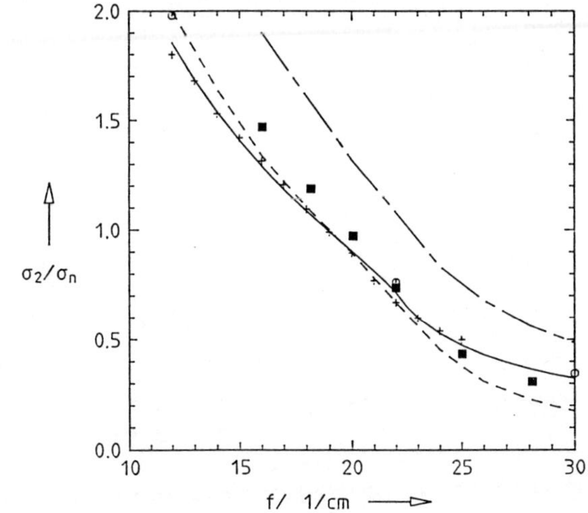

Figure 28
Imaginary part of the complex conductivity, + measured points, o correct evaluation d = 200 nm, - - - - NAM [45,46], – – – – extreme anomalous limit, ■ SHAW and SWIHART [49].

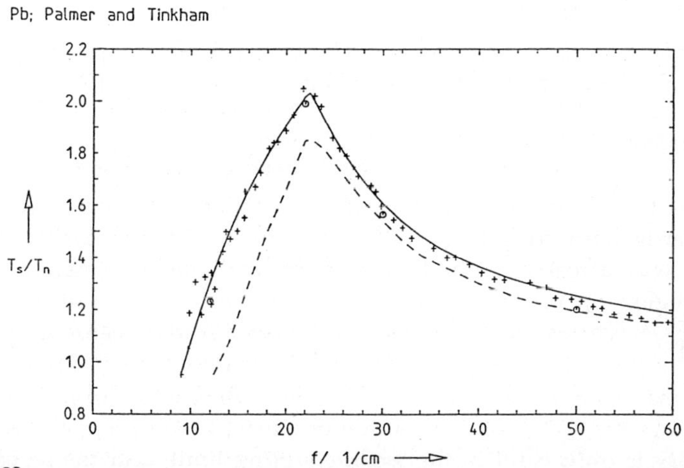

Figure 29
Transmission spectrum of PALMER and TINKHAM [47], Pb, + measured points, o correct evaluation d = 200 nm, —— correct evaluation d = 60 nm, - - - - - extreme anomalous limit.

$$\frac{T_s}{T_n} = \frac{(B+1)^2}{\left|B + \frac{\sigma_s}{\sigma_n}\right|^2} \quad \text{with} \quad B = \frac{(n+1)\,R_2}{Z_o}\,, \tag{5.3}$$

and the parameters d = 80 nm, λ_{Lo} = 38 nm, $v_o = 5.0 \times 10^5$ m/s and l = 33 nm. Here, the conductivity and transmission are nearly independent of λ_{Lo}. Again there is a small dependence of the conductivity and the transmission on the film thickness.

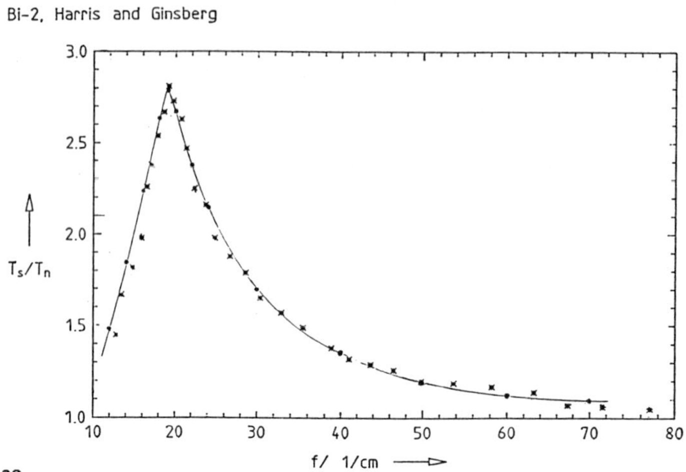

Figure 30
Transmission spectrum of amorphous bismuth probe Bi-2 from HARRIS and GINSBERG [48], * measured points, —— correct evaluation with $d = 80$ nm, $\lambda_{L0} = 38$ nm, $v_0 = 5.0 \times 10^5$ m/s and $l = 33$ nm.

Similary good results were obtained in the case of amorphous gallium and in the case of niobium nitride measured by KARECKI et. al. [50].

6. Conclusion

The equations for the anomalous skin effect in superconductors derived by Mattis and Bardeen are solved exactly for bulk conductors and for thin films. While there was already a good agreement with measurements in the microwave region on resonant cavities, now our calculations also show a very good agreement to measurements in the far infrared region above the gap frequency, where the extreme anomalous limit cannot be used.

For the first time absorption curves on bulk films with strong coupling superconductors can be described, although, strictly speaking, the Mattis-Bardeen theory is only valid in the weak coupling limit, and the results on transmission spectra are nearly identical with the results of Nam's theory for strong coupling.

Hence, the Mattis-Bardeen theory has been shown to describe the electromagnetic properties of superconductors for all frequencies, temperatures and mean free paths (as long as only nonmagnetic impurities are involved) and also for strong coupling superconductors. Five material parameters are needed: the energy gap, the critical temperature, the London penetration depth, the Fermi velocity (or the BCS coherence length) and the mean free path.

Acknowledgements

The author is indebted to Prof. J.H. Hinken and to Prof. V. Kose for helpful comments and sugestions, and to Prof. L. Bliek for reading the manuscript.

References

[1] A.B. Pippard, in Advances in Electronics, L. Marton, Academic Press, New York (1954)
[2] R.G. Chambers, Proc. Roy. Soc. (London) **A215** (1952) p. 481
[3] R. Pöpel, Dissertation University of Braunschweig, (1986)
[4] A.B. Pippard, Proc. Roy. Soc. (London) **A216** (1953) p. 547
[5] J. Bardeen, L.N. Cooper , J.R. Schrieffer, Phys. Rev. **108** (1957) p. 1175
[6] B. Mühlschlegel, Z. Physik **155** (1959) p. 313
[7] D.C. Mattis, J. Bardeen, Phys. Rev. **111** (1958) p. 412
[8] M. Tinkham, in Low-Temperature Physics, Mc Graw-Hill, New York (1966)
[9] W. Gröbner, N.H. Hofreiter, Integraltafeln, part II, Springer Verlag, Wien, New York (1966)
[10] P.B. Miller, Phys. Rev. **118** (1960) p. 928
[11] J.P. Turneaure, Dissertation Stanford University, Stanford, California (1967)
[12] R.E. Glover, M. Tinkham, Phys. Rev. **108** (1957) p. 243
[13] G.E.H. Reuter, E.H. Sondheimer, Proc. Roy. Soc. (London) **A195** (1948), p. 336
[14] J.P. Turneaure, I. Weissman, J. Appl. Phys. **39** (1968) p. 4417
[15] J. Halbritter, Externer Bericht 3/69-2, KFZ Karlsruhe (1969)
[16] P. Wilson, Int. Mitteilung SLAC-TN-70-35, KFZ Karlsruhe (1970)
[17] M. Tinkham, Introduction to superconductivity, Mc Graw Hill, New York (1975)
[18] Y. Bruynseraede et. al., Physica **54** (1971) p. 137
[19] A. Philipp and J. Halbritter, IEEE **MAG 17** (1981) p. 951
[20] P. Kneisel et. al., Advances in cryogenic engineering **22** (1977) p. 341
[21] C.M. Lyneis and J.P. Turneaure, IEEE **MAG 13** (1977) p. 339
[22] M.D. Sturge, Proc.Roy. Soc. (London) **A246** (1958) p. 570
[23] L. Szecsi, Z. Physik **241** (1971) p. 36
[24] P. Flecher et. al., Proc. of the 1969 Proton Lin. Acc. Conf., Washington
[25] J.M. Pierce, HEPL Report **No.514**, Stanfort (1967)
[26] R. Blaschke, R. Blocksdorf, Z. Physik **B49** (1982) p. 99
[27] W. Bauer et. al., J. Appl. Phys. **45** (1974) p. 5053
[28] B. Piosczyk, Thesis KFZ Karlsruhe, Externel Report **No.1991** (1974)
[29] J. Hasse, J. Lachmann, Z. Physik **258** (1973) p. 136
[30] W.H. Henkels, C.J. Kircher, IEEE **MAG 13** (1977) p. 63
[31] J. Halbritter, Z. Physik **266** (1974) p. 209
[32] M.A. Biondi, M.P. Garfunkel, Phys. Rev. **116** (1959) p. 853
[33] M.A. Biondi, M.P. Garfunkel, Phys. Rev. **116** (1959) p. 862
[34] J.D. Leslie, D.M. Ginsberg, Phys. Rev. **133** (1966) p. 362
[35] D.M. Ginsberg, Phys. Rev **151** (1966) p. 241
[36] L. Leplae, Thesis University Maryland (1962)
[37] S.L. Norman, Phys. Rev. **167** (1968) p. 393
[38] J.R. Hook, J. Low Temp. Phys. **23** (1976) p. 645
[39] M. Abramowitz and I.A. Stegun, eds., Handbook of mathematical functions, Dover, New York (1965)
[40] A.R. Barnett et. al., Computer Phys. Commun. **8** (1974) p. 377
[41] J.C. Swihart, J. Appl. Phys. **32** (1961) p. 461
[42] R.L. Kautz, J. Appl. Phys. **49** (1978) p. 308

[43] R. Sommerhaller, H. Thomas, H. P. **A. 34** (1961) p. 29
[44] R. Sommerhaller, H. Thomas, H. P. **A. 34** (1961) p. 265
[45] S.B. Nam, Phys. Rev. **156** (1967) p. 470
[46] S.B. Nam, Phys. Rev. **156** (1967) p. 487
[47] L.H. Palmer and M. Tinkham, Phys. Rev. **165** (1968) p. 588
[48] R.E. Harris and D.M. Ginsberg, Phys. Rev. **188** (1969) p. 737
[49] W. Shaw and J.C. Swihart, Phys. Rev. Let. **20** (1968) p. 1000
[50] D. Karecki et. al., Phys. Rev. **B 25** (1982) p.1565

Part II
Sensitive Detectors

High-T_C Josephson Contacts and Devices

H. Rogalla* and **C. Heiden****
* Universität Twente, NL-7500 AE Enschede
** Universität Gießen, Institut für Angewandte Physik, 6300 Gießen

1. Introduction

Ever since the discovery of the Josephson effect there has been the challenge to produce Josephson contacts using materials with high transition temperature T_C. There are among others the following reasons:

i) A more widespread use of Josephson devices can be expected if their operating temperature is raised well above liquid helium temperatures,

ii) A better performance can be expected from devices that are operated well below their transition temperature, at least if we are dealing with tunnel junctions as Josephson contacts.

Before the discovery of the new ceramic high-T_C materials [1] it was the class of the binary A15-compounds of which Nb_3Ge had the highest transition temperature near 23 K, and NbN ($T_C \approx 16$ K), that had to be considered for this purpose. The weak link of a Josephson contact can be made in different ways, the most common ones being the fabrication of either a tunnel junction or a constriction in the contact zone of the two superconducting electrodes. In the latter case we are dealing with micro- or nanobridges. Depending on the material properties, and in particular on problems connected with microstructuring processes [2] so far it has been possible to make Josephson contacts of sufficient quality using Nb_3Ge only in the form of microbridges. Good tunnel junctions on the other hand could be made with NbN.

In the following we will report on some efforts that were undertaken to make Josephson contacts on the basis of the above mentioned "classical" high-T_C materials. After dealing with basic technological issues (Sect. 2) the fabrication and properties of tunnel junctions made of NbN and also with Nb_3Ge as base electrode will be described (Sect. 3), followed in Sect. 4 by the presentation of a theoretical model for the phase slip behavior of the microbridges. Section 5 and 6 deal with devices made of the classical high-T_C materials, and the concluding Sect. 7 with first efforts to fabricate SQUIDs that work in liquid nitrogen using the new ceramic superconductors.

2. Technological Aspects

As already mentioned material and interface properties of the materials are factors that have a strong influence on the type of weak link that can be manufactured. Some materials for instance cannot be prepared in a process that is compatible with tunnel junction fabrication. Nb_3Ge needs a high preparation temperature in order to grow in the proper phase. In such a case the Josephson contacts can only be realized as superconductor-normal-conductor-superconductor junction (SNS) or as microbridges. To lay the ground for the further discussion we will focus first on some technological aspects of the preparation and the microstructuring of thin films of the model materials NbN and Nb_3Ge.

2.1 Thin Film Preparation

For the application in liquid helium and at temperatures up to 12 K the classical superconductor NbN is a material of first choice. With sputtering it can be prepared at ambient temperature and its interface to barrier materials like MgO or the oxid of NbN is so good that tunnel junctions of high quality can be fabricated.

The films, that were used here for the tunnel junction fabrication, were prepared by continuous reactive dc magnetron sputtering [3] from a Nb target in a nitrogen/argon atmosphere (see Fig. 1). The background pressure before sputtering of less than 10^{-4} Pa is obtained by an oil-diffusion pump. As sputtering gases we use argon and nitrogen of 99.999 % purity. The flow of both gases is controlled by mass flow controllers with a maximum flow of 10 ml/min for N_2 and 100 ml/min for Ar. In order to prevent too high a pressure in the diffusion pump and thus a backstreaming of oil vapor into the chamber, the cross section of the pumping line between chamber and diffusion pump can be adjusted. A large cylindrical cold trap in the sputtering chamber surrounds most of the sputtering equipment. It can be cooled to liquid-nitrogen temperature.

Nb targets with diameters of 50 mm, thicknesses of 1.5 - 3 mm, and 99.9 % purity were used in the thin film preparation. The target-substrate distance is 45 mm. As substrates we use thermally oxidized silicon of 0.28 mm thickness which are mounted on a cathode to allow for sputter cleaning of their surface. The preparation starts with a cleaning of the target by sputtering in a pure Ar atmosphere for typically 5 min. The cleaning effect can be observed by the discharge impedance. Lower impedance means a cleaner Nb surface and less oxygen in the chamber by Nb gettering. In a next step N_2 is added. After about 5 min an equilibrium is reached with regard to the nitrogen content of the target surface. For the actual sputter deposition a dc-voltage of 320 - 390 V (dependent on the target erosion) is applied to yield a current of 320 mA. This results in a deposition rate of about 50 nm/min. Superconducting films with thicknesses in the range of 7 nm - 1 µm were prepared.

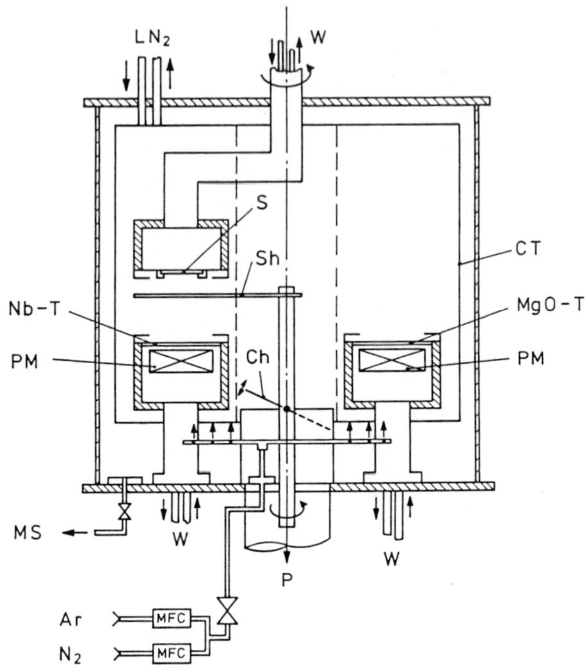

Figure 1
Schematic of the vacuum system used for NbN sputtering. Ch choke; CT cold trap; Nb-T niobium target; MgO-T magnesium-oxide target; MS mass spectrometer; P diffusion pump; PM permanent magnet; S substrate; Sh shutter; W water.

For the optimization of the preparation parameters the nitrogen partial pressure was measured as a function of the sputtering current and the injected nitrogen (see Fig. 3) by a differentially pumped mass spectrometer. In Fig. 2 the N_2 partial pressure is plotted as function of the sputtering current for different N_2 flow rates. Each curve shows three separate regimes. Starting at small currents the N_2 partial pressure decreases nearly linearly with a slope of -1.0×10^{-6} Pa/A for all curves (region A). At a characteristic current value the slope suddenly changes to -1.6×10^{-6} Pa/A (region B). Finally, at sufficiently high current, practically all nitrogen is consumed (region C). Optimum preparation conditions for the NbN-films are found at the boundaries between region A and B. Towards smaller currents the T_c of the films slowly decreases. We attribute this to the reduced sputtering rate leading to a higher inclusion of oxygen. Excess nitrogen in the film also may contribute to a T_c reduction. Towards higher currents in region B the quality of the films decreases too. Their T_c drops to below 4.2 K near the border to region C. The films reveal bad mechanical properties: the surface looks rough and they do not adhere well to the substrate. This may be due to the appearance of subnitrides and pure Nb in the films as supported by x-ray investigations.

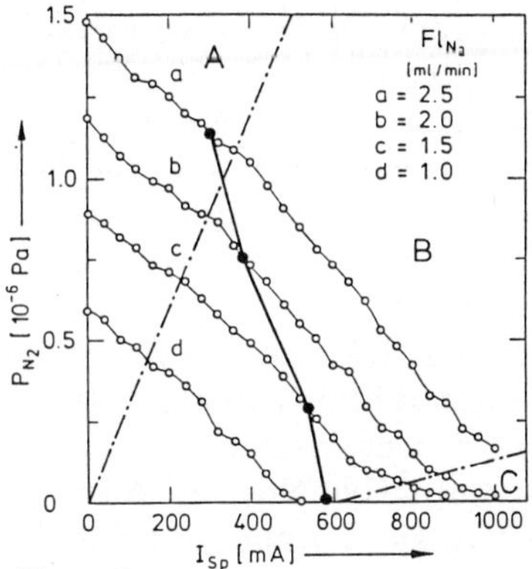

Figure 2

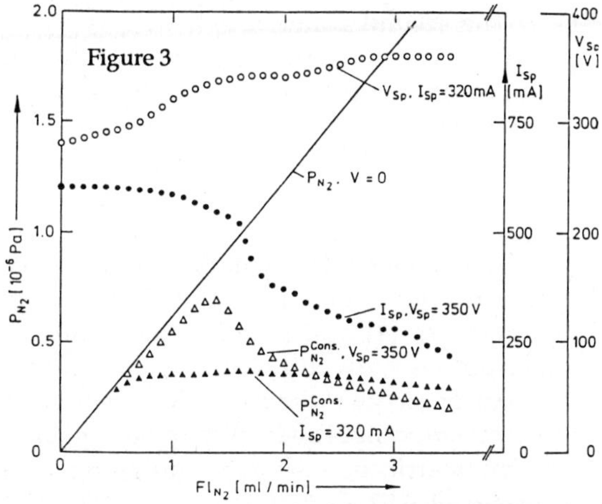

Figure 2
Dependence of the partial nitrogen pressure P_{N_2} on the sputtering current I_{sp}. Parameter is the N_2 flow Fl_{N_2} into the sputtering chamber. The bold full line corresponds to the case of constant voltage of Fig. 2.3.

Figure 3
Dependence of the partial nitrogen pressure P_{N_2} on N_2 flow Fl_{N_2} into the chamber; dependence of sputtering current I_{sp} on Fl_{N_2} for constant sputtering voltage V_{sp} and of V_{sp} on Fl_{N_2} for constant I_{sp}.

In further investigations the nitrogen partial pressure was measured as a function of the N_2 flow Fl_{N2} into the sputtering chamber for constant sputtering current I_{sp} and constant sputter voltage V_{sp}, as well as the dependence of V_{sp} and I_{sp} on FL_{N2} (see Fig. 3). The results are described in some detail elsewhere [3]. One important result is: the measurement of the sputtering current as function of FL_{N2} at constant sputtering voltage V_{sp} yields a valuable tool to determine optimum preparation parameters for the film deposition without mass spectrometer. Starting from low flow rates the sputtering current is practically constant (full circles in Fig. 3), until it changes its value significantly within only a small change in Fl_{N2}, indicating a higher discharge impedance and thus a good surface nitration of the target. At flow rates slightly higher than this step like response of I_{sp}, optimum preparation conditions are found. This result is supported by mass spectrometry.

Using the dc-sputtering technique and optimization described above, high quality NbN films with a zero resistance transition temperature T_c of up to 15.5 K could be prepared (*All T_c-values quoted further on will be those for the zero resistance downset of the resistive transition, and not those for the onset*). Starting from a thickness of 7 nm these films were superconducting above 4.2 K indicating that only very thin surface layers of the material may be destorted. These material properties are very promising for tunnel junction fabrication (c.f. Sect. 3).

The second material delt with is Nb_3Ge. Its predominant value for applications is mechanical stability and the wide temperature range in which it is superconducting. Even liquid hydrogen can be used as a coolant. Nb_3Ge, in contrast to NbN, has to be deposited at a substrate temperature of the order of 900 °C which in conjunction with the small coherence length of about 3 nm so far has prevented the fabrication of Josephson tunnel junctions. Instead, SNS-junctions or microbridges have to be used. The 23 K-phase of Nb_3Ge is a metastable compound with A15 structure. The nucleation of this phase must be forced by appropriate deposition conditions like substrate temperature [4] and kinetic energy of the sputtered atoms [5]. Otherwise other phases grow, e.g. the non-superconducting but stable Nb_5Ge_3. A small amount of oxygen during the deposition seems to play an important role in the nucleation of Nb_3Ge: it may be explained by the initial growth of a nearly stoechiometric but oxygen containing A15 phase with an enlarged lattice constant of 0.52 - 0.525 nm [6], on top of which the high-T_c phase with the right lattice constant finally is formed. Often this bottom layer is reported to have a thickness of more than 100 nm. Its properties and thickness cannot be well controlled with the consequence that the critical current of a microbridge fabricated from such a film is not very reproducible. In order to reduce the thickness of this initial bottom layer as far as possible while retaining the high-T_c of the material, the preparation was optimized for very thin films with the consequence, that the superconducting properties of the microbridges became more predictable.

The Nb_3Ge-films were dc-sputtered on sapphire substrates heated to temperatures between 700 °C and 900 °C [7]. To achieve a homogeneous and constant substrate temperature an indirectly heated niobium (or molybdenum) plate of much larger size than the substrate was used. The temperature of the plate is measured by a thermocouple mounted in a hole of the plate. An active controller keeps the temperature constant to a preset value within ± 0.1 % throughout the sputtering process. Since the surface temperature of the substrate is not well known (absorption and emissivity change during the coating with Nb_3Ge [5]) we used the heater temperature as one of the parameters to optimize. The sputtering is done from a circular water-cooled cathode of 50 mm diameter consisting of two half disks of tantalum free niobium and semiconductor quality germanium [7]. The vertical distance between substrate and cathode is about 25 mm. The Nb:Ge-ratio can be adjusted by changing the relative position between heater and cathode. During the sputtering high purity Ar gas flows through the turbo pumped vacuum chamber at a pressure of 30 - 55 Pa. Once the sputtering voltage is selected the discharge current of 40 - 100 mA may be adjusted by the gas pressure (see Fig. 4). The current density at the cathode surface is about 3 - 8 mA/cm^2.

For the fabrication of very thin films of Nb_3Ge the preparation parameters were optimized in the following way: first the onset temperature T_{on}, at which a first reduction of the film resistance towards superconductivity occurs, is measured as function of the heater temperature T_H for fixed

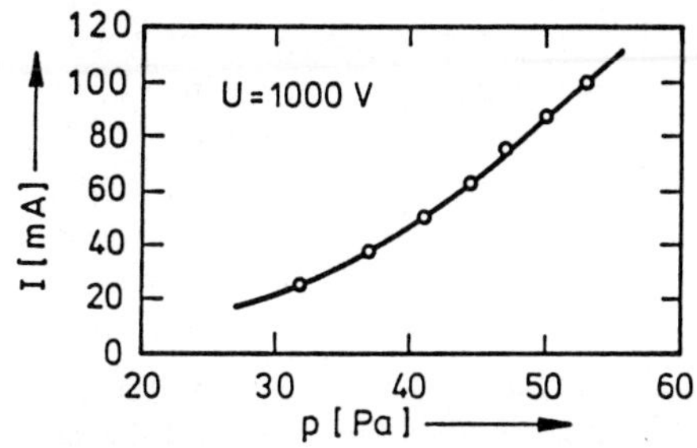

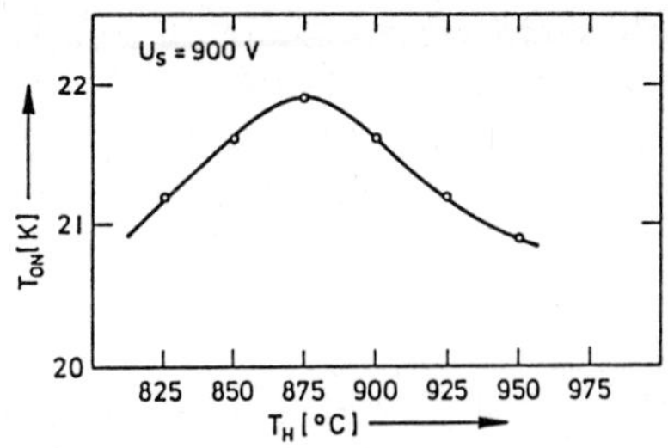

Figure 4
Dependence of the discharge current as function of Ar gas pressure for fixed cathode voltage.
Cathode diameter: 50 mm.

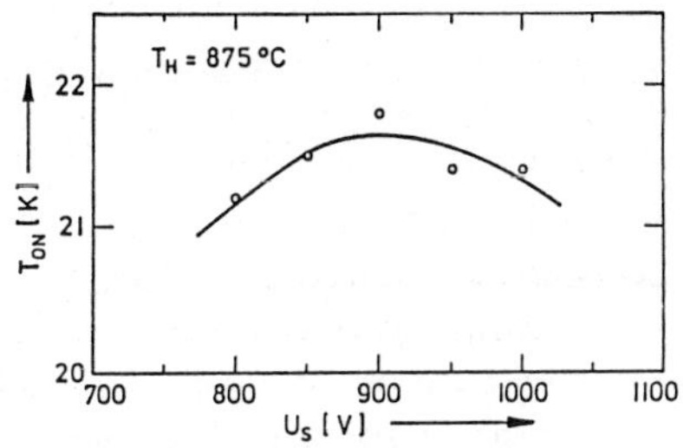

Figure 5
Dependence of the onset temperature T_{on} on the heater temperature T_H (top) and the cathode voltage U_S (bottom).

cathode voltage U_S. Next the cathode voltage is varied keeping the optimum T_H of the first step fixed. One may proceed with these steps iteratively until an optimum set for T_{on} is found (see Fig. 5).

Next the resistive transition width T_w between the superconducting and normal state is minimized using the same iterative procedure as the one to maximize T_{on}. The consequence of this two step procedure is an optimized transition temperature T_c at which the film is fully superconducting. Similar optimizations can be done for instance for the current density J_S at the cathode by varying the pressure in the chamber, or by using magnetron cathodes, and for the cathode-substrate distance d_{cs}. For the preparation chamber used here [8] we found the following optimum parameter set:

$$U_S = 850 \text{ V} \quad T_H = 875 \text{ °C} \quad p = 45 \text{ Pa} \quad J_S = 7 \text{ mA/cm}^2 \quad d_{cs} = 25 \text{ mm}.$$

These preparation parameters lead to a thermalization of the sputtered Nb- and Ge-atoms with an initial energy of about 5 eV (corresponding to 5.5×10^4 K) [9] after ca. 35 collisions to a final energy of about 0.036 eV (or 400 K). The thermalization is reached at a distance of 7 mm from the cathode. These values are calculated from the model of CADIEU and CHENCINSKI [10]. The thermalization of the sputtered niobium and germanium is important since the metastable A15 Nb_3Ge is sensitive to bombardment. Using the optimized parameters films in a wide variety of thicknesses were prepared. In Fig. 6 the dependence of T_c on the thickness of these films is shown. Down to 40 nm no change in T_c is visible. At smaller thicknesses the transi-

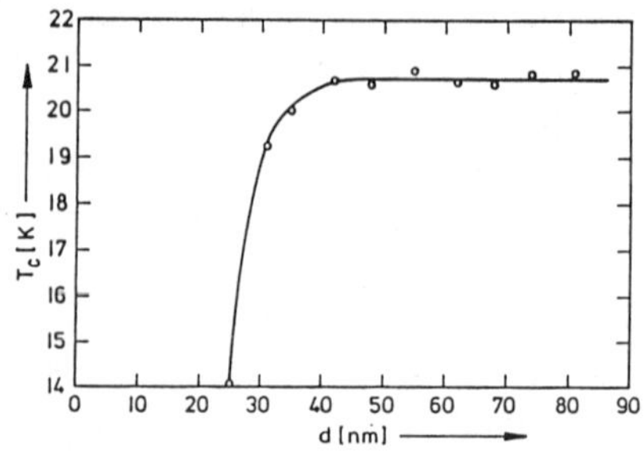

Figure 6
Dependence of T_c on thickness of Nb_3Ge-films.

tion temperature falls to about 19 K at 30 nm and below 20 nm the films become normal conducting.

Using the silicondioxid-membrane technique [11] Nb_3Ge-films were prepared on very thin and small SiO_2-membranes covering a hole in a silicon wafer [8]. At the edges of cracks in the SiO_2-films small mushroom-like protrusions of Nb_3Ge grow with a diameter of about 150 nm. Using TEM the regular structure of Nb_3Ge within grains of about 10 nm diameter can be visualized directly (see Fig. 7). A more detailed analysis reveals additionally, that the protrusion contains no hexagonal or tetragonal Nb_5Ge_3-phase. This is consistent with x-ray measurements on thicker Nb_3Ge-films

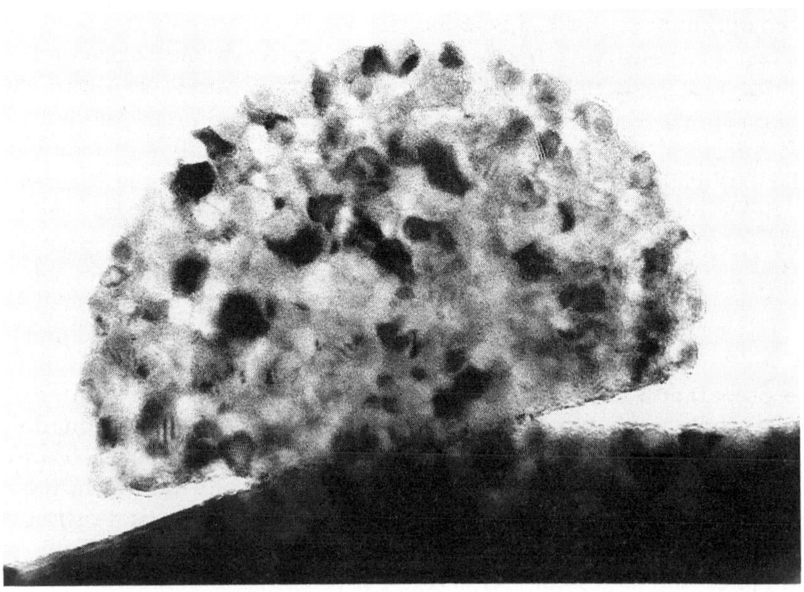

Figure 7
TEM-picture of a Nb_3Ge protrusion showing grains of about 10 nm diameter. The crystalline structure also is partially visible.

which showed no characteristic reflexes for Nb_5Ge_3. Both, TEM and x-ray measurements reveal for the lattice constant a value of 0.514 nm.

The superconducting properties of the Nb_3Ge-films are consistent with those published in the literature. The specific resistance of the films is in the range of 40 - 100 μΩcm and the resistance ratio between 1.3 and 2.2. High-T_c values correspond to low resistivity and high resistance ratio: a typical Testardi-plot [12] can be obtained from the films.

The coherence length ξ was determined from the temperature dependence of B_{c2} [13] which was measured for a 75 nm film in the temperature range between 19.0 and 21.1 K. For $\xi(0)$ a value of 2.5 nm was determined. Near T_c the typical temperature dependence of ξ in these Nb_3Ge-films is given by (see e.g. [13]) :

$$\xi(T) = 2.5\ \mathrm{nm} \left(\frac{T_c}{T_c - T}\right)^{1/2} . \tag{2.1}$$

Another important characteristic for the application of these films in microbridges is the flux flow viscosity of the Nb_3Ge-films. It can be determined from B_{c2} and the resistivity ρ_n [14]:

$$\eta = B_{c2}\, \Phi_o / \rho_n \tag{2.2}$$

assuming that $\rho_f = \rho_n$ at $B = B_{c2}$. With B_{c2} determined via $B_{c2} = \phi_o/(2\pi\xi^2)$ and with values for ρ_n measured slightly above T_c we find as a typical value for η near T_c :

$$\eta(T) = 1.4 \cdot 10^{-7}\ (1 - T/T_c)\ \mathrm{kg/ms} . \tag{2.3}$$

The magnetic penetration depth λ was determined using the method of W.H. HENKELS and C.J. KIRCHER [15]. Similar to their experiment the resonance frequency of a superconducting transmission line in stripline technique was measured in the radio frequency range. With the capacity of the meander structure determined at low frequencies its inductance is known. Using the known influence of the penetration depth on the inductance [16] the penetration depth can be determined. At 4.2 K we found a value of 460 nm for the Nb_3Ge-films. The temperature dependence of the penetration depth can be calculated with sufficient accuracy in the empirical "two-fluid" approximation near T_c :

$$\lambda(T) = \lambda(0)\,[1 - (T/T_c)^4]^{-1/2} = 460\ \mathrm{nm}\,[1 - (T/T_c)^4]^{-1/2} . \tag{2.4}$$

2.2 Microstructuring Procedures

For the fabrication of Josephson junctions from NbN- or Nb_3Ge-films microstructuring techniques have to be applied which are in accordance with the special material properties and preparation conditions. Due to the low deposition temperature NbN can easily be structured by lift-off and, since tunnel junctions can be fabricated, there is no special need for very small dimensions. A length scale of a few micrometers is sufficient for most applications. Reactive ion etching also works very well with NbN. Fluoron and chloron containing etching gases can both be applied with good success and special mixtures can be used to enhance the selectivity in the etching of different materials. Since NbN is not very sensitive to damage by ion bombardment no particular difficulties are encountered in the etching process.

In contrast to NbN, Josephson contacts made of Nb_3Ge can only be fabricated as microbridges if a high operating temperature of the resulting device is to be maintained. The necessity for small critical currents of Josephson elements used for sensor applications, a sufficient small ratio between bridge width and coherence length in order to obtain a good Josephson behavior, and the retaining of the high-T_c of the Nb_3Ge-films determine the guideline along which the structuring of microbridges out of Nb_3Ge-films has to proceed: the width of the bridges should be made as small as possible without introducing structural damage to the material in the bridge region. It is well known that the "lift-off"-process practically does not change the parameters of the deposited material. However this process does not seem to be applicable at elevated substrate temperatures. To our knowledge there exists no resist-material usable for a "lift-off"-process with characteristic dimensions of 100 nm, which withstands a deposition temperature of up to 900 °C and does not change the properties of the deposited Nb_3Ge. Tests with ZnO-films were disillusioning.

As a consequence one has to structure the films by etching. With wet chemical etching reproducible results could not be obtained since underetching becomes a severe problem for structures with dimensions less than 1 µm. Therefore we successfully tried to adapt the plasma-assisted reactive dry etching to the structuring of Nb_3Ge-films.

In a first step the Nb_3Ge-film has to be covered with a resist, which has to be structured with a resolution of better than 100 nm. Electron-beam lithography is a suitable way to fabricate versatile shaped structures on a laboratory scale. For this purpose we use an AMR 1000 SEM, into which a simple deflector for beam blanking is inserted. As resist we use PMMA which is not as sensitive as other electron beam resists but offers a high resolution and easy handling. The exposure dose is 3×10^5 C/cm^2 for a PMMA-film with a thickness of 400 nm.

The etching is done in a parallel plate reactor (see Fig. 8). It consists of a turbo-pumped vacuum chamber, which can be baked up to 100 °C and yields a background pressure better than 10^{-3} Pa, a parallel plate capacitor, whose plates can be connected either to ground or to an rf-generator, and different massflow-controlled gas inlets. SF_6 turned out to be a suitable etching gas .

SF_6 is an inert non-poisonous gas and easy to handle. The reaction products are not badly attacking the stainless steel chamber and the oil of the roughing pump and the bearings of the turbo-pump. In the plasma SF_6 is predominantly split into SF_4- and F_2-ions. The main reaction product with

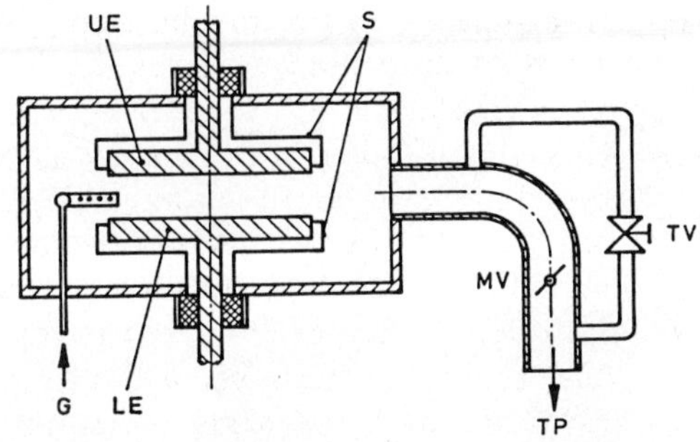

Figure 8
Schematic drawing of the plasma assisted etching reactor allowing reactive ion etching and plasma etching: LE, UE lower and upper electrode, S shield, G gas inlet, MV main valve, TV throttle valve, TP turbo pump.

Nb is NbF_5, which has a high vapour pressure at room temperature (boiling point 236 °C [17]), and with Ge is GeF_4, which is a gas at room temperature [18]. SF_4 is a gas too, the reaction products thus can easily be removed from the reaction chamber by pumping. Care has to be taken with the SF_4 reaction product since it is very poisonous. Chloride containing etching gases can also be used, but their reaction products badly influence the reliability of the chamber and pumping system and are also very hazardous.

The film to be etched rests on the lower plate of the reactor, which is ac-connected via a matchbox to the rf-source. The second plate is grounded. This reactive ion etching is highly anisotropic (see e.g [17]). The anisotropy has been verified in this chamber. It results in practically perpendicular etch corners (see Fig. 9).

The first microbridges fabricated with the described technique were a failure. The bridge width was much smaller than intended and the bridges were not superconducting. The reason for the narrowing of the bridges was the high sputtering rate of the PMMA-resist for high self-bias voltages at the cathode even if the resist was baked at elevated temperatures. The disappearance of superconductivity has its origin in the metastability of the Nb_3Ge A15-phase. It is well known that bombardment with ions of high energy leads to a damage of the material accompanied by a drastic T_c-reduction [19]. Annealing at elevated temperatures nearly restores the original properties of the material. During the etching the ion-energy however is too small to lead to a drastic damage. Instead we suppose that the ion-bombardment introduces particles into the lattice widening it or transferring enough energy into the Nb_3Ge-compound to enable its partial transition to Nb_5Ge_3. Both mechanisms lead to a decrease in T_c. To find etching parameters for a negligible decrease in T_c the influence of an ion-bombardment at low energies was studied in more detail [20, 21]. It turned out that nearly pure chemical etching with SF_6 at small ion energies, corresponding to a self-bias voltage of 20 V, gave the proper results. In this case one needs a reasonable high pressure of pure SF_6 of the order of 20 Pa. Even under these conditions

Figure 9
Example for the anisotropy of reactive ion etching with SF_6 of a Nb_3Ge-film: strips of 4 µm width were structured from a film of 1 µm thickness. The corners are nearly perpendicular. Their roughness is due to the grain structure of the film.

the etching remains fairly anisotropic (see Fig. 9). The T_c of a structured film is typically reduced by about 0.5 K due to the etching process.

Systematic variations of the bridge width and measurement of T_c and the critical current of the resulting microbridges revealed that strips of up to 50 nm width on both sides of the bridges seem to be strongly distorted by the etching process. In addition to the minimized reduction in T_c the small ion-energy has another very desirable effect: the sputtering rate for the PMMA becomes much smaller so that the etching rate ratio Nb_3Ge: PMMA is enhanced from 1:1 at $V_{csb} = 200$ V to 4:1 at $V_{csb} = 20$ V. The resist is hardened for this purpose by baking it for several hours at 170 °C. The background vacuum before the etching has to be better than 10^{-3} Pa since otherwise a very effective oxidation of the PMMA takes place. Water vapour inside the chamber seems to be responsible. It can sufficiently be removed by baking the whole chamber at about 100 °C for some hours. With these additional precautions the final structure after the etching is practically identical to the resist structure, especially since the Nb_3Ge-films can be kept significantly thinner than the PMMA-films. Using the above described etching procedure, bridges with a width as small as 100 nm were fabricated with a T_c of more than 20 K (see Fig. 10).

Figure 10
SEM-photograph of a Nb_3Ge microbridge with a width of about 200 nm and a thickness of about 60 nm.

3. Tunnel Contacts

As mentioned before NbN is a material of first choice for tunnel junction applications. Its low deposition temperature, the easy structuring of the films either by lift-off or reactive ion etching, the good interface properties and the reasonable high-T_c make it quite unique among the superconductors which could be applied to tunnel junction fabrication. For device operation in liquid helium only the $Nb/Al/Al_2O_3/Nb$-technique could be a strong competitor due to its reproducible preparation characteristics. For special applications a larger gap voltage or a higher operating temperature is needed than available with niobium or niobium-nitride tunnel junctions. In this case the A15-superconductors Nb_3Sn and especially Nb_3Ge could be applied. Unfortunately it is very difficult to fabricate tunnel junctions with both, Nb_3Ge base- and Nb_3Ge top-electrode, because Nb_3Ge needs an amorphous interface layer with a thickness of 20 nm to 100 nm to grow in the high-T_c A15-phase. Tunnel junctions with a Nb_3Ge base electrode and a counter-electrode of Pb or Nb are relatively easy to prepare. These junctions should exhibit a large gap voltage of about 5 mV and therefore be of interest for quasiparticle mixers, phonon generators, or fast switching devices. In the following we will first describe the preparation and some results on all-NbN tunnel junctions, and then deal in a second part with tunnel junctions having a Nb_3Ge base-electrode.

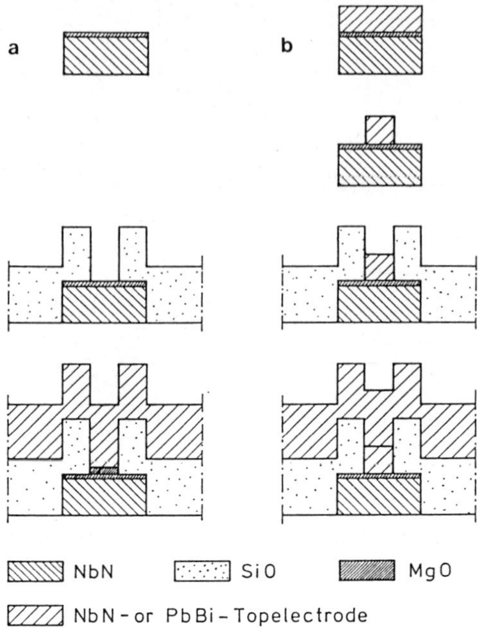

Figure 11
Schematic for the tunnel junction fabrication: structuring by (a) lift off and by (b) reactive ion-etching .

For the NbN tunnel junction fabrication we applied two methods: in the first one (see Fig 11a) successive layers of NbN, MgO and NbN were deposited without breaking the vacuum and the structuring is done by selective reactive ion-etching with SF_6; in the second method (see Fig. 11b) the base electrode and a part of the MgO barrier are prepared and structured by lift-off. Then the second part of the barrier and the topelectrode is deposited and again structured by lift-off. Both methods have their advantages and disadvantages and the decision between these techniques is more a question of available structuring techniques.

As substrates generally oxidized silicon wafers were used, cut into strips of 5 mm width and 25 mm length. The NbN is prepared onto it by the optimized reactive dc-magnetron sputtering from a Nb-target as described in Sect. 2.1. For the fabrication of the base electrode (thickness 300 nm) higher power was applied than for the counterelectrode (thickness 100 nm) in order to prevent damage to the MgO barrier. The MgO-barrier with a thickness of 0.2 nm to 1 nm was deposited by rf-magnetron sputtering from a single-crystal MgO target with an rf power density of 5 W/cm^2. As sputtering gas we used pure argon at a pressure of 1.5 Pa. The use of polycrystalline MgO targets was abandoned because the deposited MgO contained hydrocarbons as found by XPS (x-ray photoelectron spectroscopy). The hydrocarbons may have been used as binders in the target formation. Further details on the

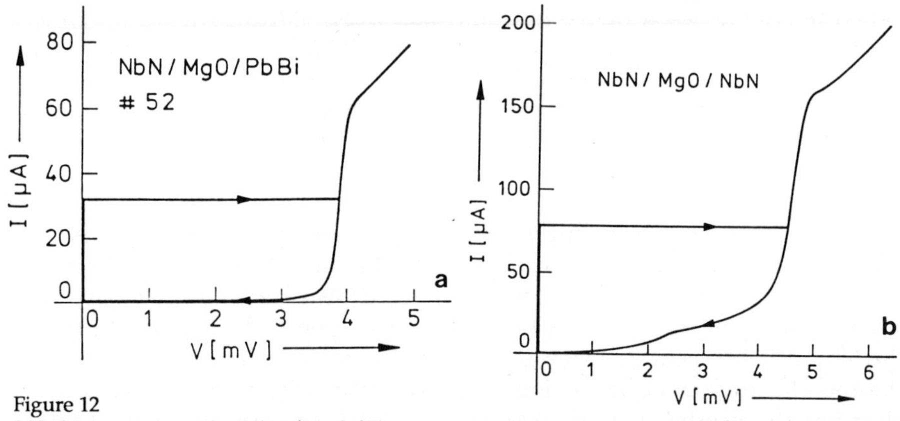

Figure 12
I-V-characteristics of a NbN/MgO/PbBi-tunnel junction (a) and a NbN/MgO/NbN-tunnel junction (b) at $T = 4.2$ K.

electrode- and barrier fabrication as well as on the insulation with SiO can be found in the literature [21].

The high quality of the NbN-base electrode and the MgO-insulator leads to very high quality tunnel junctions if evaporated PbBi-films are used as counterelectrodes (see Fig. 12a). The subgap current is very small in this case and can be explained by thermally excited quasiparticles of the superconductors. Its temperature dependence follows the theoretical predictions down to about 2 K and the quality parameter $V_m = I_c \cdot R$ (2 mV) reaches 600 mV at 2 K. The experimental data all point to the existence of a very high quality interface between the NbN base electrode and the MgO barrier.

If the PbBi counterelectrode is replaced by a sputter deposited NbN film, the quality of the tunnel junction deteriorates. The subgap current is enhanced and shows a structure at about half the gap voltage. It is nearly independent of the temperature and thus can be attributed to leakage current. Nevertheless, the resulting *I-V*-characteristic of the all-NbN tunnel junction is of reasonable quality (see Fig. 12b), good enough for most applications. The sum gap voltage is typically 4.7 mV with a transition width of 0.5 mV. Above the gap a small proximity structure is visible. All this indicates, that the barrier and the first layers of the NbN are damaged during the sputter deposition of the counter electrode. This view is supported by XPS-measurements. A similar behavior is found in the fabrication of tunnel junctions with Nb_3Ge base electrode. The tunnel junctions show very good *I-V*-characteristics, as long as the counter electrode was evaporated. Sputter deposited counter electrodes deteriorate the *I-V*-characteristics, even if they were deposited at ambient temperature. Enhanced deposition temperatures for the counter electrodes, as for instance the 900 °C for a Nb_3Ge deposition, made the tunnel junction fabrication impossible.

The Nb_3Ge-based tunnel junctions [22] were fabricated using am in-situ process. First a Nb_3Ge-film is sputtered on a heated (870 °C) sapphire substrate using the optimized deposition technique described in Sect. 2.1. These films are typically superconducting ($R = 0$) above 20 K. The next step in the sandwich preparation is the sputtering of the barrier onto the still hot substrate immediately after the Nb_3Ge deposition. For our investigations oxides like SiO_2, Al_2O_3, and MgO and fluorides like CaF_2 and MgF_2 were used and deposited by rf-magnetron sputtering at a pressure of 13 Pa pure argon. The target-substrate distance and the gas pressure have to be adjusted for full thermalization, in order to prevent a bombardment of the Nb_3Ge base electrode surface and thus a deterioration of the superconducting properties of the Nb_3Ge base electrode. After the substrate is cooled down to below 100 °C, the top electrode material is deposited. After its removal from the vacuum chamber the sandwich is structured in a SNIP process using photolithography. Depending on the top electrode material, either wet chemical etching using $FeCl_3$ for Pb or reactive ion etching with SF_6 for Nb or NbN was applied.

A series of junctions were prepared using the oxides as artificial barrier, but only contacts with SiO_2 exhibit good tunnel characteristics (see Fig. 13). The other oxidic barrier materials led to normal conducting behavior or to microshorts. Tunnel junctions with MgF_2 barrier show similar good values like those with SiO_2. As mentioned before, only evaporated top electrodes (here Pb) do not damage the barrier and the surface layer of the base electrode. Thus the junctions with sputter deposited Nb and NbN exhibit only very poor tunnel characteristics.

Altogether, junctions prepared with SiO_2 barrier exhibit V_m-values (measured at 2 mV) of up to 12 mV, those with MgF_2-barrier V_m-values of up to 50 mV. Both show gap voltages of up to 5 mV. The critical current density can be varied by the barrier thickness between 2 kA/cm^2 and 5 A/cm^2 (see Fig. 14a). The dependence of the critical current I_c on an external magnetic field exhibits a nearly ideal Fraunhofer pattern for most junctions (see Fig. 14b).

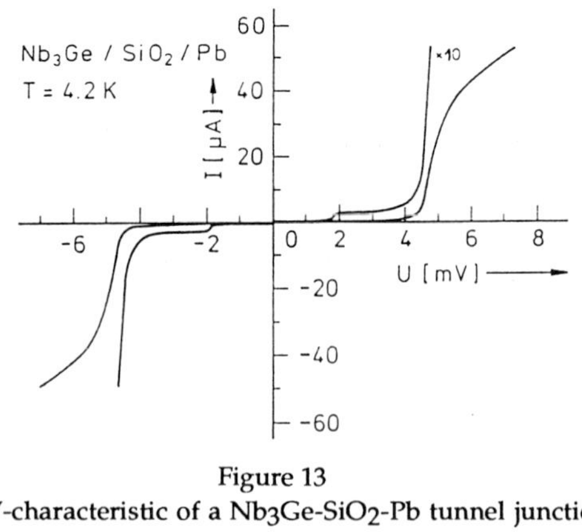

Figure 13
I-V-characteristic of a Nb_3Ge-SiO_2-Pb tunnel junction.

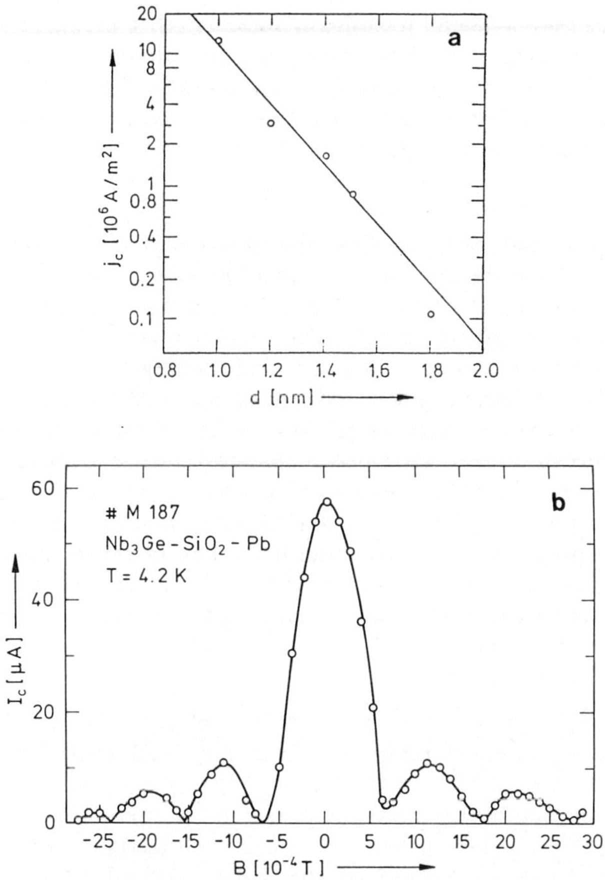

Figure 14
(a) dependence of the critical current density j_C on the barrier thickness d, (b) magnetic field dependence of the critical current I_C of a Nb_3Ge-SiO_2-Pb tunnel junction.

4. Microbridges

Even though Josephson tunnel junctions are the most accepted way of making Josephson contacts, other realizations are possible as well. SNS-junctions and microbridges often offer a better solution to application problems than tunnel junctions, which in many cases have to be artificially shunted to overcome their hysteretic behavior, e.g. for SQUID-applications. Additionally some materials cannot be fabricated in a way compatible with tunnel junction fabrication, as has been pointed out in Sect. 2 for the case of Nb_3Ge.

This material has a very small coherence length ($\xi \approx 3$ nm) which makes its application for tunnel junctions and microbridges very difficult. For microbridges dimensions of a few coherence lengths or even less are necessary to obtain a clean Josephson effect [23]. Currently such small Nb3Ge-structures cannot be fabricated with a sufficiently high-T_C. Wider microbridges thus have to be applied, which operate in the flux-flow regime, or otherwise techniques have to be developed to reduce the effective bridge length (variable thickness bridges, e.g. [24]) or to increase the coherence length in the bridge region (SNS-bridges, e.g. [25]). All three methods have their draw backs: wide microbridges of constant thickness have a large critical current which may not match to the needs of the attached superconducting circuits. Variable thickness bridges loose their high operating temperature since it is just the high-T_C top layer which is partially removed in order to decrease the critical current. The bottom layer of Nb3Ge-films possesses only a much smaller T_C if it is superconducting at all [6]. SNS-bridges show a very nice Josephson effect and can be successfully applied to many intermediate temperature range applications [26]. Their main draw back is again a T_C-reduction due to the normal material and in addition their very low resistance in the operating region of the order of μΩ to some mΩ.

In order to preserve the high-T_C and at the same time reduce the critical current I_C we optimized the Nb3Ge fabrication in order to make very thin films (30..60 nm) of the high-T_C A15-phase of Nb3Ge as described in section 2.1. The critical current was adjusted by varying the width of the bridge.

4.1 Theoretical Model

Bridges with widths between 50 nm and 200 nm work in the flux flow region [27, 28] in a parameter range, which is very well accessable to a theoretical treatment: the width and thickness of the bridges is small compared to the penetration depth, but large compared to the coherence length. In this case the phase and current distribution inside the bridge can be calculated from a potential approximation of the 2-dimensional Ginzburg-Landau equation. The bridge is assumed to be of hyperbolic shape with variable angle. In the calculation complex analytical geometry is applied and, for the ease of calculation, especially elements of conformal mapping. A point $z = \{x,y\}$ is represented by $z = x + iy$.

We first calculate the current density J_t and the phase ψ_t of the corresponding wavefunction ψ in a bridge if a current I_t is imposed. The bridge is modelled by a hyperbolic neck of width s and angle β (see Fig. 15).

The foci are located at $\{-a,0\}$ and $\{a,0\}$: $a = s/(2 \sin \beta)$. The superconducting current density in the bridge can be described by the Ginzburg-Landau equation:

$$J = \frac{2e}{2m} |\psi|^2 \left(\hbar \nabla \psi - \frac{2e}{c} A\right). \tag{4.1}$$

Since the cross-section of the bridge is very small the kinetic energy of the current is much larger than the field energy by a factor of the order of $\lambda^2/(d \cdot w)$, d being the thickness of the film. Thus the field term can be neg-

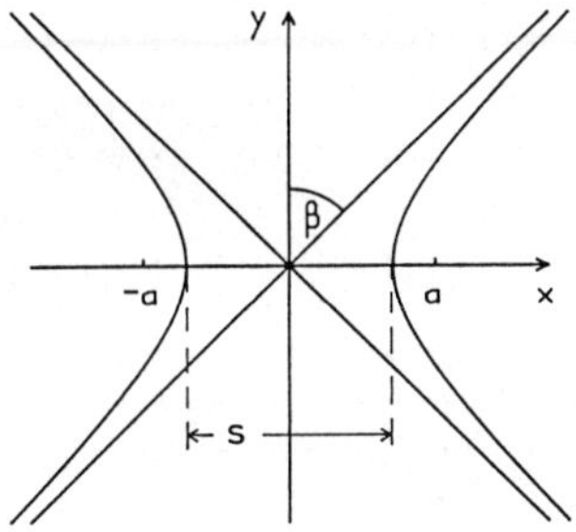

Figure 15
Model of a microbridge - a hyperbolic neck with angle β and width s.

lected. As a further simplification one can assume that the fields and currents are so weak that $|\psi|$ can be replaced by ψ_∞ everywhere [28]. This reduces the GL theory to the London theory. The result is an equation connecting the current density J and the gradient of the phase ψ of the wavefunction:

$$J = \frac{c^2 \hbar}{8 \pi e \lambda^2} \nabla \psi . \tag{4.2}$$

The next step in the calculation of the current density J is the determination of the phase from:

$$\nabla^2 \psi = 0 \qquad \text{and} \qquad \partial\psi/\partial n = 0, \qquad \text{since} \qquad \operatorname{div} J = 0 .$$

Solutions for such boundary value problems in conjunction with Laplace's equation can be calculated using the standard methods of potential theory. One of these methods, conformal mapping, is especially well suited to our problem since conformal transformations of hyperbolic necks onto strips and planes are well known. Applying this technique yields the phase ψ_t and current density J_t in a hyperbolic bridge (without vortices) due to an imposed current I_t:

$$\begin{aligned} \psi_t(z) &= \frac{I_t}{I_o}\frac{\pi}{2\beta} \operatorname{Re}\left\{\ln\left(z/a + \sqrt{z^2/a^2 - 1}\right)\right\} \\ J_t(z) &= i\,\frac{I_t}{d\pi}\frac{\pi}{2\beta}\left(a^2 - z^2\right)^{-1/2} . \end{aligned} \tag{4.3}$$

The influence of β on the current distribution in the hyperbolic neck is clearly visible in the 3-dimensional representation of the coordinate dependence of $|J_t|$ in the bridge (see Fig. 16). With increasing β the current concentration in the edges of the bridge becomes more and more pronounced.

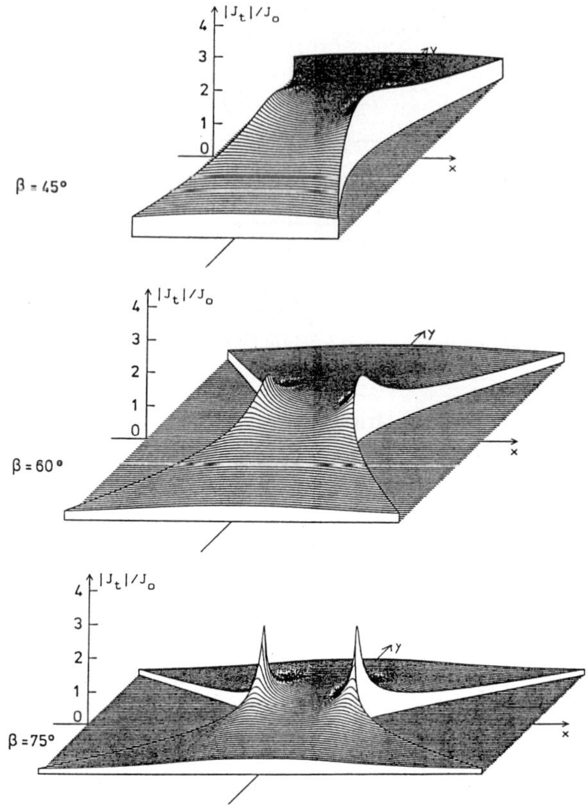

Figure 16
Distribution of the absolute value of the current density J_t in three bridges of different angle β. The current and the width of the bridges are all the same.

In an infinite area a vortex may be represented by a phase field which is only dependent on the angle:

$$\psi_v(z) = q \arg(z) = q\,\mathrm{Im}\{\ln z\}\,. \tag{4.4}$$

q is +1 or -1 depending on the orientation of the vortex. This picture is valid for the high- -limit where a vortex can be considered of radius ξ, $\xi << \lambda \cdot A$ solution will be obtained in the radial area $\xi < r < \lambda$. As mentioned before the films under consideration are thin: $d << \lambda$. In this case we can assume a constant current density throughout the thickness of the film. The current density of a vortex in these limits is then given by 11:

$$J_v(r) = \frac{c^2 \hbar}{8 \pi e \lambda^2} \nabla \psi_2(r)\,. \tag{4.5}$$

The phase of a system of vortices at positions z_1 - z_n results from linear superposition as long as their distance is sufficient:

$$\psi_v(z) = \sum_{j=1}^{n} q_j \, \mathrm{Im}\left\{\ln\left(z - z_j\right)\right\}. \tag{4.6}$$

Making again use of properties of conformal mappings we finally get a closed expression for the current density J_v and phase ψ_v in a current-carrying hyperbolic bridge with vortices at locations z_j, j = 1 - n:

$$\psi(z) = \frac{I_t}{I_o}\frac{\pi}{2\beta}\,\mathrm{Re}\left\{\ln\left(z/a + \sqrt{z^2/a^2 - 1}\right)\right\} + \sum_{j=1}^{n} q_j \,\mathrm{Im}\left\{\ln \frac{w^{\pi/2\beta} - w_j^{\pi/2\beta}}{w^{\pi/2\beta} + \left(\overline{w_j}\, e^{i\pi}\right)^{\pi/2\beta}}\right\} \tag{4.7}$$

$$J = \frac{I_o}{\pi d}\frac{\pi}{2\beta}\,\overline{\left(a^2 - z^2\right)^{-1/2}}\left\{ i\frac{I_t}{I_o} + \sum_{j=1}^{n} q_j\left(\frac{\overline{w}^{\pi/2\beta}}{\overline{w}^{\pi/2\beta} - \overline{w_j}^{\pi/2\beta}} - \frac{\overline{w}^{\pi/2\beta}}{\overline{w}^{\pi/2\beta} + \left(w_j\, e^{-i\pi}\right)^{\pi/2\beta}}\right)\right\} \tag{4.8}$$

with $w = z/a + \sqrt{z^2/a^2 - 1}$ and $w_j = z_j/a + \sqrt{z_j^2/a^2 - 1}$.

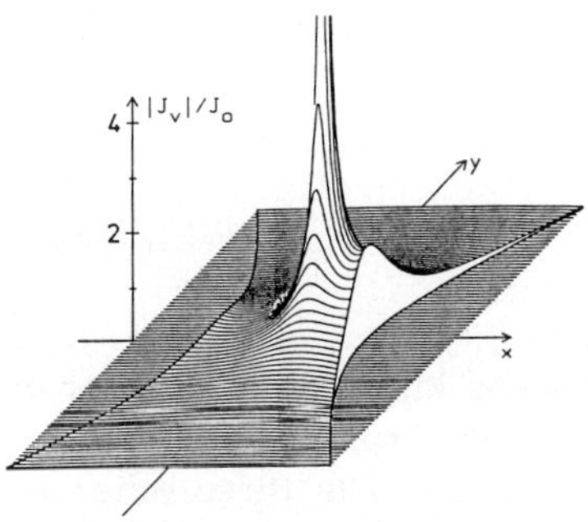

Figure 17
Current density J_v in a 45°-bridge resulting from a vortex at (0.3a, 0). No external current is imposed.

In Fig. 17 the modulus of the current density J_v in a 45°-bridge is plotted in the form of a 3-dimensional representation. No external current is imposed. As in the case of the current-carrying bridge, a current concentration is visible at the edges of the neck. Compared to the current density J_t resulting from an external current $I_t = I_o$ (see Fig. 16) the vortex current density J_v is significant. This has the consequence, that after the nucleation of one vortex the nucleation of a complementary vortex at the other side of the bridge is likely as the first vortex travels through the bridge, even if there is some asymmetry in the bridge.

The kinetic energy of currents within a bridge can be calculated from:

$$E = \frac{2\pi}{c^2}\int_V \lambda^2 J^2 \, d\tau = \frac{2\pi}{c^2}\, d \int_S \lambda^2 J^2 \, da\,, \tag{4.9}$$

neglecting the small contribution by the core of the vortices and assuming a constant current density throughout the thickness of the film ($d << \lambda$). The total current density J is the sum of the external current density J_t and the vortex current density J_{vi}, i = 1 ... n:

$$J = J_t + \sum_{i=1}^{n} J_{vi} \, . \tag{4.10}$$

In this case the kinetic energy can be written as:

$$\begin{aligned} E = \frac{2\pi}{c^2}\lambda^2 d \Bigg\{ & \int\!\!\int_S |J_t|^2 \, da && \text{external current} \\ & + 2\sum_{i=1}^{n} \int_S J_t J_{vi} \, da && \text{vortex-current interaction} \\ & + \sum_{\substack{j,k=1 \\ j\neq k}}^{n} \int_S J_{vj} J_{vk} \, da && \text{vortex-vortex interaction} \\ & + \sum_{i=1}^{n} \int_S |J_{vi}|^2 \, da \Bigg\} && \text{vortex selfenergy} \end{aligned} \tag{4.11}$$

The first term represents the kinetic energy E_t of the transport current in a bridge containing no vortices. From this term the kinetic inductance of the microbridge can be calculated. The three other terms contain the vortex dependent parts. Since all the current densities are known in a closed form, an explicit expression for the energy can be given and via differentiation explicit expressions for the forces $F_v(z_j)$ acting on vortices at positions z_j, j = 1 ... n, within a current-carrying bridge. For ease of calculation we again use complex notation: $F = F_x + i\,F_y$, where F_x and F_y are the x- and y-components of F. $F_v(z_j)$ can be calculated from the vortex energy E_v:

$$\begin{aligned} F_v(z_j) &= - \nabla_j E_v \\ &= - \nabla_j E_t + \nabla_j E_1 + \nabla_j E_w \\ &= F_t(z_j) + F_1(z_j) + F_w(z_j) \end{aligned} \tag{4.12}$$

With F_t we denote the vortex-current interaction force, with F_l the lineforce, and with F_w the vortex-vortex interaction force. In explicit form the total force acting on a vortex at rest in a hyperbolic bridge is:

$$F_v(z_j) = E_o \frac{\pi}{\beta} \overline{\left(a^2 - z_j^2\right)}^{-1/2} \Bigg\{ q_j \frac{I_t}{I_o} \qquad \text{vortex-current force}$$

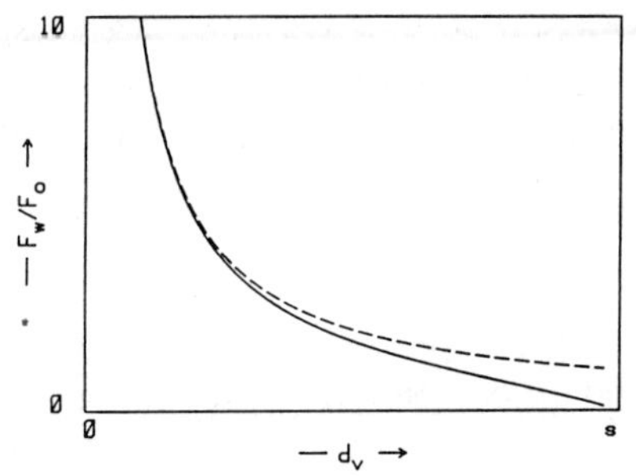

Figure 18
Vortex-vortex interaction force F_W as function of distance d_V for two symmetrically located vortices at $x_1 = -x_2$ and $y = 0$ with different orientation. The full line corresponds to $\beta = 45°$, the dashed line to $\beta = 90°$.

$$+ \frac{\beta}{\pi}\overline{z_j}\left(a^2 - z_j^2\right)^{-1/2} + \frac{1}{2}\tan\frac{\pi\left(\pi/2 - \theta_j\right)}{2\beta} \qquad \text{line force} \qquad (4.13)$$

$$+ i \sum_{\substack{k=1 \\ k\neq j}}^{n} q_j q_k \left(\frac{\overline{w_j}^{\pi/2\beta}}{\overline{w_j}^{\pi/2\beta} - \overline{w_k}^{\pi/2\beta}} - \frac{\overline{w_j}^{\pi/2\beta}}{\overline{w_j}^{\pi/2\beta} + \left(w_k\, e^{-i\pi}\right)^{\pi/2\beta}}\right) \qquad \text{vortex-vortex force}$$

As an example the vortex-vortex-interaction force F_W for two vortices of opposite orientation is plotted in Fig. 18 as function of the vortex distance d_V for $y = 0$. For small dv the interaction force F_W is practically independent of the bridge angle β. In this case F_W is proportional to $(x_j - x_k)^{-1}$ and represents a strong contribution to the total force acting on a vortex.

From the expression for the forces acting on vortices in a current carrying bridge we can derive the dynamic behavior of the vortices: if we apply a current larger than the critical current I_c the vortex begins to move against the electromagnetic damping force F_c. Thus the sum of forces acting on a single vortex in the bridge is:

$$F_t(z) + F_l(z) + F_c(z) = 0\,. \qquad (4.14)$$

This is a first order nonlinear differential equation in z, which can be separated into two coupled equations for the real and imaginary part. For a system of n vortices we have to include the interaction force F_W. This leads to a system of 2n coupled first order differential equations, j = 1 ... n:

$$-\eta\frac{dx_j}{dt} = \mathrm{Re}\left\{F_t(z_j) + F_l(z_j) + \sum_{\substack{i=1 \\ i\neq j}}^{n} F_w(z_j, z_i)\right\}$$

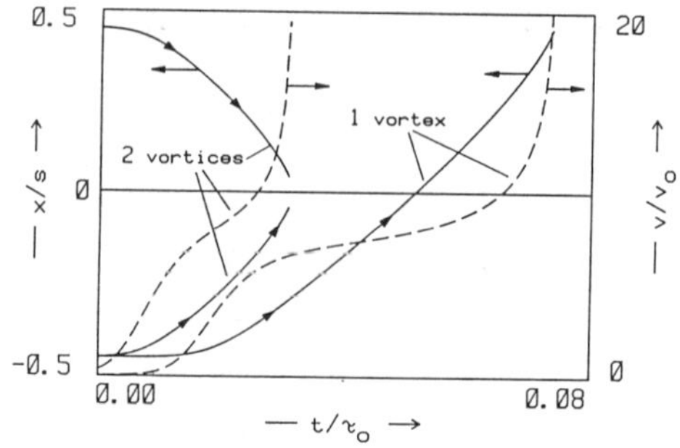

Figure 19
x-t-dependence (full) and v-t-dependence (dashed) for a single vortex and a vortex pair moving along the x-axis in a 45°-bridge. Diameter of the vortex core is $2\xi = 0.1$ s.

$$- \eta \frac{dy_j}{dt} = \mathrm{Im} \left\{ F_l(z_j) + F_l(z_j) + \sum_{\substack{i=1 \\ i \neq j}}^{n} F_w(z_j, z_i) \right\} \tag{4.15}$$

The effect of pinned vortices can be included: as long as their position is fixed they only appear additively in the interaction terms. If one wants to observe their time dependent motion too, one has to include counteracting pinning forces in these equations.

The set of equations can easily be solved by standard methods of numerical mathematics as for instance the Runge-Kutta-Fehlberg algorithm [29]. Some simple cases for the equation of motion can even be solved in closed form. As an example for the motion of a single vortex and a vortex pair in a 45°-bridge, the equation of motion was solved numerically and the resultant x-t-dependence is plotted in Fig. 19. The diameter $2\ \xi$ of the vortex is assumed to be 10 % of the width of the bridge. Additionally the diagram includes the time dependence of the velocity v. Without surface barrier the vortices start with a finite velocity if a nonzero transport current is applied. In the presence of pinning the dynamics of the vortex transport in the bridge is again determined by $I - I_c$.

In the presence of a surface barrier we have to apply a current larger than its critical current to free the vortex from the barrier. As long as the current is only slightly larger than the critical current, the velocity of the vortex is very low and only slowly increasing as the vortex moves through the bridge until it is fully unbound. The velocity rapidly increases as the vortex crosses the center of the bridge since it is then attracted by the surface barrier of the opposite side of the bridge. The vortex velocity is very sensitive to a small current change in the low-velocity region.

The experimentally most easily available electrical information on microbridges is their I-V-characteristic. It can be separated into two principle parts. The first one is the current range up to the critical current I_c. This of course does not mean, that there are no vortices in the bridge, but the vortices are pinned either at the surface barrier or at pinning centers inside the bridge. The magnitude of the critical current is determined by the surface barrier and the strength and distribution of the pinning centers. Both depend strongly on the preparation technique and material of the bridge.

If we apply a dc-current larger than the critical current a voltage appears at the terminals of the bridge, meaning that vortices start to move across the bridge: between to points z_1 and z_2 a voltage arises, which can be calculted via the phase change at z_o and z_1 and the second Josephson equation:

$$V_m = \frac{\hbar}{2e}\left[\dot{\phi}(z_1) - \dot{\phi}(z_o)\right] \tag{4.16}$$

with:

$$\dot{\phi}(z) = \frac{d}{dt}\phi_t(z) + \frac{d}{dt}\sum_{j=1}^{n}\phi_v(z, z_j),$$

$$\dot{\phi}(z) = \frac{\dot{I}_t}{I_o}\frac{\pi}{2\beta}\,\mathrm{Re}\left\{\ln\left[z/a + \sqrt{z^2/a^2 - 1}\right]\right\},$$

and:

$$\dot{\phi}(z, z_j) =$$

$$\frac{\pi}{2\beta} q_j\,\mathrm{Im}\left\{\frac{\dot{z}_j}{\sqrt{z_j^2 - a^2}}\frac{w_j^{\pi/2\beta}}{w^{\pi/2\beta} - w_j^{\pi/2\beta}} + \frac{\dot{\bar{z}}_j}{\sqrt{\bar{z}_j^2 - a^2}}\frac{\left(\overline{w_j}\, e^{i\pi}\right)^{\pi/2\beta}}{w^{\pi/2\beta} + \left(\overline{w_j}\, e^{i\pi}\right)^{\pi/2\beta}}\right\}$$

$$+\, \mathrm{Im}\left\{\ln\frac{w^{\pi/2\beta} - w_j^{\pi/2\beta}}{w^{\pi/2\beta} + \left(\overline{w_j}\, e^{i\pi}\right)^{\pi/2\beta}}\right\} q_j \frac{1}{\tau_v \cosh^2(2t/\tau_v - 1)}\,.$$

In the actual measurement of the *I-V*-characteristics we apply a dc-current of slowly increasing magnitude. Therefore we can neglect the dI_t/dt-term. Additionally the voltage probes are far away from the center of the bridge. For simplicity we assume that they are located at positions $\pm\infty$ on the y-axis symmetrical to $x = 0$. With these assumptions we can derive a simple expression for the voltage, dependent on the position z_j and velocity $d/dt\, z_j$:

$$V_m = \frac{\hbar}{2e}\frac{\pi}{\beta}\sum_{j=1}^{n}\mathrm{Re}\left\{\frac{\dot{z}_j}{\sqrt{a^2 - z_j^2}}\right\}. \tag{4.17}$$

This equation is valid in bridges of any angle and for any distribution of vortices in the bridge.

From the equation of motion and the equation for the instantaneous voltage we can calculate the mean voltage numerically for different externally imposed currents. In Fig. 20 the resulting *I-V*-characteristic is plotted for vortex pairs, created simultaneously with opposite spin on opposite sides

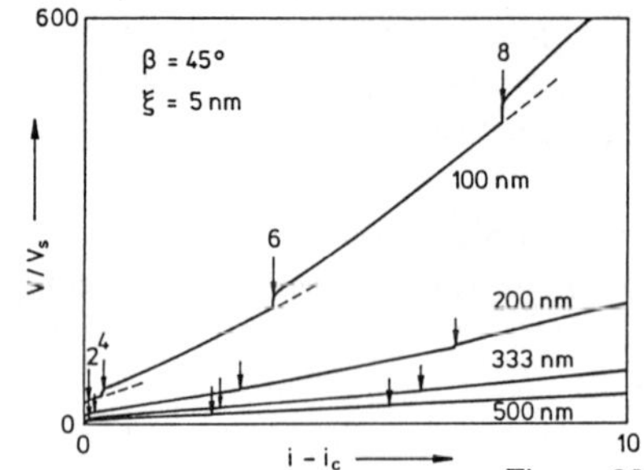

Figure 20

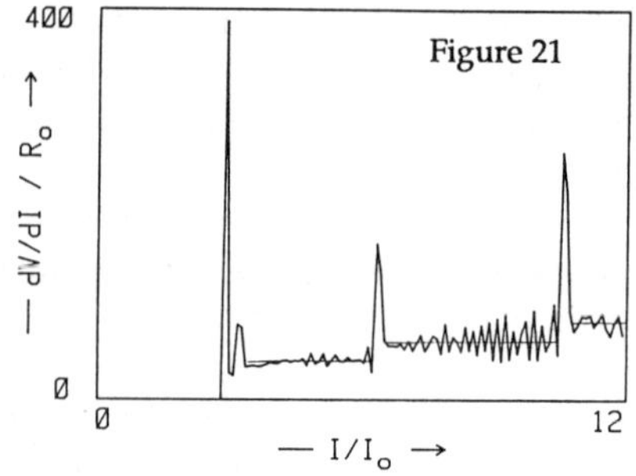

Figure 20

Calculated I-V-characteristics for varying microbridge widths s and fixed coherence length ξ. Notice that the voltage is referred to $V_S = V_{OS2}$ thus putting the width dependence back into V_S. The appearance of new vortex pairs is indicated by arrows.

Figure 21

Derivative dV/dI of the I-V-characteristic of Fig. 20 for $s = 100$ nm ($\xi/s = 0.05$). The curve is obtained by pointwise numerical differentiation. The oscillations within the plateaus are due to the digitizing.

of 45°-bridge. For small bridges ($< 50\ \xi$) the appearance of new vortex pairs can clearly be seen in the steps in the I-V-characteristics.

For the calculation of the I-V-characteristic of Fig. 20 we take into account the surface barrier, the driving vortex-current force and the vortex-vortex interaction. The time to create a vortex is assumed to be negligible against the characteristic period of the vortex motion. This is reasonable if the vortex nucleation current is significantly smaller than the current necessary to free a vortex from the surface barrier. The critical current I_c is suppressed.

The structure in the experimental curves is more obvious if dV/dI-plots are measured. The derivative dV/dI of the I-V-characteristic for $s = 100$ nm plotted in Fig. 20 is shown in Fig. 21. The step-like structure with an overshoot at the beginning of each step can be observed both in theoretical curves and experimental ones.

The curve is found by direct numerical pointwise differentiation of the I-V-characteristic. The oscillations within the plateaus stem from the digitization.

The height of the steps increases with increasing ratio ξ/s and vice versa. For instance with $\xi/s = 0.01$ the steps are practically invisible and the I-V-characteristic is mainly made up by a series of straight lines. The same effect can be observed if the vortex motion inside the microbridge is more stochastic: the small steps vanish and we again find an I-V-characteristic made up of a series of straight lines connected by smooth curves. Such characteristics were observed for wide bridges (e.g. [30]).

The computer evaluation was performed for currents up to $I - I_c = 10\ I_0$. This corresponds to 8 moving and interacting vortices. For higher currents or larger vortex diameters the vortices will not necessarily be lined up in one row: we have to take into account parallel paths of the vortex motion. The calculations are in principle the same as used here in the single path model but one has to include the two-dimensional form of the equation of motion.

With this model a number of I-V-characteristics can be explained which were found for microbridges with a width of a few times the coherence length up to a width of the order of the penetration depth [30, 31]. The main restriction in the theory is the assumption that the calculation can be done using potential theory, thus neglecting the vector potential in the Ginzburg-Landau equation and non-linear terms. The experimental results support the applicability of this model.

4.2 Experimental Results

The methods described in Sect. 2 and 3 allow to fabricate small microbridges from thin Nb_3Ge-films with a width of less than 100 nm and a T_c of more than 20 K. Even though the lithography and film preparation are sufficiently reproducible the electrical characteristics of the microbridges are not so well controlled. The reason is, that the combination of a very small width of the bridges and a grain size of about the film thickness leads to only few grains in the neck of the bridge. Thus the electrical characteristics are influenced too by grain boundary pinning, which may differ significantly from bridge to bridge. This is especially true for very small bridges.

The I-V-characteristic of wide bridges (500 nm and more) is made up of a polygon (see Fig. 22) which approximates a parabolic behavior [24], [32]. In such I-V-characteristics a step structure is usually invisible even if the region is magnified where the straight lines meet. However one often finds a slight overshoot in the dV/dI-curves differentiated with the lock-in method as the current proceeds from one line segment to the next one. This indicates a small step rounded by noise. Reasons for the rounding may be external and thermal noise and an additional stochastic component in the vortex motion due to pinning and nucleation. From the calculations in Sect. 4.1 it is clear that the step height becomes small for wide bridges (see Fig. 20).

If one further reduces the width of the bridges to less than 200 nm a step structure in the I-V-characteristic is often clearly visible (see Fig. 23) in accordance with the results of Sect. 4.1 . The increasing height of the steps with increasing step number was also found in the simulations. The distribution of the steps along the I-axis can be attributed to some asymmetry in the bridge.

For larger currents (above 160 μA in Fig. 23) the structure changes. The step height decreases and a rounded slope like the one of Fig. 22 appears. Both, heating and parallel paths for the vortex motion can be the reason. The agreement between the theoretical model and the experimental data is very good and the results indicate, that these microbridges have unique features and are well adapted to Josephson contact applications, as will be shown in Sect. 5. and 6.

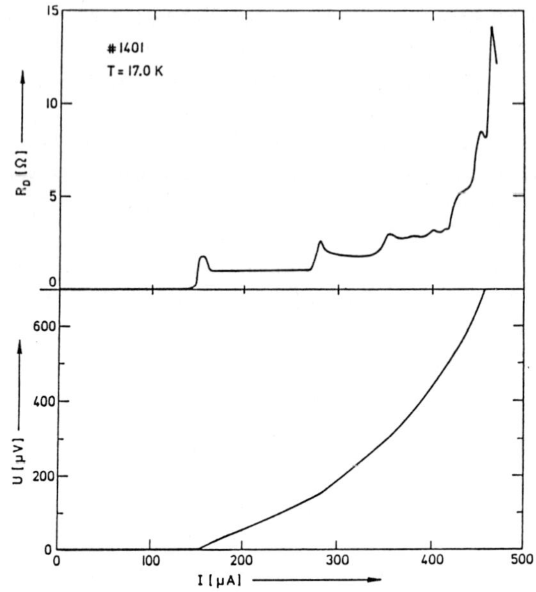

Figure 22
I-*V*- and d*V*/d*I*-characteristic of a typical wide Nb3Ge-bridge. The width is about 500 nm and the thickness about 60 nm.

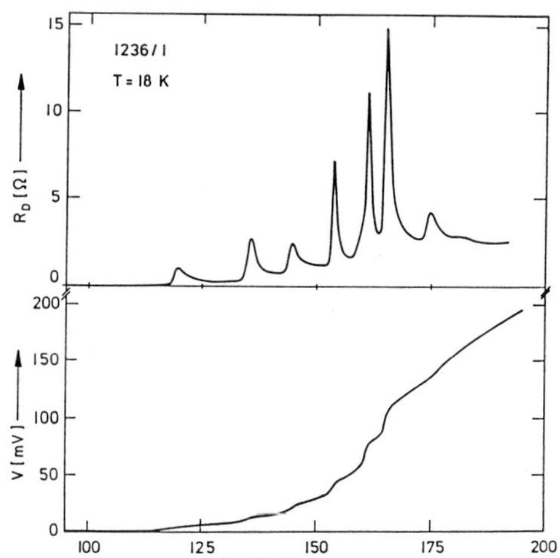

Figure 23
I-*V*- and d*V*/d*I*-characteristic of a small Nb3Ge-bridge. The width is less than 200 nm and the thickness is about 60 nm.

5. High -T_c SQUIDs

In this chapter we will deal with the application of Nb_3Ge-nanobridges in superconducting quantum interference devices (SQUIDs). In the first section the preparation and operating characteristics of single layer Nb_3Ge dc-SQUIDs will be presented. In the second one we describe a Nb_3Ge multilayer technique for the integration of more complex superconducting circuits. As an example we present a dc-SQUID with integrated input coil, whose superconducting parts consist of high-T_c Nb_3Ge and which operates at temperatures up to about 20 K.

5.1 Single Layer Nb_3Ge-DC-SQUIDs

The properties of small Nb_3Ge-nanobridges described above led to their use in dc-SQUIDs [33]. The fabrication of single layer dc-SQUIDs is similar to the preparation of individual nanobridges. In a first step the two bridges and the SQUID-ring (see Fig. 24) are structured by electron beam lithography and reactive ion etching from a single Nb_3Ge-film. In a second step the contact pads are prepared by conventional photolithography using AZ 1350 resist. Connections to the measuring system are made by bonded aluminum wires and the measurements were performed using the same equipment as for single nanobridges.

The temperature dependence of the critical current of these dc-SQUIDs corresponds to the one of single bridges, typically $I_c \sim (1-t)^{3/2}$ where $t = T/T_c$.

Figure 24
SEM-picture of a small area Nb_3Ge dc-SQUID

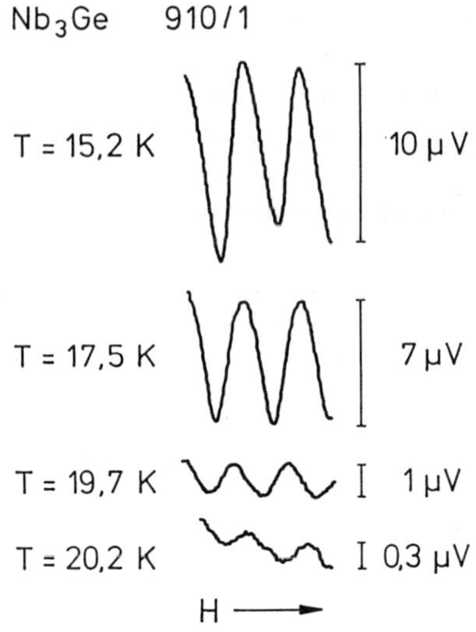

Figure 25
Voltage modulation pattern of a Nb_3Ge dc-SQUID for different temperatures. The bias current was optimized for each temperature.

The *I*-*V*-characteristic exhibits linear regions approximating a parabolic behavior just like single nanobridges. If the dc-SQUIDs are made from very narrow bridges, again a pronounced step-structure in the *I*-*V*-characteristics appears. If an external magnetic field is applied to the dc-SQUID periodic modulation in the *I*-*V*-characteristic is observed up to temperatures of about 1 K below the transition temperature of the unstructured Nb_3Ge-film (see Fig. 25). At low temperatures (typically below 10 K) a thermal hysteresis appears. For most devices a total operating temperature range of about 10 K is obtained with a maximum temperature of 19.5 K. One sample showed clearly visible modulation even at 20.2 K [34].

If, as hypothesis, we assume the applicability of the RSJ-model [35] on the nanobridges of a dc-SQUID a high modulation depth in the *I*-*V*-characteristic of the dc-SQUID could be achieved if the parameter β, $\beta = LI_c/\phi_o$, is of the order of one [36]. L is the SQUID-inductance and I_c the critical current of each individual bridge which is assumed to be the same for both bridges. Since the critical current varies over the operating temperature range from typically 1 mA at low temperatures to 100 μA near T_c the inductance of the SQUID-structure should be only some pH. Thus we performed our first experiments with a SQUID having a small area A of about 5 μm*5 μm.

From the approximation $L = \mu_o \cdot \sqrt{A}$ for the geometrical inductance L of the SQUID we found an inductance of about 6 pH. Thus we should observe a modulation depth in the current of about 40 % for $I_c = 500$ μA. But typically we found a modulation depth of only a few percent practically independent of the area A of the SQUID as A was varied between 25 μm^2 and 250 μm^2. Two reasons may be responsible for this behavior. The first is that the validity of the RSJ-model is quite questionable for nanobridges whose behavior is dominated by flux motion. And effects of pinning and of the vortex dynamics are not included in the basic RSJ-model.

The second more obvious reason is the large kinetic inductance of the microbridges. The kinetic inductance of a hyperbolic nanobridge (bridge-angle $\beta = 45°$, width 100 nm, film thickness 50 nm, length 500 nm, $\lambda = 500$ nm) can be calculated to be 16 pH. Thus we observe a total kinetic inductance of 32 pH of the two bridges corresponding to an equivalent magnetic inductance of a ring of 28 µm diameter. Even without any further magnetic inductance of the SQUID-structure we would observe a modulation depth of only 10 % for $I_C = 500$ µA corresponding to a SQUID-parameter $\beta = 2$. For large values of the parameter β the modulation depth is not very sensitive to changes in β. Thus a drastic change in the modulation depth can not be expected if the SQUID-area is varied.

Apart from one sample where a 50 % modulation in the critical current was observed most of the other SQUIDs exhibited only a small modulation depth of one to five percent in I_C. With differential resistances of the order of 0.1 Ω to 1 Ω this small I_C-modulation nevertheless leads to a sufficient voltage modulation.

Voltage modulation appears only at currents slightly above I_C in a range of about 5 % to 20 % of I_C (see Fig. 26). The quite sharp limitation to higher currents is another indication that the RSJ-model is not applicable. Instead the modulation should be discussed in terms of vortex motion, its synchronization in the two bridges, and the influence of the surface barrier.

In the non hysteretic mode of operation the maximum modulation voltage for a given SQUID is a linear function of the temperature (see Fig. 27) with a typical value of 10 µV at 18 K. The relative current modulation

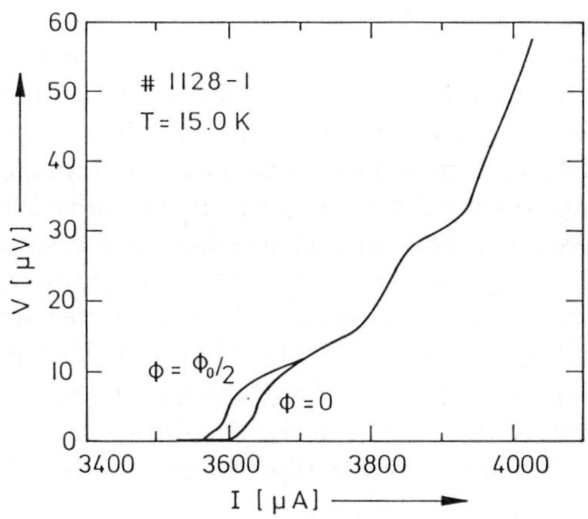

Figure 26
I-V-characteristic of a dc-SQUID with and without applied flux.

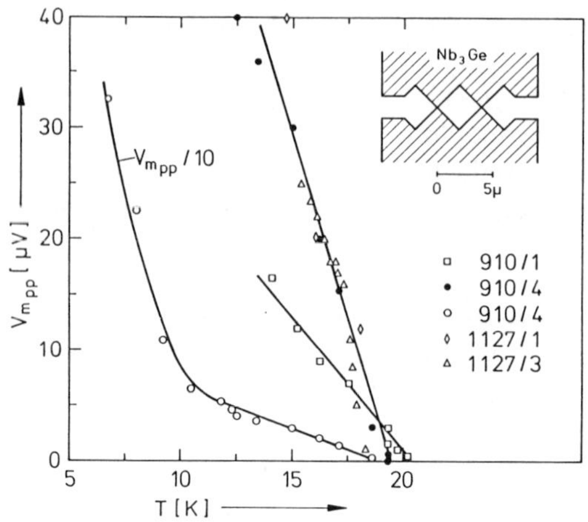

Figure 27
Voltage modulation for same dc-SQUIDs as function of the temperature. The dc-bias current was readjusted at each temperature point to obtain optimum modulation.

$c_m = \Delta I_c / I_c$ is nearly independent of temperature. The voltage modulation is determined by $\Delta V(t) = c_m \cdot R_d(t) \cdot I_c(t)$. Here $R_d(t)$ is the temperature dependent differential resistance and t the reduced temperature. Typically R_d has values between 0.1 Ω and 10 Ω. Since R_d often behaves like $(1\text{-}t)^{-0.5}$ and I_c like $(1\text{-}t)^{1.5}$ this results in the observed linear temperature dependence of $\Delta V(t)$.

Many periods of modulation are always observed without change in the amplitude. This is due to the very small sensitivity of the individual bridges to an external magnetic field. Depending on the current bias point either sinusoidal or sinus-square like modulation is obtained.

The magnetic flux sensitivity $S_b = \phi_o / \Delta B$ is defined by the change ΔB in the external flux density necessary to increase or decrease the flux in the SQUID by one flux quantum. Calculations of this value from the geometrical data of the SQUID lead to a much lower value than was measured. This is due to flux compression by the superconducting banks on both sides of the SQUID. A bank width of 3 mm and a length of each bank varying from 100 μm to 1.5 mm lead to an increase of S_b by a factor of ten. To optimize the geometry of the banks the dependence of the flux sensitivity on the length b of the banks (see Fig. 28) was studied. We found the dependence $S_b \sim b^{0.83}$ for a bank width of 3 mm. This is consistent with computer simulations [37]. For most SQUIDs an overall size of the banks and SQUID-structure of 3 mm by 3 mm was chosen, which leads to a sensitivity S_b of 1 ϕ_o per 4×10^{-7} T.

Some dc-SQUIDs were operated in the flux-locked-loop mode using an external feed-back coil with a mutual inductance to the SQUID of about 1 pH. We used conventional 100 kHz dc-SQUID electronics [38] which needs about 2 μV as minimum input voltage. This limits the maximum operating temperature to about 19.3 K. Over the entire operating temperature range

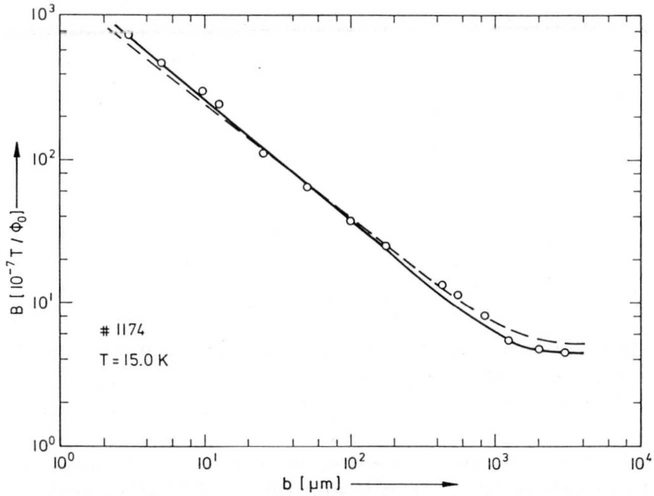

Figure 28
Influence of the flux compression by superconducting banks around a SQUID. With increasing length b of the banks the flux density B needed to change the flux in the SQUID by ϕ_0 decreases like $b^{-0.83}$. Dashed line: Result of numerical simulation [37].

flux-lock-loop operation was easily obtained. Due to the weak temperature dependence of I_c and V_m small changes in the operating point of the dc-SQUID by temperature variations did not destabilize the loop.

Preliminary noise measurements in flux-locked-loop operation yielded values of some 10^{-4} $\phi_0/\sqrt{Hz}$ independent of temperature. The noise obviously originates from external sources and the dc-SQUID electronics itself. A more extensive study of the noise properties of such SQUIDs could not be performed due to limitations of the cryostat available for this work.

To increase the sensitivity of the SQUIDs one can try to increase the area of the SQUIDs and thus their inductance. Various geometries were realized and finally a slit configuration as shown in Fig. 24 was chosen [39]. This type of SQUID has the advantage of a good utilization of the flux compression. The slit length can be made quite long to provide good coupling. Increasing the width of the slit does not significantly increase the sensitivity of the SQUID. Devices with a slit length of up to 60 μm and widths of 4 μm exhibit a current modulation depth of about 1 % and a voltage modulation of some μV.

5.2 Nb_3Ge Multi-Layer Technique

For practical high-T_c Josephson devices like Nb_3Ge dc-SQUIDs a single layer technique as described in Sect. 5.1 is not sufficient in most cases. One often needs links between different parts of the electronic circuits which cannot be reduced to a two-dimensional structure without crossings. Therefore a versatile Nb_3Ge multilayer technology based on SiO_2 as insulating material [40] was developed which we will describe in this section.

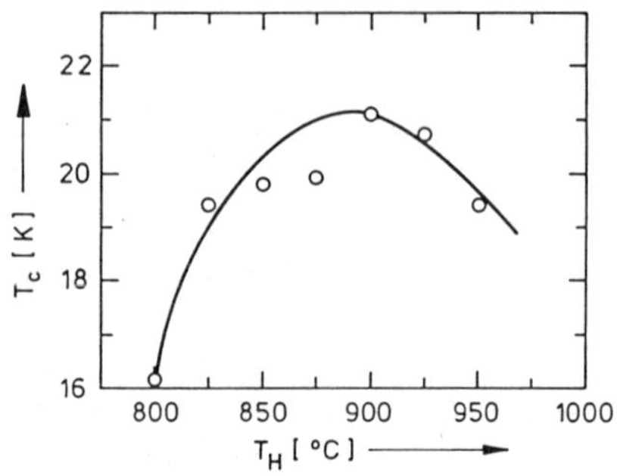

Figure 29
Dependence of the zero resistance temperature T_c on the heater temperature for Nb_3Ge-films sputtered on top of SiO_2-covered Nb_3Ge-films.

The main difficulty for such a technique is the high substrate temperature of 800 °C - 900 °C needed for the preparation of A15 Nb_3Ge. Insulating materials separating different Nb_3Ge-layers should withstand these temperatures without chemically reacting significantly. Due to the lack of thermal stability this excludes for instance the use of ZnO-films, which otherwise would have been interesting for a 'lift-off'-process.

Beside its thermal stability one has to take into account the properties of the insulator with regard to etching. Some materials withstand elevated temperatures very well like Al_2O_3 but are quite difficult to etch. Also the etchant itself may not be selective in the sense that it attacks the Nb_3Ge-films too. We will discuss in the following some materials which were tested for this application (see also table 1):

SiO is normally used as insulator in the Pb-In- and Nb-technology. It offers many good properties but cannot be used here since it decomposes into silicon and silica at temperatures above 600 °C [41].

Si_3N_4 is of great practical importance as an insulating material in semiconductor electronics. We prepared silicon nitrite films by reactive sputtering of silicon in a pure nitrogen atmosphere. Deposition at ambient substrate temperature results in soft and porous films. Sputtering onto substrates heated to 800 °C leads to hard and pinhole free films. Unfortunately the transition temperature T_c of the covered Nb_3Ge-films suffered a drop of about 4 K: the Nb_3Ge-film seems to react with the nitrogen at these elevated temperatures.

SiO_2 and Al_2O_3 possess excellent insulating and thermal properties. Both materials are rf-magnetron sputtered in an argon atmosphere (7 Pa) using a fused quartz target for SiO_2 and a sintered Al_2O_3-plate for the aluminum oxide. Deposition rates of 12 nm/min for SiO_2 and 20 nm/min for Al_2O_3 were realized. To minimize stress during the following deposition of the Nb_3Ge top-layer the substrate is held at a temperature of 600 °C during the insulator deposition. Films of about 200 nm thickness are very hard, pinhole free and stable against thermal cycling.

Table 1
Deposition method and etching conditions used in testing some insulating materials. The upper limit for thermal stability is included.

material	formation	etching	thermal stability
SiO	evaporation	reactive on etching with SF_6	600 °C
SiO_2	rf-magnetron sputtering	$HF+NH_4F$ wet etching reactive ion etching with SF_6	> 1 000 °C
Si_3N_4	reactive sputtering	conc. HF wet etching	900 °C
Al_2O_3	rf-magnetron sputtering	reactive ion etching with BCl_3, CCl_4	> 1 000 °C

These tests show that both SiO_2 and Al_2O_3 are of sufficient thermal stability and do not react with Nb_3Ge if one uses the deposition parameters mentioned above. The final decision towards SiO_2 is determined by the etching behavior: for Al_2O_3 there exists to our knowledge no suitable wet chemical etchant of sufficient etch rate and selectivity. Reactive ion etching with BCl_3 or CCl_4 is also not selective and imposes quite severe security measures for the etching chamber.

By contrast SiO_2 can easily be wet etched with buffered HF. This method has the advantage against reactive ion etching that the wet etching stops on top of the Nb_3Ge since the etch rate for Nb_3Ge is small in this case. In the case of reactive ion etching however the etch rates for Nb_3Ge and SiO_2 are practically the same. The wet etching method has the further benefit that it results in oblique edges of the insulator which is of advantage for covering by a possible top-layer.

The preparation conditions for the Nb_3Ge-films deposited on top of SiO_2 were optimized in the same way as described earlier. The conditions are very similar to those used for the sputtering onto sapphire substrates. Most of the films prepared with a thickness of 100 nm have a T_c of about 21 K (see Fig. 29).

In addition to the high-T_c of the Nb_3Ge-layers below and on top of the SiO_2-film superconducting contacts between these layers are needed (see Fig. 30). The T_c of these contacts should be nearly the same as that of the superconducting layers and its critical current density j_c should be high enough so that the contacts remain superconducting under all working conditions.

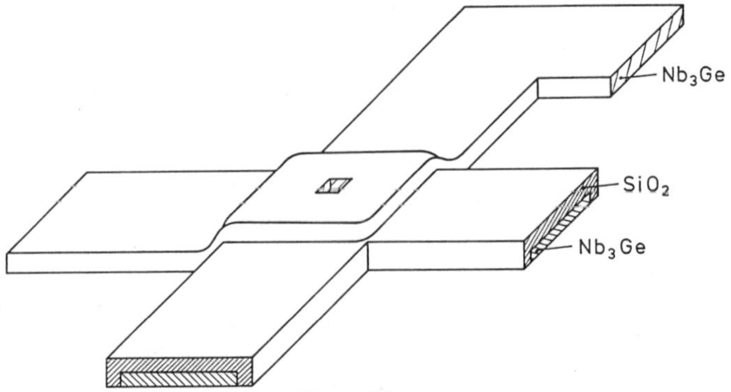

Figure 30
Scheme of a test contact between two Nb_3Ge-layers separated by a SiO_2-film.

The fabrication process begins with the deposition of a thin Nb_3Ge-film (100 nm thickness or less). Then this layer is structured using photo- or

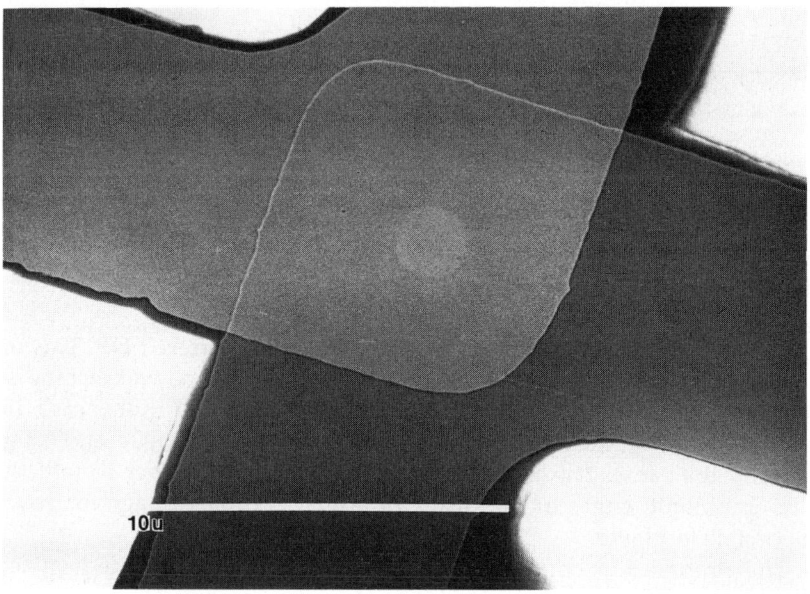

Figure 31
SEM-picture of a manufactured test-contact between Nb_3Ge-layers.

electron-beam-lithography and reactive ion etching with SF_6. Then a 200 nm SiO_2 layer is deposited on top at a substrate temperature of 600 °C. In the next step the contact window is etched into the SiO_2-film via photolithography and wet chemical etching as described above. As last step the top Nb_3Ge-layer is sputterd and structured under the same conditions as the bottom Nb_3Ge-film. An SEM-picture of a manufactured contact is shown in Fig. 31.

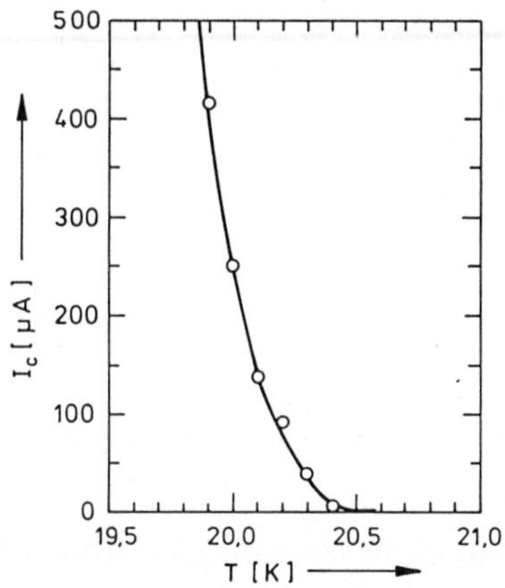

Figure 32
Dependence of the critical current I_C on temperature for a superconducting Nb_3Ge-Nb_3Ge contact. The contact area is 400 μm^2.

The critical current of such contacts and its temperature dependence were measured by a standard four probe method with a 2 μV voltage criterion. Typical contacts have a T_c of about 20 K and a critical current density of about 50 A/cm^2 at 19.5 K. Some contacts even offer a T_c of 20.5 K and a critical current density of 60 A/cm^2 at 20 K (see Fig. 32). These values indicate that the interface between the top and bottom Nb_3Ge-layer was not severely disturbed by the structuring of the SiO_2-film in the contact area. In addition possible damage in the contact area after etching might have been annealed by the high substrate temperature during the deposition of the counter electrode. The contact shows no Josephson tunnel junction behavior: no Fraunhofer pattern is observed in the critical current if a magnetic field is applied.

With the described technique multilayer Nb_3Ge integrated circuits can be realized. As an example for a two-layer technique we will present a dc-SQUID with integrated input coil in the next section.

5.3 Nb_3Ge Multi-Layer DC-SQUID

The techniques described in the previous sections can be applied to fabricate an all Nb_3Ge dc-SQUID with integrated input coil [37, 39]. The preparation starts with structuring the dc-SQUID and its contact pads from a Nb_3Ge bottom film (see Fig. 33) prepared on a sapphire substrate as described before. In the next step a 200 nm SiO_2-film is sputtered on top of the bottom film and a contact window is etched into this insulating layer. The window is needed for a superconducting link between the inner contact of the coil and the SQUID-body. In the next step a Nb_3Ge -layer is sputtered on top of the insulator. The coil is structured from this top layer by conventional photolithography and reactive ion etching with SF_6. The linewidth used for the coil is 3 μm and the spacing between the lines also 3 μm.

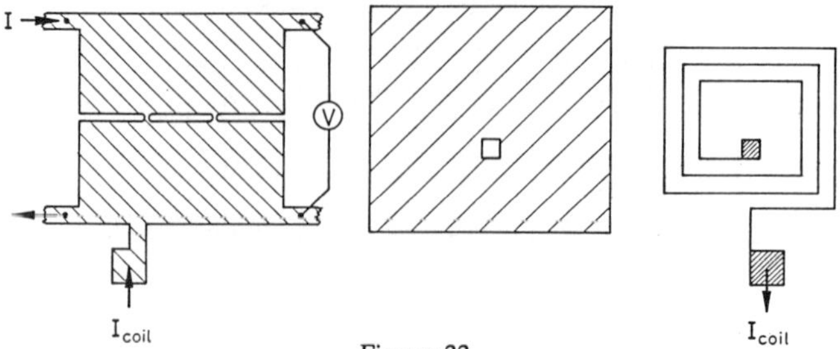

Figure 33
Preparation steps for a Nb3Ge dc-SQUD with integrated input coil.

Various SQUIDs with input coils ranging from 10 to 50 turns were fabricated (see Fig. 34). These dc-SQUIDs including the input coil operate at temperatures up to about 20 K with a total working temperature range of about 10 K. The mutual inductance between the SQUID and an input coil is proportional to the number of turns : for a coil with 10 turns we find 100 pH and with 50 turns about 500 pH. The coupling coefficient $k = M/\sqrt{L_1L_2}$ between the SQUID and input coil is of the order of 0.2 to 0.4. Here L_1 is the SQUID inductance, L_2 the inductance of the input coil, and M their mutual inductance.

Noise measurements were performed on these multilayer dc-SQUIDs in both flux-locked and free running operation. In the latter case the noise of

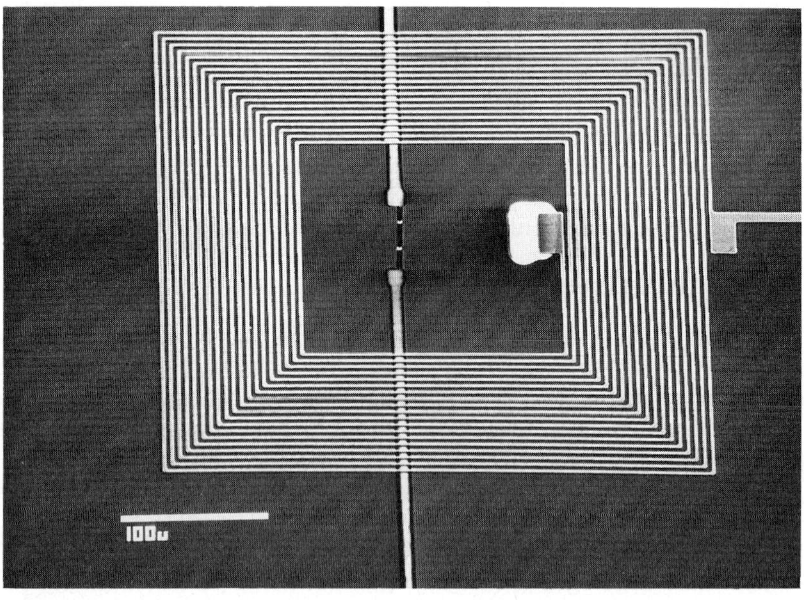

Figure 34
SEM-picture of a Nb3Ge dc-SQUID with an integrated Nb3Ge input coil of 20 turns.

the SQUIDs was estimated by measuring the output noise voltage of the SQUID. Knowing the slope $dV/d\phi$ (dV is the differential change of output voltage resulting from a differential change $d\phi$ of the flux inside the SQUID, $dV/d\phi$ is typically some $\mu V/\phi_0$), the gain and the bandwidth of the amplifier, the equivalent input flux-noise was calculated. To improve the coupling between the low-impedance SQUID and the high-impedance amplifier input a series tuned *LC*-circuit of known *Q* was connected between SQUID and amplifier as impedance transformer. The sensitivity of this method to measure the noise again was limited in our experiment by amplifier noise and other external noise sources. The measured temperature independent equivalent input noise had a value of some $10^{-5}\ \phi_0/\sqrt{Hz}$. We are convinced that this value does not represent the intrinsic SQUID-noise but is of external origin.

Flux-locked-loop operation was performed using conventional dc-SQUID electronics operating at a frequency of 100 kHz. Again a series tuned *LC*-circuit was used to match the low impedance SQUID (10 Ω) to the high impedance preamplifier circuit (1 kΩ). Stable operation was achieved with all tested dc-SQUIDs, some of them still working at temperatures above 19 K, the highest temperature limited by the sensitivity of the electronics. The spectral density of the equivalent input flux noise was measured in the flux-locked-loop mode too. In this mode of operation it has a value of $6 \times 10^{-5}\ \phi_0/\sqrt{Hz}$ at 17 K again practically independent of temperature.

To make an ordinary flux-transformer at these elevated temperatures is somewhat difficult since Nb_3Ge can be used only as a thin film on top of a more ductile material. Thus one either has to integrate the whole flux-transformer onto the same substrate on which the SQUID was fabricated or one has to use a Nb_3Ge ribbon. The one we used has a T_c of above 20 K and was fabricated on steel ribbon using a CVD process [42]. Despite the hardness of Nb_3Ge, the ribbon can be easily bent to radii as small as 1 cm.

The ribbon has to be connected carefully to the input-transformer pads to yield contacts with high-T_c and large critical current density. After carefully cleaning the pads by polishing or etching the ribbon was clamped to the pads of the input coil by a small spring. The critical current of the whole input loop was larger than 1 mA at 19 K. This limit is given by the planar contact inside the SQUID.

6. High Frequency Applications

Nb_3Ge nanobridges exhibit under microwave irradiation constant voltage steps as they are typical for Josephson contacts (c.f. Fig. 35) [43]. And, for sufficient high frequency, the zero voltage critical current I_0 of the bridge shows an oscillatory dependence on the microwave power as shown in Fig. 36. In principle one could use such contacts like Josephson tunnel junctions for a voltage standard. In this chapter we will report on some other high frequency applications of the nanobridges, i.e. on their use for micro-switches and as relaxation oscillators.

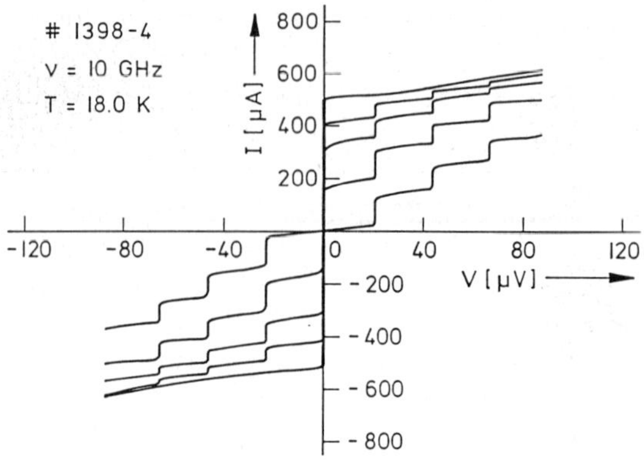

Figure 35
I-V-characteristics of Nb3Ge nanobridge under microwave irradiation at different power levels.

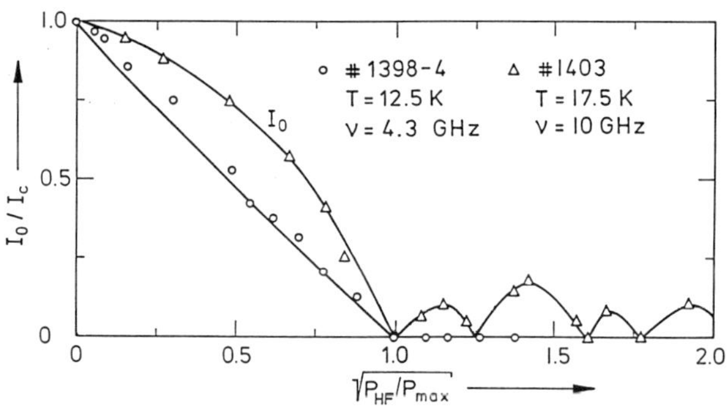

Figure 36
Zero voltage critical current I_0 as function of microwave power for two different microwave frequencies.

6.1 Microwave Driven Switching Device

As shown also in Fig. 36 however there is at lower microwave frequencies (here 4.3 GHz) an almost linear reduction of I_0 with increasing rf-current, and no recovery after I_0 becomes zero. Due to this behavior, which has been explained by the vortex flow model [44, 45], nanobridges can be used as switching devices that by application of microwave power switch from zero resistance to a resistive state.

Fig. 37a shows the time behavior of voltage drop across a Nb3Ge nanobridge and of the envelope of the applied microwave pulse, Fig. 37b the employed measuring scheme.

Microwaves of 2 GHz are coupled to the nanobridge in this experiment using a strip line placed immediately on top of the bridge. The microwave power needed to drive the switch into the resistive state is in the range of 1 - 10 nW. According to Fig. 37a the switching times (4 ns) for the microwave

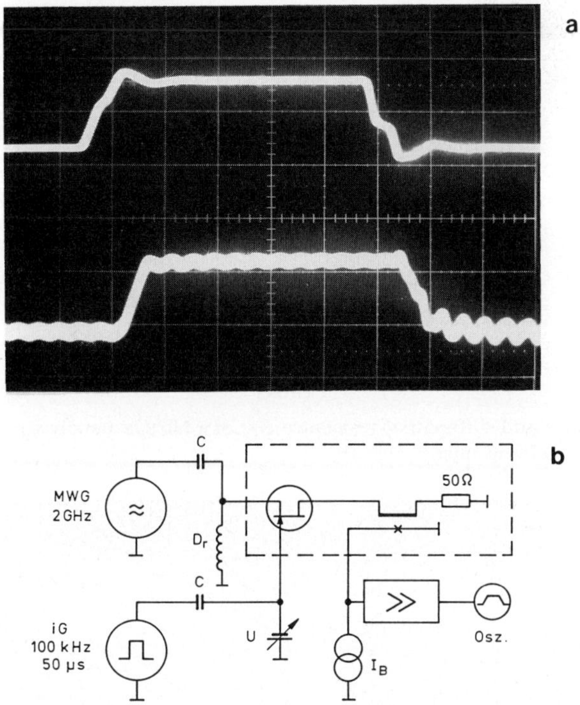

Figure 37
a) Switching behavior of single Nb3Ge-nanobridge (lower trace) and envelope of applied microwave pulse (upper trace). One horizontal division = 5 ns. b) Corresponding measuring scheme.

pulse and for the switch are equal, and are determined by the measuring circuit. Shorter switching times therefore can be expected for the microswitch. Since frequencies up to 5 GHz could be used to drive the nanobridges it appears that switching times less than a ns should be obtainable. Fig. 38 shows the *I-V*-characteristics of a Nb_3Ge nanobridge both with microwaves applied (on state, curve a) and without microwaves (off state, curve b). The differential resistance R_N in the on state also is shown. It is found to be practically constant for dc-transport currents up to $I = 2\,I_c$.

With a single nanobridge, resistance values R_N of up to 100 Ω have been achieved. Higher resistances can be obtained by an array of bridges connected in series, or also simply by a narrow strip line. Fig. 38 shows such a device consisting of a meander like strip line. It has a total length of 115 μm, a width of 250 nm, and a resistance R_N near 1 kΩ [46].

The microswitch can be used to transfer dc-power to a resistor R_L connected for instance in parallel to the nanobridge. Maximum power transfer results under matching conditions, i.e. $R_N = R_L$. Since the dc-current through the switch has to be limited to values $I \leq I_c = I_o$, the maximum power that can be transferred to R_L is $P_{max} = R_N \cdot I_c^2/4$. For a bridge with

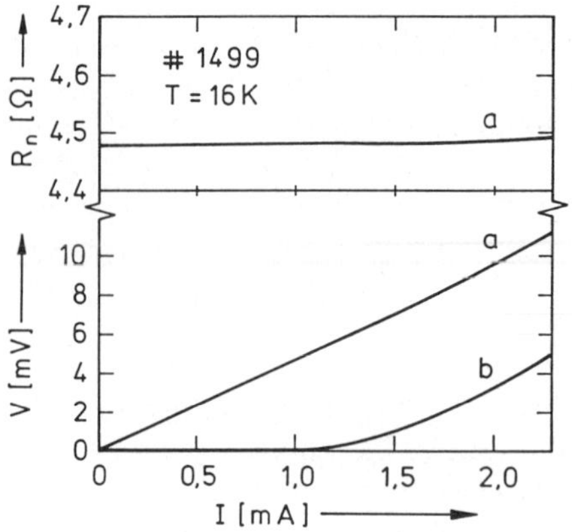

Figure 38
I-V-characteristics and differential resistance R_N of a Nb_3Ge nanobridge with microwaves applied (a), and without microwaves (b).

Figure 39
SEM picture of microswitch made of a Nb_3Ge strip line meander.

I_c = 1 mA, $R_N \approx 4.4\ \Omega$ (c.f. Fig. 39), this results in P_{max} = 1.1 μW. This power can be switched by the application of some nanowatts of microwave power!

6.2 Nanobridges as Relaxation Oscillators

Nanobridges with a sufficient high critical current I_c exhibit hysteresis in their *I-V*-characteristics, i.e. the transition into the resistive state occurs at a higher critical current I_{cmax} than the return at I_{cmin} to the fully superconducting state. In contrast to the case of capacitively shunted tunnel junctions we are dealing here mainly with a thermal effect. After going resistive the power dissipation in the bridge leads to a local temperature rise which in turn lowers the critical current of the device. This effect can be used to create a rather simple relaxation oscillator [47] by connecting the nanobridge to a *RL* network. There are two basic possibilities, the series and the parallel circuit (c.f. Fig. 40).

The devices start to oscillate if bias voltage or bias current meet the restrictions given in Fig. 40. The output voltage of such an oscillator, which was simply made by shunting the nanobridge with a 3 mm long aluminum bond wire of 25 μm diameter, is shown in Fig. 41 together with the resulting dependence of frequency on bias current .

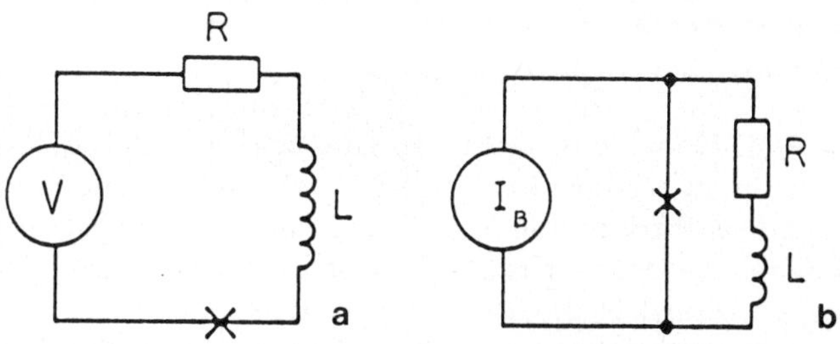

$\nu \approx R/L + (R+R_N)/L$

$R \cdot I_{cmax} < V < (R+R_N) \cdot I_{cmin}$

$\nu \approx R/L \cdot (\ln\{I_B - I_{cmin}\}/\{I_B - I_{cmax}\})^{-1}$

$I_{cmax} < I_B < \{(R_N+R)/R\} \cdot I_{cmin}$

Figure 40
Basic schemes for a relaxation oscillator using a nanobridge: a) series circuit, b) parallel circuit.

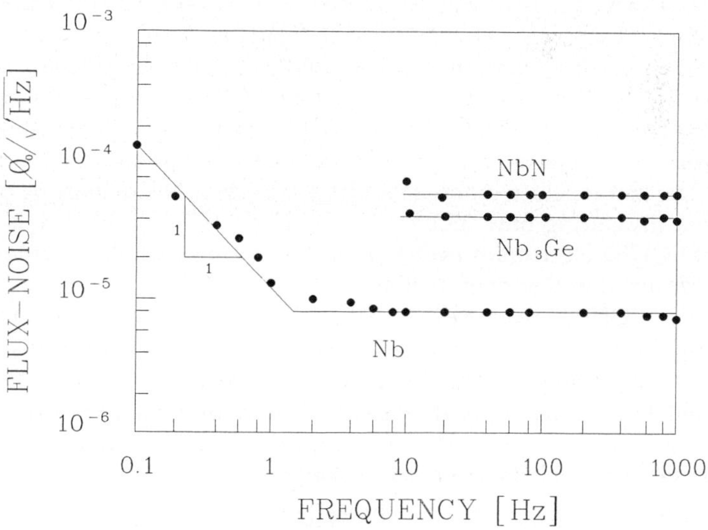

Figure 41
Flux noise spectra for three different SQUIDs measured in the flux-locked-loop mode.

Pulse amplitudes of several 10 mV and frequencies up to 500 MHz have been obtained with such devices. The frequency spectrum contains many harmonics. Microwave power in the range of μW at frequencies of several GHz can be produced with such oscillators. Due to the temperature dependence of critical current the oscillation frequency is not very stable. Fortunately the oscillators can be synchronized easily by a more stable oscillator like a quartz crystal.

The relaxation oscillators can be used in applications where rf-power in the range of μW is needed without high requirements with regard to spectral purity. They may for instance be used as microwave generators to drive the above mentioned microswitches. A suitable local coupling scheme between oscillator and switch however has to be developed for this combination to become attractive. Another application provides a simple fm-read-out scheme for dc-SQUIDs which will be discussed in the following.

6.3 FM-Read-Out Scheme For DC-SQUIDs

Shunting a dc-SQUID structure with two nanobridges (c.f. Fig. 24) by a *RL*-combination results in a relaxation oscillator, whose frequency is dependent on the magnetic flux through the SQUID. Instead of using the amplitude of the SQUID voltage as signal it is now the information in the time domain which is being processed in the associated SQUID electronics. As in an earlier scheme also based on the relaxation oscillator [48, 49] this has the additional advantage, that the SQUID produces a rather strong output signal. Special low noise preamplifiers and matching networks therefore are not needed.

Using the frequency of the relaxation oscillation as information a very simple and compact read out electronics was developed consisting only of two integrated circuits [50]. The first IC contains a limiter amplifier and a quadrature type frequency to voltage converter. A ceramic filter is used as phase shifter leading to a conversion factor of 20 mV per kHz frequency variation at the output of the IC. The second IC is a dual operational amplifier which serves as integrator for flux-locked-loop-operation and as buffer amplifier. The entire electronics, including a 14.4 V NiCd battery fits into a metal box measuring only 35 by 35 by 30 mm^3, which probably is the most compact SQUID electronics ever built so far. Using surface mounted devices should reduce this size even further.

Planar SQUID structures similar to Fig. 24, but with a square hole of 40 μm by 40 μm were made of thin Nb, NbN, and Nb_3Ge films. The *RL*-shunt is made either by an aluminum bond wire or by thin 100 μm wide and 1 mm long evaporated Au-Cu-strip. For a flux change of $\phi_0/2$ the frequency variation for such devices typically is of the order of 1 %, which amounts to about 50 kHz at a center frequency of 5.5 MHz that was chosen by adjusting the bias current to the appropriate value.

The result of noise measurements on three devices made of different materials is given in Fig. 41. For ease of temperature stabilization the measurements were made at 4.2 K in a helium storage container. The observed noise appears to be strongly dependent on the SQUID material, the lowest values being obtained in the case of niobium. Nb_3Ge SQUIDs also were operated at a temperature of 17 K with a noise level slightly above that at 4.2 K, but still below $10^{-4}\ \phi_0/\sqrt{Hz}$ at frequencies above 10 Hz. A more detailed study of the noise mechanism inherent in these devices, which probably is related to pinning of vortices, is needed.

7. Emerging Developments: SQUIDs at 77 Kelvin

Immediately after the discovery of the high-T_c ceramic superconductors attempts have been made to use these materials for quantum interference devices [51]. The first determination in 1987 of the flux quantum $h/2e$ in these materials [52] demonstrated the potential of the materials for making SQUIDs. Since then both thin film dc-SQUIDs [53 - 56] as well as rf-SQUIDs made of bulk specimens [57 - 61] have been made with different success concerning their performance at liquid nitrogen temperature. While the spectral flux noise density at frequencies above 100 Hz approaches that of conventional devices at 4.2 K, strong excess noise with a $1/f$-like behavior is observed at lower frequencies. Its origin probably is associated with the granular nature of the material used so far for the devices. In the following we describe a particular version of a bulk rf-SQUID [60].

The most common type of weak link used so far in devices made of $YBa_2Cu_3O_{7-x}$ consists of a constriction of comparatively large cross section, a "millibridge", with a diameter of the order of 1 mm, which in the case of bulk specimens is prepared mechanically by drilling, sawing, grinding, etc. It contains a network of many Josephson contacts. Among the prevalent problems with these devices two appear to be most urgent: i) the already mentioned low frequency excess noise, and ii) the as yet unsatisfactory stability with regard to thermal cycling and prolonged storage at room temperature under ambient atmosphere which leads to corrosion. While the solution to the first problem may have to wait until better materials are available a simple way to overcome the corrosion problem is to make a monolythic device, in which the millibridges are hermetically encapsuled.

Following this idea rf-SQUIDs were manufactured from bulk $YBa_2Cu_3O_{7-x}$ cylinders that were produced using the citrate method [61]. Zero resistivity transition temperature T_c was near 92 K. Two kinds of devices were made:

i) a two hole SQUID as shown in Fig. 42a, and

ii) a four hole SQUID according to Fig. 42b.

After drilling the holes the weak links were made in the central part of the cylinder using thin saw blades, that could be inserted through the holes. After achieving a good SQUID signal in liquid nitrogen, the units are sealed

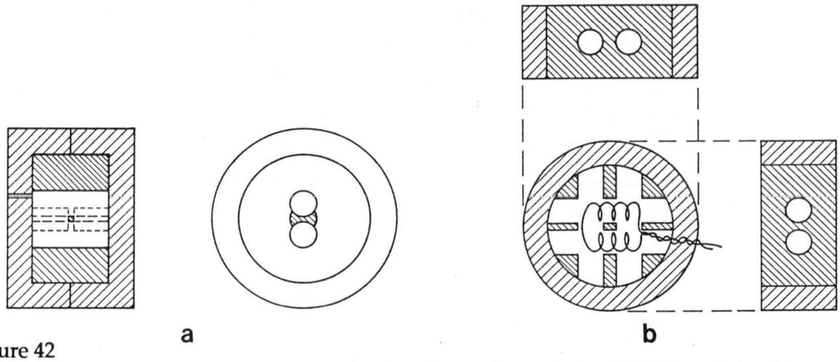

Figure 42
RF-SQUID structures made of bulk $YBa_2Cu_3O_{7-x}$: a) two-hole SQUID sealed with caps, b) four-hole SQUID sealed with a ring.

either by two caps (two-hole SQUID) or by a ring (four-hole SQUID). The sealing structures are also made of $YBa_2Cu_3O_{7-x}$, and are glued to the SQUID-cylinder using a mixture of epoxy and quartz powder in order to match expansion coefficients. In this way rf-SQUIDs are made in which the weak link is part of a monolithic structure, and is at the same time hermetically sealed from ambient atmosphere. Coating the whole SQUID structure by a suitable material can serve as barrier against diffusion of gas or moisture protecting further the device against degradation. For our SQUIDs this was done using again a quartz epoxy mixture.

Prior to sealing the unit a coil made of insulated copper wire is inserted into the holes (c.f.fig 42b) for read out and coupling purposes. The SQUIDs are operated with standard read out electronics with 20 MHz oscillator for the rf-current, and a modulation signal (ca. 1 kHz) for flux-locked-loop operation. Fig. 43 shows response signals obtained with a SQUID immersed in liquid nitrogen at 77 K at four different rf-levels.

The noise performance was analyzed with the SQUID immersed in liquid nitrogen at 77 K, the dewar being surrounded by a double Mu-metal shield reducing the earth magnetic field to a level of less than 10^{-5} T. Fig. 44 shows flux noise spectra of some devices determined in the frequency range of 1 to 300 Hz. There is always a pronounced increase of the noise level with decreasing frequency below ca. 100 Hz. The "white" noise region at higher frequencies differs from SQUID to SQUID demonstrating that the observed noise is dominated by the properties of the particular device rather than by the read out electronics. Lower noise levels might be achieved during the course of further development. The employed method of adjusting the weak link of course is a rather crude one which only by chance will lead to optimum performance. This can be seen by the effect of a magnetic field on the SQUID response. The magnetic field generated by a dc bias current I_b applied to the coupling coil of the SQUID affects the critical current of the weak link, optimum SQUID signals being generated for certain values of I_b.

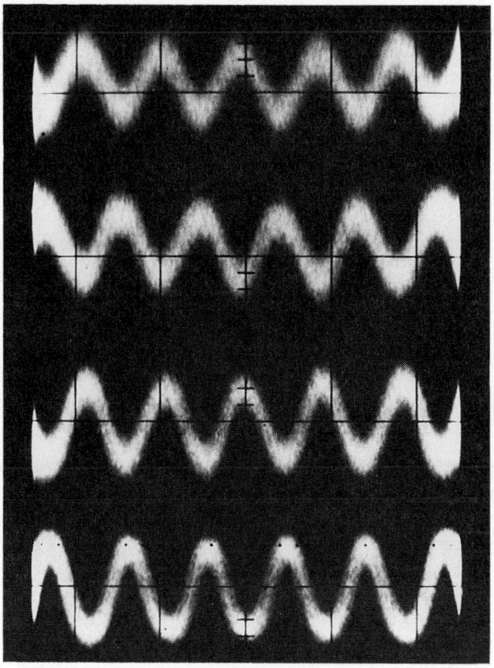

Figure 43
SQUID response to external flux at four different rf-levels.

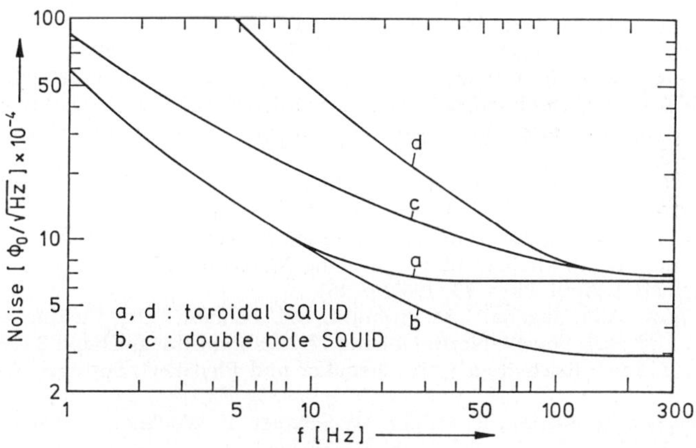

Figure 44
Flux noise spectra of four different SQUIDs.

The stability of the devices with respect to thermal cycling was tested in a straightforward manner by repeatedly immersing them into liquid nitrogen followed by warming up to room temperature. For specimens that were sintered at least 6 days no degradation in performance was detected after more than 30 cycles. Samples with a shorter sintering time were destroyed after 2 to 3 cycles caused by differential contraction and expansion during temperature changes that broke the weak links.

Sealing the devices offers a possible solution to the corrosion problem. A major remaining task is the reduction of the strong excess noise at low frequencies which most likely involves a systematic material developement such that Josephson contacts can be manufactured in a better defined way. It may well be that thin film devices have a better chance towards this goal. Results obtained recently with TlBaCaCuO dc-SQUIDs [62] point in this direction.

References

[1] J.G. Bednorz, and K.A. Müller, Z. Phys. **64** (1977) p. 189

[2] M.R. Beasley, C.J. Kircher, in "Superconductor Materials Science. Metallurgy, Fabrication, and Application" S. Foner and B.B. Schwartz, Eds., Plenum Press, N.Y. (1981) p. 605

[3] H.-J. Hedbabny, H. Rogalla, J. Appl. Phys **63** (1988) p. 6

[4] J.R. Gavaler, M.A. Janocko, C.K. Jones, J. Appl. Phys. **45** (1974) p. 3009

[5] R. Somekh, Proc. 7th Int. Conf. on Vacuum Metallurgy 1982 in Tokio, The Iron and Steel Institut of Japan, Tokyo, 17 (1982)

[6] J.R. Gavaler, M. Ashkin, A.I. Braginski, A.T. Santhanam, Appl. Phys. Lett. **33** (1978) p. 359

[7] H.F. Braun, PhD Thesis, Justus-Liebig-Universität, Giessen (1977)

[8] B. David, PhD Thesis, Justus-Liebig-Universität, Giessen (1986)

[9] G.K. Welmer, G.S. Anderson, "Handbook of Thin Film Technology", Chapt. 3, Ed. L.I. Maissel and R. Glang, McGraw-Hill Book Company (1970)

[10] F.J. Cadieu, N. Chencinski, IEEE Trans. Mag. **MAG-11** (1975) p. 227

[11] C. Csepregi, K. Kühl, R. Nießl, H. Seidel, Forschungsberichte BMFT-FB-T 84-209 und BMFT-FB-T 83-089 des Bundesministeriums für Forschung und Technologie (1984)

[12] L.R. Testardi, R.L. Meek, J.M. Poate, W.A. Royer, A.R. Storm, J.H. Wernick, Phys. Rev. **B11** (1975) p. 4303

[13] P.G. de Gennes, "Superconductivity of Metals and Alloys", W.A. Benjamin Inc., New York (1966)

[14] Y.B. Kim, C.F. Hempstead, A.R. Strnad, Phys. Rev. **131** (1963) p. 2486

[15] W.H. Henkels, C.J. Kircher, IEEE Trans. Mag. **MAG-13** (1977) p. 63

[16] J.C. Swihart, J. Appl. Phys. **13** (1961) p. 461

[17] T.P. Chow, A.N. Saxena, L.M. Ephrath, R.S. Bennett, "Dry Etching for Microelectronics", Ed. R.A. Powell, North-Holland Physics Publishing, Chapt. 2 (1984) p. 72

[18] D'Ans, E. Lax, "Taschenbuch für Chemiker und Physiker", Springer-Verlag, **1** (1967) p. 347

[19] C. Nölscher, H. Adrian, R. Müller, W. Schauer, F. Wüchner, G. Saemann-Ischenko, Proc. LT17 Karlsruhe, Elsevier Science Publishers, part I (1984) p. 491

[20] Y. Kato, B. David, H. Rogalla, Jpn. J. Appl. Phys. **23** (1984) p. 1881

[21] H.-J. Hedbabny, H. Rogalla, Proceedings of the ASC '88, San Francisco, IEEE Trans. Mag., to be published

[22] M. Mück, H.-J. Hedbabny, H. Rogalla, IEEE Trans. Mag., **MAG-23** (1987) p. 1493

[23] K.K. Likharev, Sov. Phys. JETP **34** (1972) p. 906

[24] T. Goto, H. Tanihara, J. Appl. Phys. **54** (1983) p. 3291

[25] A. de Lozanne, M.S. Di Iorio, M.R. Beasley, Appl. Phys. Lett. **42** (1983) p. 541

[26] A. de Lozanne, "High Critical Temperature SNS Josephson Microbridges", Promotional Thesis, Stanford University (1982)
[27] K.K. Likharev, Sov. Phys. JETP **34** (1972) p. 906
[28] L.G. Aslamazov, A.I. Larkin, Sov. Phys.-JETP **41** (1975) p. 381
[29] E. Fehlberg, ZAMM **44** (1964) p. T17
[30] S. Kuriki, A. Yoshida, H. Konishi, J. Low Temp. Phys. **51** (1983) p. 149
[31] M. Mück, H. Rogalla, B. David, Phys. Stat. Sol. **87** (1985) p. K105
[32] S. Kuriki, A. Yoshida, H. Konishi, J. Low Temp. Phys. **51** (1983) p. 149
[33] H. Rogalla, B. David, J. Rühl, J. Appl. Phys. **55** (1984) p. 3441
[34] M. Mück, H. Rogalla, B. David, Phys. Stat. Sol. (a) **87** (1985) p. K105
[35] W.C. Stewart, Appl. Phys. Lett. **12** (1968) p. 227
[36] C. Tesche, J. Clarke, J. Low. Temp. Phys. **29** (1977) p. 301
[37] H. Rogalla, B. David, M. Mück, SQUID'85 "Superconducting Quantum Interference Devices and their Applications", H.D. Hahlbohm and H. Lübbig, Eds., W. de Gruyter, Berlin 1985, p. 671
[38] J. Clarke, W.H. Goubeau, M.B. Ketchen, J. Low. Temp. Phys. **25** (1976) p. 99
[39] H. Rogalla, B. David, M. Mück, Y. Kato, IEEE Trans. Mag. **MAG-21** (1985) p. 536
[40] B. David, M. Mück, H. Rogalla, "Advances in Cryogenic Engineering" **32** (1986) p. 543
[41] L. Holland, in "Vacuum Deposition of Thin Films", Chapman and Hall LTD., London (1958) pp. 438-491
[42] A. Müller, Zeitschrift für Metallkunde **71** (1980) p. 507
[43] M. Mück, H. Rogalla, B. David, C. Heiden, Z. Phys. **B 61** (1985) p. 81
[44] K.K. Likharev, Sov. Phys. JETP **34** (1972) p. 906
[45] A.I. Golovaskin and A.N. Lykov, Sov. Phys. JETP **47** (1978) p. 110
[46] M. Mück, H.J. Hedbabny, H. Rogalla, IEEE Trans. Mag. **MAG-23** (1987) p. 1493
[47] Relaxation oscillators using Josephson contacts habe been reported in the literature, see e.g.:
J.E. Zimmerman, A.H. Silver, Phys. Rev. Lett. **19** (1967) p. 14
F.L. Vernon, R.J. Pedersen, J. Appl. Phys. **39** (1968) p. 2661
Y. Taur, P.L. Richards, J. Appl. Phys. **46** (1975) p. 1793
N. Calander, T. Claeson, S. Rudner, Appl. Phys. Lett. **39** (1981) p. 504
[48] P. Gutmann, in "SQUID, Superconducting Interference Devices and their Applications", H.D. Hahlbohm and H. Lübbig, Eds., W. de Gruyter, Berlin 1976, p. 501
[49] P. Gutmann, Electronic Lett. **15** (1979) p. 373
[50] M. Mück, C. Heiden, IEEE Trans. Mag. **MAG-25** (1989)
[51] J. Clarke, Nature **333** (1988) p. 47
[52] C.E. Gough et al., Nature **326** (1987) p. 885
[53] R.R. Koch et al., Appl. Phys. Lett. **51** (1987) p. 200
[54] H. Nakane et al., Jap. Journ. Appl. Phys. **26** (1987) p. L1925
[55] C.X. Fan, et al., Physica **C 153-155** (1988) p. 1413
[56] B. Häuser, M. Diegel, H. Rogalla, Appl. Phys. Lett. **52** (1988) p. 844
[57] J.E. Zimmerman et al., Appl. Phys. Lett. **51** (1987) p. 617
[58] S. Harrop et al., Supercond. Sci. Technol. **1** (1988) p. 68
[59] N.V. Zavaritsky, V.N. Zavaritsky, Physica **C 153-155** (1988) p. 1405
[60] Y. Zhang, M. Diegel, C. Heiden, IEEE Trans. Mag. **MAG-25** (1989)
[61] D.H.A. Blank, H. Kruidhof, J. Flokstra, J. Phys. **D21** (1988) p. 226
[62] R.H. Koch et al., (1989)

Biomagnetic Sensors

H. Koch
Physikalisch-Technische Bundesanstalt, Institut Berlin, D-1000 Berlin 10

1. Introduction

Measurements of physiologically originated magnetic fields form the basis of a diagnostic discipline commonly termed biomagnetism. Biomagnetic investigations are often closely related to the better known electrophysiological methods, eg. electro-cardiography (ECG) and electro-encephalography (EEG). Similarly, the methods dealing with the magnetic field generated by heart or brain activities are termed magneto-cardiography (MCG) and magneto-encephalography (MEG) respectively. The main aim of biomagnetism is to obtain, with respect to standard diagnostic methods, additional information on the patients condition and to be able to determine the site of electrophysiological functions (eg. a particular brain activity etc). The strengths of biomagnetic signals (e.g. the amplitude of the magnetic field strength) range from nT (magnetic lung contaminants) to less than a few fT (evoked cortical activity) with intermediate values for a typical MCG ($\approx$ 0.1 nT) and MEG ($\approx$ pT). Their bandwidths are from dc up to 100 Hz (see [1]). For comparison: the steady component of the earths magnetic field strength at the surface near the equator is 1.3×10^{-5} T. A general overview on biomagnetism is given by [1], [2] and [3]. More recent advances are dealt with in [4], [5], [6] and [7].

In most applications a biomagnetic sensor is a SQUID sensor in principle, and SQUID sensors were the subject of the previous chapter. Up to recently most experiments in biomagnetism have been performed with a "general purpose" SQUID sensor like the one shown in Fig. 1. The only degree of freedom left for the experimentalist seems to be the design of the detector coil to be connected to the 4 screw contacts leading to the input coil of the sealed SQUID sensor package. One would therefore expect that nothing needs to be added here except a few application-related remarks.

But as it turns out, applying this kind of SQUID sensor to biomagnetic experiments *successfully* requires a very thorough system-oriented approach. Some of the main obstacles that have to be dealt with are:

- A SQUID is extremely sensitive to magnetic fields over an extremely wide frequency bandwidth. Thus special care must be given to the problem of separating strong unwanted signals such as urban noise from the very weak wanted signals, such as signals due to brain activi-

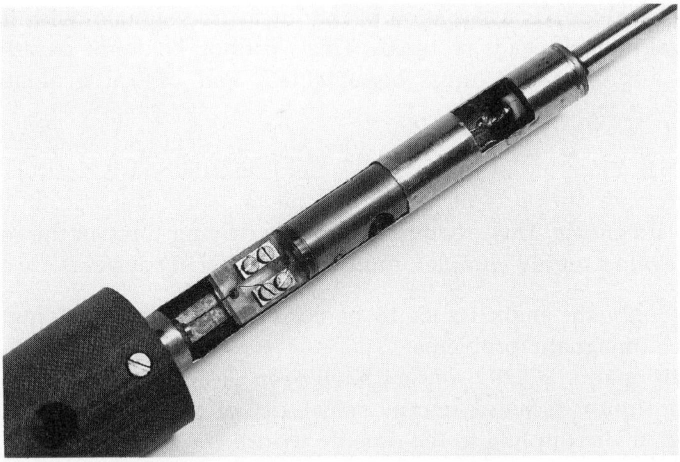

Figure 1
Typical SQUID sensor package (at the center). The detector coil for magnetic field pick-up is connected to the input coil contained inside the package via the 4 screw contacts visible in front.

ty, which seldom exceed magnetic field strengths in the order of pT. Very advanced shielding, noise discrimination and filtering techniques must be implemented.

- The sources of the magnetic fields to be investigated in biomagnetism are within the human body. The body must be kept at room temperature for obvious reasons, whereas (conventional) SQUID sensors operate only at temperatures close to 4.2 K, the temperature of liquid helium. Thus a severe temperature gradient must be maintained. For this purpose, specially designed liquid helium dewars are used.

- The detectability of physiological magnetic fields decreases drastically with increasing source-detector distances therefore one of the main aims of biomagnetic instrumentation design is to keep this distance as small as possible, for instance by an optimized dewar wall design.

- When applying biomagnetic sensors to patients tight safety measures have to be considered, in particular the liquid helium has to be contained securely.

- Different diagnostic requirements call for differently adapted detector coil designs: utmost sensitivity calls for large coils whereas precise localization of magnetic sources calls for small ones; discrimination between different types of sources calls for different gradiometer coil designs etc.

- An unambiguous association of the measured signals with the physiological phenomena is required. This cannot be done exactly, so the best approximation must be obtained, and critical evaluation is necessary.

- A quick mapping of the magnetic field distribution is desirable in order to relieve the patient and to achieve short turn-around time for the diagnosis. This requirement is the driving force in the recent race to develop highly complex multichannel SQUID systems.

These are only the main issues to be considered when applying a SQUID sensor to biomagnetic problems.

Traditionally, SQUID sensors have been designed first and foremost to achieve optimum *intrinsic* energy sensitivity of the SQUID and only then to consider their coupling to the outside world.

For biomagnetic applications this line of thought should be reversed to obtain the best performance: What is the physiological phenomenon to be investigated? How precise should the source localization, that is the spatial discrimination, be? What degree of magnetic field sensitivity is required? What type of detector coil configuration is best adapted to the physiological problem, to the mapping algorithm, and to the multisensor configuration used? The question then arises of how to couple the SQUID effectively to this specific coil configuration.

In conclusion: a SQUID sensor becomes a biomagnetic sensor only if the specific diagnostic problem and the interdependence of all components of the sensor environment have been considered in order to yield results of any value.

The purpose of the following sections is to illuminate this background for a successful biomagnetic sensor system design exemplifying a typical system- oriented approach equally necessary for any other cryoelectronic application.

2. The Biomagnetic Method

The aim of a typical biomagnetic experiment is to determine the temporal evolution of the spatial distribution of the normal component of the magnetic field within a so-called *mapping plane,* a surface as close as possible to the object examined. As the method is non-invasive this mapping plane is usually more or less equidistant to the body's surface close to the object of interest, i.e. for heart investigations a plane parallel to the chest, for brain research a spherical surface surrounding the head.

The procedure of the biomagnetic method is exemplified below in the case of magnetocardiology. The biomagnetic signals are usually detected and recorded for each coordinate of a well-established *standard grid* [8] like the one shown in Fig. 2a for a sample biomagnetic heart investigation. For each position of the standard grid a signal is recorded resembling an ECG signal

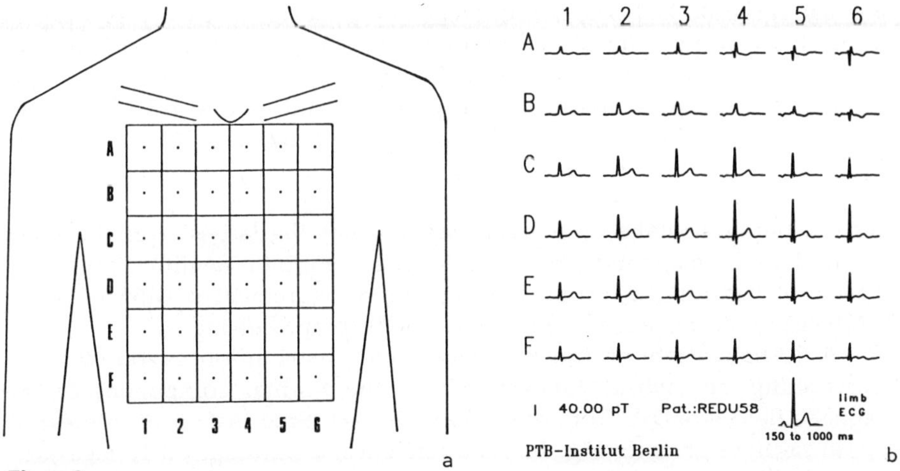

Figure 2
a) Standard grid for biomagnetic heart investigations. b) MCG signals obtained at each position of the standard grid (bottom right: corresponding ECG signal).

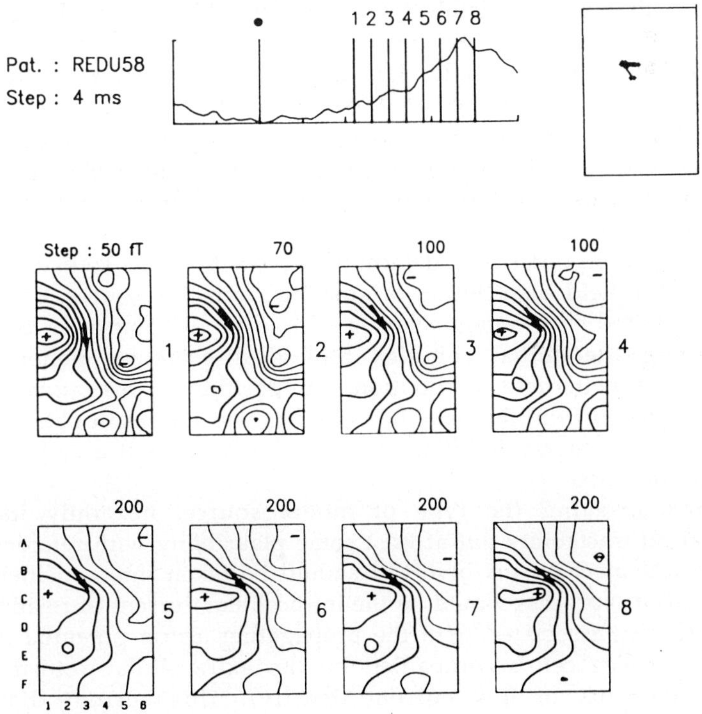

Figure 2
c) Isofield contour maps obtained from the data in Fig. 2b at time instants marked in the top of the figure at the simultaneously recorded ECG-signal.

(Fig. 2b). It should be borne in mind that biological signals hardly ever repeat in exactly the same manner and with the same period. Each signal in Fig. 2b is actually the result of a sophisticated correlation and averaging technique for obtaining a representative stationary signal and getting rid of artifacts. Some of these techniques are described in more detail in [2] and [5]. The signals are also frequently conditioned by digital filtering procedures (cf. [9]). Obviously, there is a great deal of data to be handled and signal processing and computation make up a large part - if not the largest - of the work involved in biomagnetic experiments. The computer facilities must be capable of handling a typical biomagnetic measurement session of about 20 Mbyte raw data to be reduced to 200 kbyte by preprocessing.

A visualization of the magnetic field distribution is possible by constructing an *isofield contour map* of the normal component of the magnetic field strength from these data for each time instant of interest. A typical result is given in Fig. 2c. As a reference, a simultaneously measured ECG signal is added at the top of Fig. 2c and the time instants are marked for which the maps displayed have been derived from the data in Fig. 2b.

Furthermore one is often interested in the type and precise location of the source that is responsible for generating the measured signals. The procedure to try to retrieve information about the source of the magnetic field from the magnetic field data alone is associated with the notion: *inverse problem* (see [10]). Field theory states that a field can be exactly calculated if the type and position of the field sources are known, but that no unique solution exists for the inverse problem of determining in general the accurate type and position of the sources from the given field distribution data.

What *can* be done, is to determine the position of a source of a *known type* from the field distribution originating from this source. This is commonly exploited in biomagnetism: a simple model source is considered, representing the actual complex biological structure generating the measured magnetic field distribution. The position of this *equivalent source* is then calculated from the data. It is obvious that the result obtained in this manner is to be interpreted with great caution because it may be too crude and oversimplified.

When selecting the type of model source, not only its simple mathematical tractability but at least some plausibility with reference to the physiological mechanisms involved should be taken into consideration. A basic bioelectric process would be the intracellular current of relatively high current density associated with the propagating action potential along the axon of a nerve cell accompanied by the extracellular return currents (volume currents of low current density). This current distribution intuitively leads to the widely used *current dipole model* (Fig. 3). In reality, a huge number of such events occur in parallel. For example a whole bundle of nerve cells is involved. This gives rise to measurable biomagnetic fields [10]. The current dipole is, however, still the model of choice in biomag-

netism in view of the lack of any better description and for practical computational reasons [11].

3. Current Dipole Model

A current dipole model consists of a current distribution with a high current density concentrated along a short linear stretch as shown in Fig. 3. The associated return currents flowing back through a relatively large conducting volume are usually neglected.

The current dipole $\vec{Q}$ itself in ampere meters is given by the product of a volume element ΔV with the space dependent current density $\vec{J}(\vec{r}\,')$ inside:

$$\vec{Q} = \vec{J}(\vec{r}\,')\,\Delta V \tag{3.1}$$

The magnetic flux density $\vec{B}$ is obtained by applying Biot Savart´s law:

$$\vec{B} = (\mu_o/4\pi\, R^3)\,\vec{Q} \times \vec{R} \tag{3.2}$$

with $\vec{R} = \vec{r} - \vec{r}\,'$ and μ_o the permeability of vacuum.

A more detailed discussion on the current dipole model, in particular its modifications due to inhomogeneities and boundaries of the surrounding conducting medium, is to be found in [12]. There and in [13] it is shown that the return currents do not contribute to the value of the magnetic field measurable at the field point (Fig. 3).

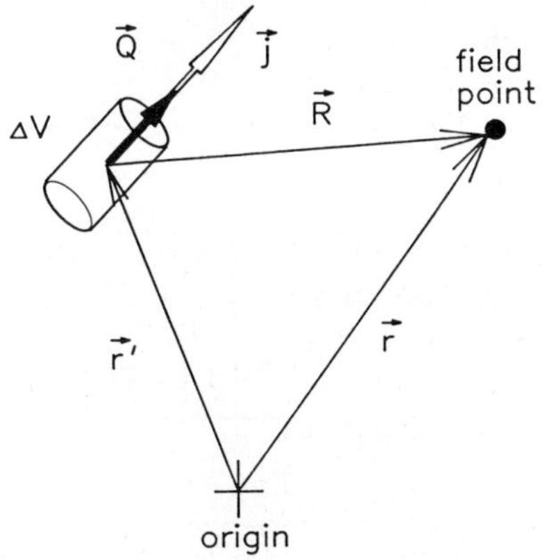

Figure 3
Current dipole model.

In biomagnetic experiments the *normal* component of the magnetic field $\vec{B}$ is measured within the mapping plane (c.f. Fig. 4) that is situated outside the body close to the body's surface, ie, the current dipole is no longer surrounded by an infinite, homogeneously conducting environment. Taking boundaries and inhomogeneities into account complicates the problem considerably because now the return currents also have to be taken into consideration. Again, simplified models of the human body have been introduced in biomagnetism: the conducting half space is mainly used for magnetocardiology and the conducting sphere for brain research. The interested reader is referred to [2] and [5]. Only within the limit of these models does the spatial distribution of the *normal* component of the magnetic flux density $\vec{B}_n(x,y)$ in a mapping plane parallel to the boundary surface of the conducting volume allow a correct solution to the problem of localizing the effective current dipole. The other components of the magnetic field are considerably more affected by boundaries and inhomogeneities and are therefore of little diagnostic value. However, recently considerable effort is devoted to complex inverse problem calculations [10] including the other components of the magnetic field, the actual boundary

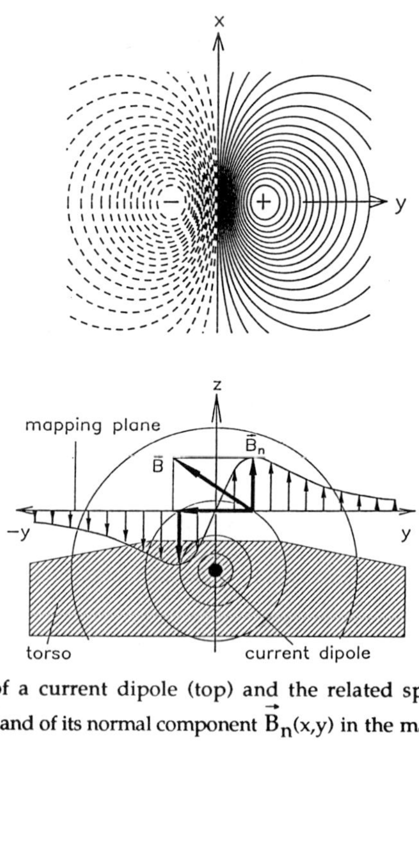

Figure 4
Isofield contour map of a current dipole (top) and the related spatial distribution of the magnetic flux density $\vec{B}$ and of its normal component $\vec{B}_n(x,y)$ in the mapping plane.

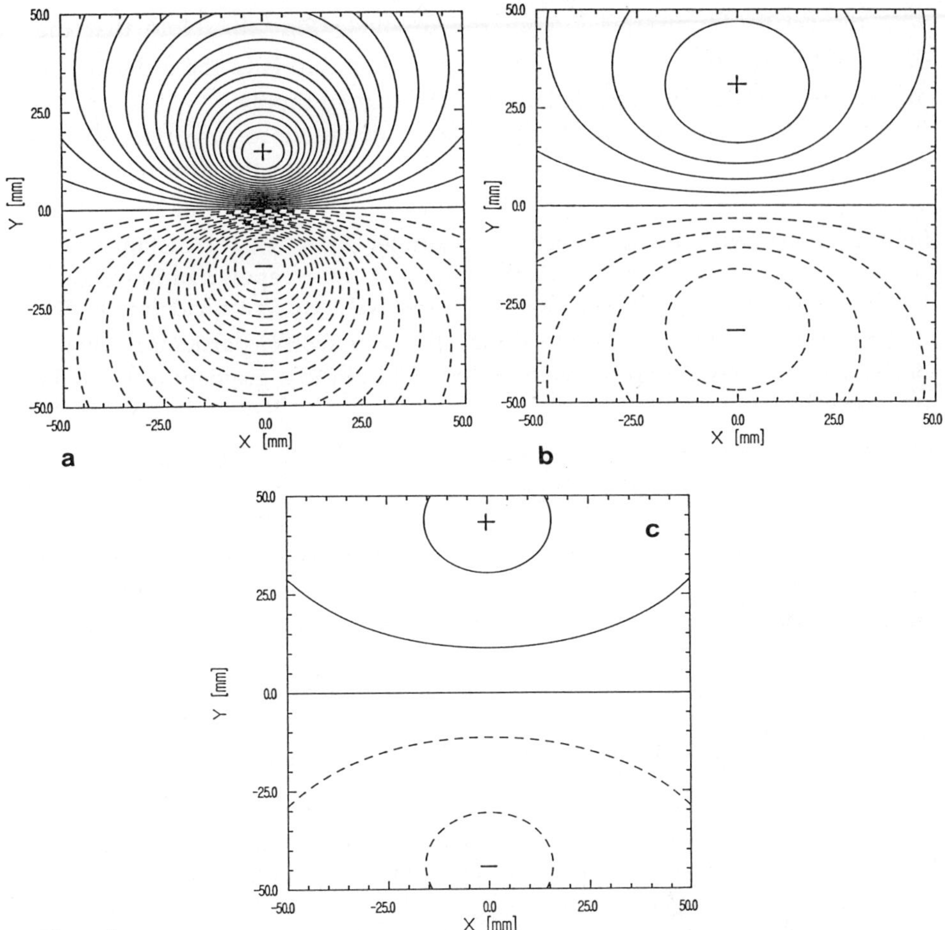

Figure 5
Isofield contour maps for three detector-to-source distances: a) 20 mm, b) 40 mm, and c) 60 mm; contour spacing 5×10^{-13} T for $|\vec{Q}| = 1 \times 10^{-7}$ Am.

conditions and inhomogeneities as derived from advanced NMR image processing techniques.

It is not possible to localize the source of the field with non-invasive electro-physiological methods and the ability to do this is therefore a unique feature of biomagnetism.

Fig. 4a shows schematically the type of spatial distribution of the normal component $\vec{B}_n(x,y)$ of the magnetic field originating from a current dipole $\vec{Q}$ as detected in the mapping plane. This distribution is obvious if one considers how the normal components $\vec{B}_n$ are derived from the field vectors $\vec{B}$ of the radially symmetrical field with decreasing field strength with respect to the distance $\vec{R}$ from the dipole $\vec{Q}$ (Fig. 4b). Note that the distance between the nodes (maximum and minimum) is in the order of the distance between the mapping plane and the current dipole.

Figs. 5a-5c demonstrate the importance of the close proximity of the detector to the source: the values of $\vec{B}$ decrease drastically with increasing distance $\vec{R}$ and the nodes become far less distinct. The true power law for the field decrease with distance may be stronger than R^2, for instance, if a propagating depolarization signal is followed by a repolarization signal, forming two opposing dipoles yielding a distorted current quadrupole. The detector response with respect to $\vec{R}$ may be even worse, particularly if the detector is not sensitive to the value of the field but to a gradient of the field (see Sect. 4).

4. Detection Coil Configurations

4.1 Wire-Wound Flux Transformers

Most users of SQUID sensors in biomagnetic research have to rely on commercially available SQUIDs. These usually consist of a hermetically sealed package like the one shown in Fig. 1. The housing, usually made of a metallic, non-magnetic wall material (e.g. BeCu-alloy) and sometimes with an additional superconducting cylindrical cover, protects the inner components from mechanical and thermal shock and from electrical and magnetic interference. It contains the SQUID ring itself (either a thin film or a bulk version), the input coil, the modulation coil, the tank circuit and/or the matching circuit and sometimes filter elements (see the previous chapter). The screw contacts visible in Fig. 1 are made of superconducting material and are connected to the superconducting input coil contained inside the package. The input coil is inductively coupled to the SQUID ring. Thus, if a field pick-up coil (detector coil) made of superconducting wire is connected to these contacts, together with the input coil it forms a superconducting *flux transformer* (Fig. 6).

A severe design constraint is imposed by the impedance matching condition of a flux transformer $L_d = L_i$ with L_d detection coil inductance and L_i input coil inductance. Only in this case is an optimum signal transfer to the SQUID provided. As L_i is fixed for a given SQUID, L_d is more or less fixed, too. Thus the design of the geometric configuration of the detector coil is the only freedom left to those users, who do not have access to SQUID design and fabrication facilities. Under such circumstances wire wound detector coils are usually used. The simplest detection coil configuration is either a single superconducting loop or a single coil of several turns of superconducting wire. This detection coil in conjunction with the SQUID sensor acts as a *magnetometer*. Given the constraint $L_d = L_i$ the following

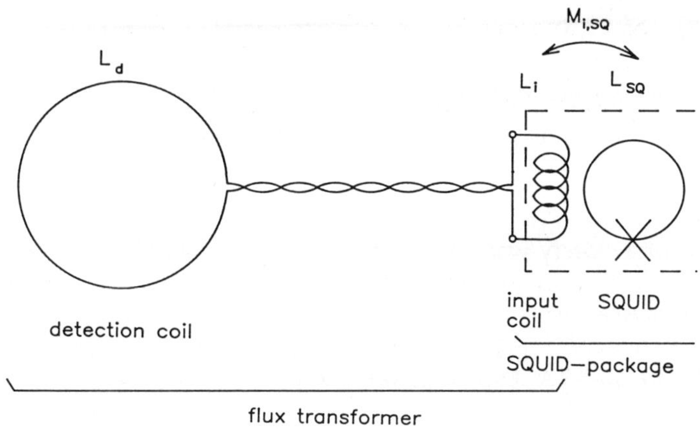

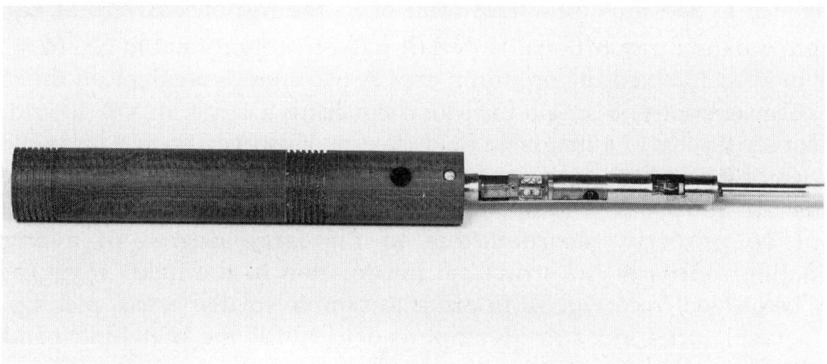

Figure 6

Flux transformer coupled to a SQUID (schematic diagram and real device).

four parameters of the detection coil can be varied in order to optimize the magnetometer response: the detection area A surrounded by the coil, the number of turns N, the cross section of the wire a and the pitch of the coil. Maximum response is achieved for a given SQUID input inductance L_i if the change ΔI of the screening current I flowing through the flux transformer due to an applied flux change in the detection coil $\Delta\Phi = (N\ \Delta B_n\ A)$ is maximized. The relation between the magnetic flux applied to the detection coil and the screening current of the superconducting flux transformer is given by:

$$N \Delta B_n\ A = (L_d + L_i)\ \Delta I \qquad (4.1)$$

Hence for $L_d = L_i$:

$$\Delta I/\Delta B_n = N\,A/2L_d \tag{4.2}$$

The inductance L_d can be related to the inductance of one of its single turn loops L_1 by:

$$L_d = N^{2\,\alpha}\,L_1 \tag{4.3}$$

with $\alpha \to 0$ for a closely wound coil and $\alpha \to 1$ for a large pitch.

L_1 may be expressed as:

$$L_1 = 4 \times 10^{-7}\,\pi^{1/2}\,\sqrt{A}\left[\ln\left(8\,\sqrt{A}\,/\,\sqrt{a}\right) - 1.75\right] \tag{4.4}$$

with A and a in m^2 and L_1 in H. Then for a given SQUID with a fixed input inductance L_i and thus for a *fixed value* of L_d the response $\Delta I/\Delta B$ of Eq. (4.2) is approximately proportional to N^{-3} ($\alpha = 0$) or proportional to N^{-1} ($\alpha = 1$). In order to keep L_d fixed the optimum area A becomes dependent on the choice of N. Consequently, a single-turn loop covering a large area A would give the best sensitivity to a magnetic field change. However, such a large pick-up coil area of a response-optimized magnetometer may be too large for source localization purposes. The spatial variation of the magnetic field strength cannot be properly recorded due to the large degree of averaging, eliminating details of the structural information in the field maps, as can easily be judged from Fig. 4. In order to sample smaller areas, pick-up coils with several turns must to be employed to fulfill the matching condition $L_d = L_i$. Here a coil with a large pitch ($\alpha = 1$) is advantageous because of less response degradation compared with that of closely wound coils. According to Eq. (4.4) a thicker wire would give better results, but in practice thin wires ($a \approx 0.01\ mm^2$) are preferred because they trap less magnetic flux and produce less noise [14]. A more extensive description of inductance calculations for wire-wound detection coil configurations may be found in [15] and [16].

Due to the high sensitivity of a magnetometer sensor to wanted and unwanted signals (noise), every effort must be directed towards the suppression of the interfering noise. This can only be achieved by performing biomagnetic experiments with magnetometers in electrically and magnetically shielded rooms.

Another efficient discrimination between wanted and unwanted signals is possible with the use of special detection coil configurations called *gradiometers*. Gradiometers take advantage of the fact that unwanted signals originate from sources much further in distance than the distance between the physiological source and the detector. At the point of measurement, therefore, the distant noise sources produce a spatially rather homogeneous field, whereas the field of physiological origin is accompanied by a strong gradient. The spatial discrimination performed by gradiometers is thus more effective the closer they are positioned to the signal source.

The simplest gradiometer comprises two identical coils of opposite winding sense at a certain distance from each other and sharing a common axis (Fig. 7a). The coil closest to the source acts as the signal pick-up coil (L_p), the other, the compensating coil (L_c), cancels all signals resulting from a spatially homogeneous field. The distance between the two coils is termed the *baseline.* A good choice of baseline ensures that less than 10 % of the signal flux threading the pick-up coil is gathered by the compensating coil. Then, to a first approximation, the signal output provided by a gradiometer sensor is comparable to a magnetometer output with good noise rejection.

Two other types of gradiometers are in common use:

i) A configuration called an *asymmetrical gradiometer* is frequently used (Fig. 7b). This geometry with $L_p > L_c$ has been designed to keep the overall inductance $L_d = L_p + L_c$ of the gradiometer low (for easier impedance matching to L_i), but the sensitivity of the pick-up coil high.

ii) Another very popular configuration is a *second-order gradiometer,* an extension of the first-order gradiometers previously described. In practice the unwanted signals are often not spatially homogenous to a tolerable degree but exhibit a gradient. The reason may be a field distortion due to the steel structures of adjacent building elements or other causes. A problem more intrinsic to biomagnetism is the need to discriminate the strong heart signals while examining the faint signals originating from brain activity. Under such circumstances it is necessary to spatially discriminate against a constant field gradient, too. This is done by a coil configuration as shown in Fig. 7c, called a second-order gradiometer.

The gradiometer performance is primarily determined by the art of flux balancing. In a perfect spatially homogeneous magnetic field the compensated signal of a first order gradiometer should be a few parts per

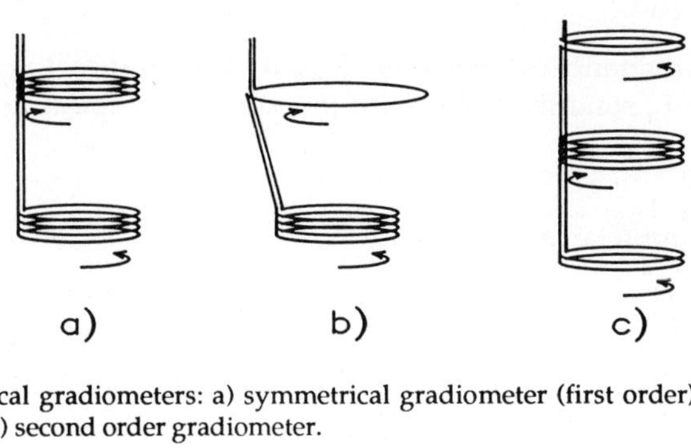

Figure 7
Wire wound vertical gradiometers: a) symmetrical gradiometer (first order), b) asymmetrical gradiometer and c) second order gradiometer.

million of the signal that the pick-up loop alone would provide. A second order gradiometer should be balanced with respect to a spatially homogeneous magnetic field and for a magnetic field of constant gradient, too. This degree of balancing cannot be achieved with wire-wound gradiometers without a final balancing after the sensor is in place and immersed in the liquid helium bath. The fine tuning is usually done from outside the dewar by means of rods that position small superconducting tabs close to the coils.

4.2 Thin Film Flux Transformers

Modern SQUID sensors are fabricated in an "all-Nb" thin film technology. The term "all-Nb" indicates that the base electrode, the counter electrode of the Josephson junctions, and the stripline wires of the circuit are all made of niobium. Nb is a superconducting material very well suited for cryoelectronic circuits due to its relatively high critical temperature of 9.2 K, its compatibility with barrier materials yielding excellent junction characteristics and because devices made of Nb are stable against multiple thermal cycling between room temperature and liquid He temperature. Thus a logical extension of fabricating all-Nb thin film SQUIDs is to couple an all-Nb thin film flux transformer to the SQUID by integrated circuit fabrication.

A common design for such a scheme has been introduced by Wellstood et al. [17]. Figs. 8a and 8b show a schematic circuit diagram and a schematic layout of the thin film configuration of a so-called washer-type DC SQUID with an integrated flux transformer with a single turn pick-up loop forming a magnetometer. The scanning electron microscope picture displays the washer-SQUID with the input spiral coil on top. The coupling concept - thin film input spiral coil and washer-SQUID - was developed by Dettmann et al. [18] and refined by Jaycox and Ketchen [19]. The DC-SQUID "ring", shaped like a washer, acts as a ground plane to the input coil, making it possible to achieve very high coupling coefficients of $k > 0.9$.

For such a configuration the SQUID inductance is mainly determined by the hole dimension d of the washer:

$$L_{SQ} = 1.25\,\mu_o d \tag{4.5}$$

The mutual inductance M between SQUID inductance L_{SQ} and the input coil inductance L_i roughly scales with the number of spiral turns N:

$$M \approx N\,L_{SQ} \tag{4.6}$$

The input coil inductance scales with N^2:

$$L_i \approx N^2\,L_{SQ} \tag{4.7}$$

Thus the matching condition between the detection coil configuration inductance L_p and the input coil inductance $L_i = L_p$ is easily maintained: the

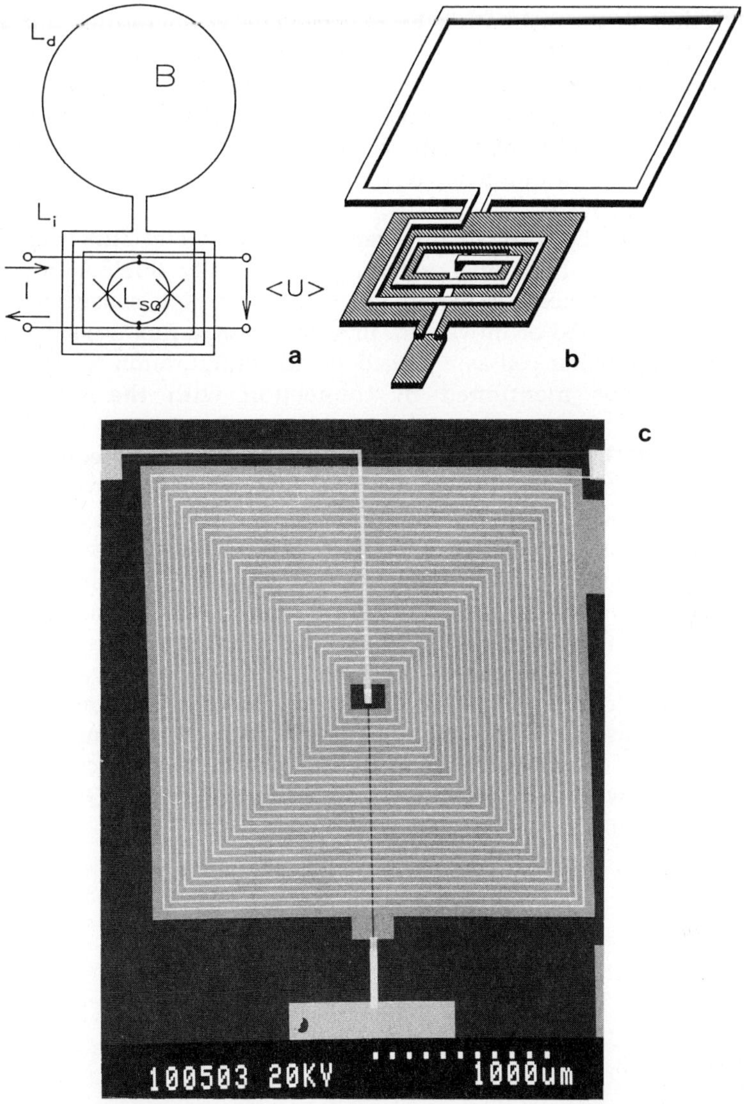

Figure 8
Thin film DC SQUID magnetometer with an integrated flux transformer. a) circuit diagram, b) schematic drawing, and c) scanning electron microscope picture showing a detail: the washer SQUID with the spiral input coil.

SQUID loop inductance L_{SQ} is chosen to fulfill the optimization conditions $\beta_L = I_c 2L/\Phi_o \approx 1$ and $\beta_C = 2\pi I_c R^2 C/\Phi_o < 1$ with I_c, R, C being the critical current, the normal resistance and the capacitance of the junctions respectively and $\Phi_o \approx 2 \times 10^{-15}$ Wb the flux quantum (see [20]). The number of spiral coil turns is then chosen accordingly to yield $L_i = L_p$. One drawback of this design is its inclination to LC resonances degrading the SQUID performance, especially if the number of turns becomes too large. Incorporation of carefully designed damping resistors or filter elements may help [21]. Alternatively, a double transformer concept has been introduced in order to keep the number of spiral turns small [22]. It is often desirable to let the SQUID act as a gradiometer to reduce its sensitivity to environmental interference [18], [21] [23].

The DC SQUID with an integrated magnetometer flux transformer shown in Fig. 8 has been operated in a magnetically shielded room. In a more noisy environment, it would fail due to its high sensitivity to any signal, whether wanted or unwanted. In such a case, a flux transformer with a thin film gradiometer coil as a detection coil configuration would help for the same reasons mentioned in connection with the wire wound gradiometers.

In contrast to wire wound gradiometers thin film configurations require planar geometries. In Fig. 9a - 9d several shapes proposed in the literature are shown. For linear first-order gradiometers it has been shown that the separation area between the two pick-up loops should have the same area as each pick-up loop covers [24]. The parallel-connected gradiometer has the advantage of a low inductance, making matching to the SQUID an easier task. On the other hand, a strong homogeneous magnetic field would cause a large circulating current. The series-connected gradiometer has an extremely small circulating current for the almost spatially homogeneous noise fields but its inductance is larger. Preference is given to the latter lay-out in most practical designs. Second-order configurations have been dealt with mainly in theory, as matching to the SQUID becomes even more difficult [25], [26], [27], and [28].

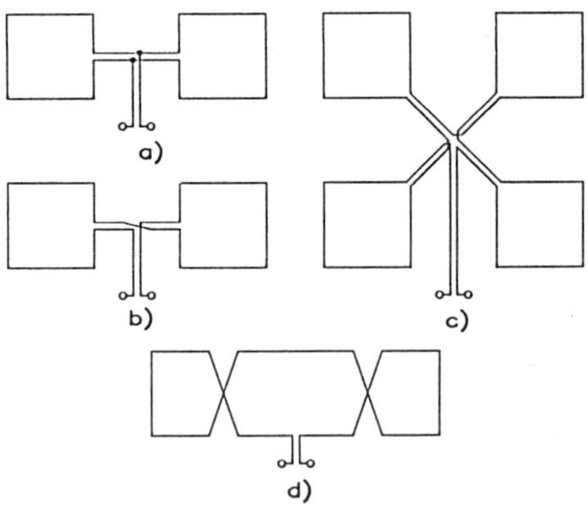

Figure 9
Planar gradiometer configurations: a) parallel connected, sensitive to $\partial B_n/\partial x$, b) series-connected, sensitive to $\partial B_n/\partial x$, c) series connected, sensitive to $\partial^2 B_n/\partial x \partial y$, and d) sensitive to $\partial^2 B_n/\partial x^2$

Planar gradiometers made in thin film technology have a major advantage over wire-wound ones: the precision achievable by microfabrication produces gradiometers that are already quite well balanced and only small corrections, for instance done by laser trimming, are needed. This is of importance for multisensor systems where a balancing procedure is a laborious task, when all sensors are in place within the liquid He bath.

However, planar gradiometers introduce a further complication to the interpretation of the data in comparison with the straightforward results obtained with magnetometers and vertical gradiometers. The output of a vertical gradiometer scarcely differs from the output of a magnetometer, as the pick-up loop generally collects much more signal flux than the compensation coil. The difference with *planar* gradiometers is evident for the case of the simple current dipole model if one considers the two oppositely wound loops of a series-connected planar gradiometer to be moved along the mapping plane in Fig. 4. Obviously, a completely different isocontour map will be derived for different orientations of the current dipole relative to the gradiometer axis. Fig. 10a and 10b show the resulting iso*gradient* contour maps for parallel and orthogonal orientation. The size of the gradiometer baseline and/or the depth of the source also give very different responses and isocontours. The maps in Fig. 10 would actually have been obtained as iso*field* contour maps with a magnetometer or vertical gradiometer as a response for a distorted current quadrupole source, which means two current dipoles in opposition and a short distance apart. Source localization from isogradient maps may be as precise as from isofield maps if ideal model sources have produced the data. But for realistic i.e. distorted and noisy signals the maps may be less unambiguously interpreted when the data were obtained with planar gradiometers. In particular, the

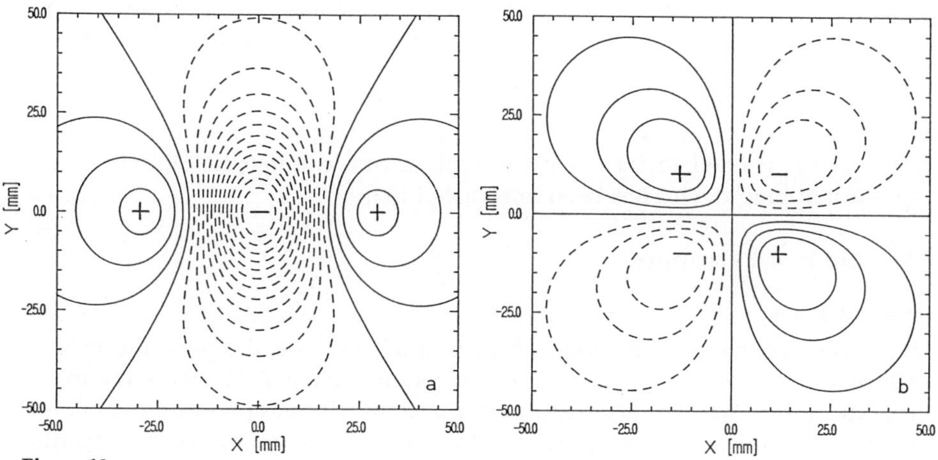

Figure 10
Isogradient contour maps obtained with a planar first-order gradiometer for a current dipole source oriented a) parallel and b) orthogonal to the gradiometer axis.

depth resolution may be worse for planar gradiometers compared with vertical gradiometers. Research concerning the questions of optimally designing planar gradiometers and the data evaluation problems is in progress but as yet, only preliminary results have been obtained.

4.3 Multisensor Configurations

Another up-to-date topic is the development of multisensor systems. A truly parallel simultaneous read-out of several SQUID channels could reduce the measurement time from typically hours to minutes and would allow single events and propagation effects to be detected. Such systems are therefore in great demand and much effort is being put into developing multichannel biomagnetic instrumentation in industry and research laboratories. Several systems of a few channels (in the order of 7) are commercially available or have been set up in some laboratories, but these few-channel systems allow only a small reduction in the measurement time in comparison with single-channel systems. They still have to be scanned mechanically over the subject in order to obtain a useful isocontour map. In this case the data "stitching" and the comparability of data obtained with different systems in different laboraties becomes a problem, because no standard grid exists for such configurations. Only multichannel instruments of 24 channels or more deserve the name: *mapping devices*, as these do not need to be scanned mechanically and allow single-event detection and localization. Beside the cross-talk problem between the channels, the reduction of the heavy heat leakage caused by the huge number of electrical connections becomes the main obstacle in developing such systems. Schemes to reduce the cross talk have been proposed by [29] and [30]. The heat leakage problem could be minimized if less electric wire connections per channel were needed. This requirement has triggered the development of novel read-out schemes in SQUID electronics. Complete digital read-outs integrated on the same chip that contains the SQUID and the flux transformer would be the ultimate goal. The first developments in this direction have been introduced by [23] and [31], leading in the latter case to two wires per channel only. However, further work devoted to higher integration and/or better performance is needed though. Very recently a first commercial SQUID multisensor mapping device has been introduced [39].

5. Sensor Periphery

5.1 Dewars

Superconducting cryoelectronic circuits must be cooled below the critical temperature T_c of the material used, preferably below $T_c/2$ where the material and Josephson junction parameters become less temperature dependent than for an operation closer to T_c. The SQUID sensor including the flux transformer is usually immersed in a bath of liquid He (4.2 K) where temperature fluctuations and mechanical vibrations are orders of magnitude smaller than in any alternative cooling system. With a few exceptions [32] cryocoolers have not yet become established in biomagnetic applications because of the production of magnetic noise signals.

Instead of cryocoolers, dewars are invariably used in biomagnetism, like the one shown schematically in Fig. 11 containing the liquid He with the SQUID sensors immersed. The choice of an epoxy-fiberglass composite as the wall material is dictated by the strict requirements concerning mechanical strength, low magnetic noise due to magnetic or magnetizable impurities or to eddy currents that would be generated in normal metals by Johnson noise currents. A vacuum is established between the double walls to reduce heat conduction and convection. The vacuum is maintained by a getter material placed inside and by continuous liquid He refill. As (fiber) glass is permeable to gaseous helium at non-cryogenic temperatures, the vacuum would be considerably degraded within a few weeks if gaseous He were stored in the dewar at room temperature. Heat leakage due to radiation is kept to a minimum by superinsulation (multilayered, metallized plastic foils with high reflectivity coefficients). Evaporating He is forced by baffles to flow along the neck of the dewar and thus cool a shroud attached to it. This shroud is thermally connected to various parts of the superinsulation by heat-conducting wires fixed to it. The baffles act as radiation shields, too. Further details on dewar construction may be found in [33] and [34].

The detection coil of the flux transformer is situated in a tail section of the dewar. Here, the wall thickness is minimized to the limits permitted by considerations of mechanical security and heat leakage. In practice the dewar wall may well be the major noise source in very sensitive biomagnetic measurements. It is in particular the normal metal layers of the

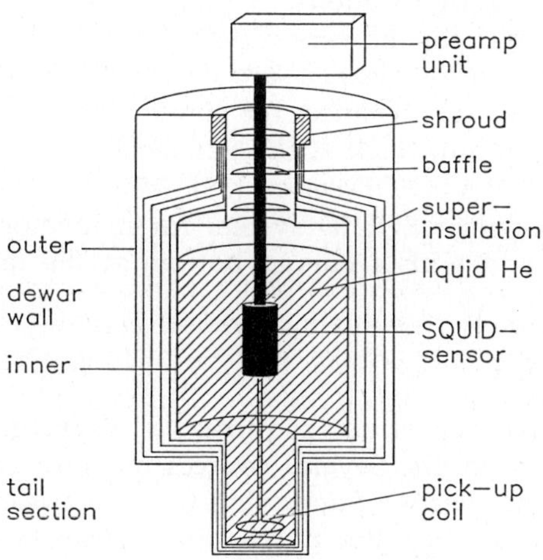

Figure 11

Schematic cross section of a dewar suited for biomagnetism.

Figure 12

Dewar insert for a 37-channel SQUID multisensor system [36].

superinsulation which generate noise, therefore the number of layers is reduced to a tolerable amount in the tail region.

Such requirements as minimum detection-coil-to-physiological-source distance, low noise level, low heat leakage, sufficient mechanical strength call for demanding design options. Optimum results cannot be obtained with a standard dewar for the very different applications in biomagnetism, particularly if multisensor systems are employed. Various dewar designs have been presented, for instance, for single nerve investigations, with a dewar wall only one mm thick at the tail section fitted for a conductively cooled pick-up loop of a SQUID sensor [35], with a flat bottom accomodating a 37-channel multisensor SQUID system for magnetocardiac experiments [36] (see Fig. 12) and with a curved bottom at the tail section for brain research [7].

5.2 Shielded Rooms

Biomagnetic instrumentation requires a very great reduction of environmental noise levels. Several sources of noise and their magnitude are listed in [1]. One means of achieving this has already been introduced: the concept of gradiometry. But this is often unsatisfactory and additional shielding against electrical and magnetic interference becomes unavoidable. A solution is provided by shielded rooms with walls containing several layers of either Cu, Al and/or mu-metal [37].

Figure 13
Shielded room in Berlin.

One of the rooms most advanced in construction for biomagnetic purposes is situated at the PTB in Berlin (Fig. 13). This cube-shaped room measures 4 m x 4 m x 4 m outside and 2 m x 2 m x 2 m inside, and its wall consists of one Cu shell and 6 mu-metal shells.

6. Possible Implementation of High-T_c Superconductors in Biomagnetic Instrumentation

Three areas of high critical temperature superconductors (HTS) applications seem possible:

i) Shielding against magnetic interference

ii) Hybrid systems, consisting of an HTS flux transformer coupled to a Nb-SQUID and/or

iii) All-HTS SQUID sensors.

HTS thick-film fabrication techniques like plasma spraying or the sol-gel method allow large wall surface areas of shielding containers or rooms to be covered with an uninterrupted layer of HTS. Cooled by liquid N_2 and well insulated, such walls should provide a shield against magnetic interference due to the Meißner-Ochsenfeld effect. The extremely large demagnetizing factors of a room shielded against magnetic interference would require additional mu-metal walls and cooling is a considerable maintainance factor. However, one of the biggest problem to date that rules out such an application is the low critical current density of the granular HTS material. This property allows for a large amount of flux creep caused by moving magnetic flux generating strong noise signals, particularly 1/f noise.

The importance of the close proximity of the flux transformer pick-up coil to the physiological source has already been pointed out. HTS flux transformers cooled with liquid N_2 to 77 K would permit a considerably thinner dewar wall. The appropriate flexible HTS wire for fabricating flux transformers has not yet been invented. Again, inherent 1/f noise may be a problem here, too. At present, only truly single-crystal HTS thin-films seem to exhibit tolerable 1/f noise levels [38]. HTS thin film flux transformers would inevitably require that the coupled SQUID also be made of HTS thin film.

All-HTS SQUID sensors have been described in the preceding chapter. For most biomagnetic applications their field sensitivities are still far too low when operated at 77 K. It should be noted that cooling is not a major cost factor in operating biomagnetic systems. There is some hope for a competitive performance of all-HTS SQUID sensors operating at 4.2 K. However, intensive long-term research and development activities seem to be necessary to develop a mature multilayer epitaxial HTS thin-film technology that could compete with all-Nb cryoelectronics. Of course, superconductors with appreciably higher T_c or any other important invention, could change the scene again.

7. Conclusion

The central point of biomagnetic sensor design is the layout of the detection coil configuration. The constraints on the design procedure are:

- the input parameters of the SQUID used,
- the type of physiological source to be investigated,
- the physiological questions to be solved such as localization, source discrimination, single events, signal fine structure , etc.
- the source to detector distance (dewar wall design),
- multichannel considerations (cross talk, measurement grid),
- environmental noise discrimination (shielding, filtering, gradiometer type),
- material properties (types of superconductor and their quality, low noise properties).

As the signals to be investigated are extremely small, only by considering all these points and finding specialized solutions to various diagnostic problems can successful results be obtained. The research related to these questions is still at an early stage. The first multi-channel systems have been introduced commercially, true mapping devices only very recently. Therefore an evaluation of their performance is not possible yet.

Acknowledgements

The author would like to thank the members of the research group "Biomagnetism" of Physikalisch-Technische Bundesanstalt (PTB), Institut Berlin, Prof. H.D. Hahlbohm, and Dr. S.N. Erné, Mr. R. Stehr and Dr. L. Trahms for their helpful contributions to this manuscript. In particular the calculations for Fig. 2, 5, and 10 performed by S.N. Erné, A. Schnabel and R. Stehr are highly appreciated.

References

[1] G.L. Romani, S.J. Williamson, and L. Kaufmann, Rev. Sci. Instr. **52** (1982) pp. 1815-1845

[2] Biomagnetism - An interdisciplinary approach, eds. S.J. Williamson, G.L. Romani, L. Kaufmann, and I. Modena, Plenum, New York, 1983

[3] Biomagnetism: Applications & theory, eds. H. Weinberg, G. Stroink, and T. Katila, Pergamon, New York, 1985

[4] SQUID'85 - Superconducting quantum devices and their applications, eds. H.D. Hahlbohm and H. Lübbig, de Gruyter, 1985

[5] Biomagnetism'87, Proc. 6th Int. Conf. Biomagn.,Tokyo, eds. K. Atsumi, M. Kotani, S. Ueno, T. Katila, and S.J. Williamson, Tokyo, Denki Univ. Press, 1987

[6] Advances in cryogenic engineering, Vol. 33, ed. R.W. Fast, Plenum, New York, 1988

[7] J. Knuutila, S. Ahlfors, A. Ahonen, J. Hallström, M. Kajola, O. Lounasmaa, V. Vilkman, and C. Tesche, Rev. Sci. Instr. **58** (1987) pp. 2145-2156

[8] M. Saarinen, P. Siltanen, P.J. Karp, and T.E. Katila, Ann. Clin. Res. **10** (S21) (1978) pp. 1-43

[9] H.P. Lehmann, in [2], pp. 591-624

[10] J. Sarvas, Physics in Medicine and Biology **32** (1987) pp. 11-22

[11] S.N. Erné, G. Curio, L. Trahms, Z. Trontelj, and P. Aust, in [5]; and L. Trahms, S.N. Erné, Z. Trontelj, G. Curio, and P. Aust, Biophys.J. **55** (1989)

[12] J.H. Tripp, in [2] pp. 101-139

[13] D. Cohen, and H. Hosaka, J. Electrocardiology **9** (1976) p. 409

[14] V. LeTourneau, J.H. Claasen, S.A. Wolf, D.U. Gubser, T.L. Francavilla, and R.A. Hein, Advances in Cryogenic Engineering, Vol. 32, eds. R.P. Reed, and A.F. Clark, Plenum, New York (1986) pp. 579-584

[15] P. Carelli, I. Modena, and G.L. Romani, in [2] pp. 85-100

[16] H.J.M. ter Brake, SQUID Magnetometers, thesis, Technische Hogeschool Twente, The Netherlands

[17] F. Wellstood, C. Heiden, and J. Clarke, Rev. Sci. Instrum. 55 (1984) pp. 952-957

[18] F. Dettmann, W. Richter, G. Albrecht, and W. Zahn, Phys. stat. solid (a) **51K** (1979) pp. K-185-188

[19] J.M. Jaycox, and M.B. Ketchen, IEEE Trans. Magn. **MAG-17** (1981) pp. 400-403

[20] C.D. Tesche, and J. Clarke, J. Low Temp. Phys. **29** (1977) pp. 301-331

[21] J. Knuutila, M. Kajola, H. Seppä, R. Mutikainen, and J. Salmi, J. Low Temp. Phys. **71** (1988) pp. 369-393

[22] B. Muehlfelder, W. Johnson, and M.W. Cromar, IEEE Trans. Magn. **MAG-19** (1983) pp. 303-307

[23] D. Drung, E. Crocoll, R. Herwig, A. Knüttel, M. Neuhaus, and W. Jutzi, Ext. Abstr. 1987 Int. Supercond. Electronics Conf., Tokyo (1987) pp. 21-24
[24] C.M. Pegrum and G.B. Donaldson, SQUID ´80, eds. H.D. Hahlbohm and H. Lübbig, de Gruyter, Berlin (1980) pp. 535-548
[25] P. Carelli, and V. Foglietti, J. Appl. Phys. **54** (1983) pp. 6065-6067
[26] S.N. Erné, and G.L. Romani, in [4] pp. 951-961
[27] M.B. Ketchen, J. Appl. Phys. **58** (1985) pp. 4322-4325
[28] C.D. Tesche, in [4] pp. 933-937
[29] J.H. Claassen, J. Appl. Phys. 46 (1975) pp. 2268-2275
[30] H.J.M. ter Brake, F.H. Fleuren, J.A. Ulfman, and J. Flokstra, Cryogenics **26** (1986) pp. 667-670
[31] N. Fujimaki, H. Tamura, T. Imamura, and S. Hasuo, IEEE Int. Solid State Circ. Conf. (1988) pp. 40-41, 291
[32] D.S. Buchanan, D. Paulson, and S.J. Williamson, in [6] pp. 97-106
[33] W.S. Goree, SQUID applications to geophysics, ed. Weinstock (Society of Exploration Geophysicists, 1981) pp. 85-91
[34] D. Crum, in [3] pp. 21-28
[35] J.P. Wikswo, in [6] pp. 107-116
[36] S.N. Erné, K.P. Matthies, and A. Pasquarelli, private communication
[37] S.N. Erné, in [2] pp. 569-578
[38] M.J. Ferrari, M. Johnson, F.C. Wellstood, J. Clarke, P. Rosenthal, R. Hammond, and M.R. Beasley, IEEE Trans. Magn. **25** (1989) pp. 806-809.
[39] G. Daalsmans, H. Seifert, W. Folberth, F. Bömmel, H.E. Hoenig, and S. Schneider, Ext. Abstr. 1989 Int. Supercond. Electronics Conf., Tokyo (1989) pp. 40-45

The Josephson Junction as a Spectral Detector

J. H. Hinken
Hans Kolbe & Co., FUBA-Forschungszentrum
D-3202 Bad Salzdetfurth

1. Introduction

At microwave frequencies and in the far infrared, receivers can employ Josephson junctions in various functions [1], [2]. These also include such functions which in principle can be provided by semiconductor devices, at least in the microwave range: broadband detectors, mixers, amplifiers, and local oscillators. But beyond this, Josephson junctions offer further functions which are not usual or not possible with standard semiconductor devices. These functions are the self-oscillating mixer and the spectral detector, which provide not only amplitude information but also frequency information on the signal received without the need for a separate local oscillator. The spectral detector can possibly be used for the spectroscopic analysis of gases which are located on earth and in space, but also the use as a surveillance receiver can be imagined. In the following sections, the working principle and a few measurement results of the spectral detector will be described.

In order to operate the Josephson junction at high frequencies, the influence of the parasitic elements of its equivalent circuit should be as small as possible. This is especially valid for the structure-dependent junction capacitance. Point-contact or microbridge Josephson junctions are therefore normally used for high frequency receiver application. To a good approximation, these junctions can often be modelled in an equivalent circuit with the resistively-shunted-junction (RSJ) model, see Fig. 1. R is the normal resistance of the junction, I_c is the critical current and the symbol J represents the so-called sinφ-term or ideal Josephson junction, whose voltage U and current I_J are given in parametric form by the Josephson equations

$$I_J = I_c \sin \varphi \tag{1.1}$$

$$\frac{d\varphi}{dt} = \frac{2e}{\hbar} U \tag{1.2}$$

(e - elementary charge; $\hbar = h/2\pi$; h - Planck's constant). The parameter φ can be understood as the quantum mechanical phase difference between the

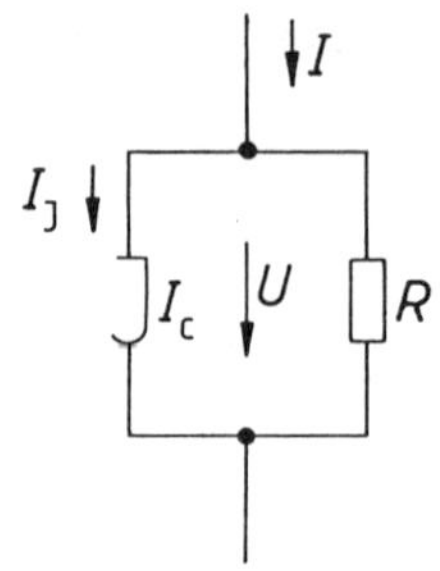

Figure 1

RSJ model for a Josephson junction.

Cooper pairs in the two superconducting electrodes of the Josephson junction.

Devices which satisfy Eqs. (1.1) and (1.2) have a dc current voltage characteristic which is changed characteristically if an rf signal is injected in addition to the dc current. The special form of the deviation gives information on the amplitude and frequency of the signal. Details are discussed in Sect. 2.

There is, however, as in all practical receivers, a lower threshold of rf power which can be detected. This is determined by internal and external noise. This influence is treated in Sect. 3.

In Sect. 4, experimental set-ups, measurements, and results are described. The final Sect. 5 presents a survey on what might be improved in the further development of spectral detectors employing Josephson junctions.

2. Current and Voltage Sensitivity

In this section the dc current voltage characteristic of a junction according to Fig. 1 is described. A distinction is made between a pure dc current bias (autonomous case) and an additionally injected small rf current. We will see that at constant dc current, the dc voltage difference from both cases depends partly on the internal noise of the junction. This dependence can be described by the spectral linewidth of the internal Josephson oscillation. The influence of the external-circuit impedance will also be considered.

2.1 Autonomous Junction

In this section the junction is assumed to be biased with only a dc current, i. e. $I(t) = I_0$ in Fig. 1. Kirchhoff's law then gives

$$\frac{U}{R} + I_c \sin \varphi_0 = I_0 \tag{2.1}$$

where the phase difference φ is written as φ_0, thus indicating the autonomous case without ac current injection. Using normalized quantities $i_o = I_o/I_c$, $u = U/(RI_c)$, $\tau = t \cdot \omega_c$, $\omega_c = 2eRI_c/\hbar$ and the Josephson eq. (1.2) we can write

$$\frac{d\varphi_o}{d\tau} + \sin \varphi_o = i_o . \tag{2.2}$$

If $d\varphi_o/d\tau = 0$, then $\varphi_o = \arcsin i_o$ is time independent and the voltage $u = d\varphi_o/d\tau$ is zero.

If $d\varphi_o/d\tau \neq 0$, a solution can be found by separating the variables φ_o and τ:

$$\frac{d\varphi_o}{i_o - \sin \varphi_o} = d\tau. \tag{2.3}$$

Integration of (2.3) gives the function $\tau(\varphi_o)$. Taking the inverse function one obtains $\varphi_o(\tau)$. The derivation with respect to τ is performed, which gives $\dot{\varphi}_o(\tau) = u(\tau)$, i. e. the voltage across the contact. This can be written as

$$\dot{\varphi}_o(\tau) = \begin{cases} 0 & \text{for } |i_o| \leq 1 \\ \dfrac{\overline{u}^2}{i_o - \cos(\overline{u}\tau)} & \phantom{\text{for }} |i_o| > 1 \end{cases} \tag{2.4}$$

with an appropriate choice of the integration constant and with the time averaged voltage

$$\overline{u} = \begin{cases} -\sqrt{i_o^2 - 1} & i_o < -1 \\ 0 & \text{for } -1 \leq i_o \leq 1 \\ \sqrt{i_o^2 - 1} & i_o > 1 \end{cases} \tag{2.5}$$

The characteristic (2.5) is shown as curve N in Fig. 2. In regions with $\overline{u} \neq 0$ the differential resistance r_d of the dc characteristic can be written as

$$r_d = \frac{d\overline{u}}{di_o} = \frac{i_o}{\overline{u}} . \tag{2.6}$$

With the time-averaged voltage $\overline{u}$, the function $\varphi(\tau)$ can be given in the following compact form:

$$\sin(\varphi_o(\tau)) = \begin{cases} i_o & \text{for } 0 \leq i_o \leq 1 \\ \dfrac{1 - i_o \cos \overline{u}\tau}{i_o - \cos \overline{u}\tau} & \phantom{\text{for }} i_o > 1 \end{cases} \tag{2.7a}$$

For the phase φ_o the following is valid

$$\cos \varphi_o = -\frac{\ddot{\varphi}_o}{\dot{\varphi}_o} = -\frac{d \ln \dot{\varphi}_o}{d\tau} \tag{2.7b}$$

which follows from differentiation of (2.2).

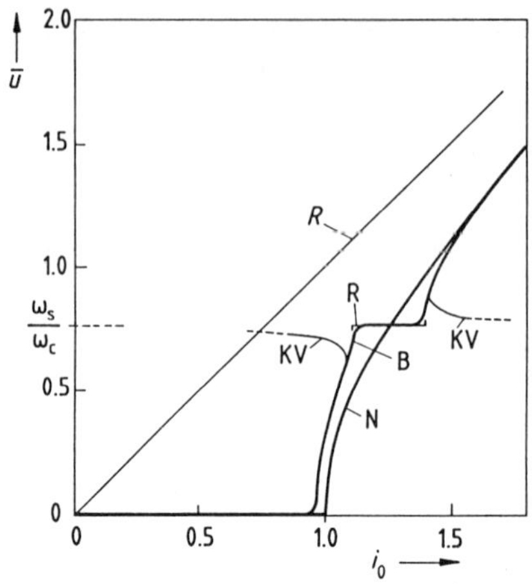

Figure 2

Voltage current characteristic of a Josephson junction obeying the RSJ model. N: $\hat{i}_s = 0$ without noise rounding; otherwise $\hat{i}_s = 0.3$ and $\omega_s/\omega_c = 0.77$; KV-according to Eq. (2.31); R - after [3]; B - experimental characteristic with noise rounding.

2.2 Impressed RF Current

In this section we assume that in addition to the dc current, there is a small sinusoidal rf current impressed into the junction. In the case of rf currents with large amplitudes, we would expect the appearance of Shapiro steps in the $\overline{u}(i_0)$ curve. The calculation of the deformation caused by small rf amplitudes is the subject of this section.

In the case of dc and ac current bias, the current through the junction is written as

$$I(t) = \overline{I} + \tilde{I} \ . \tag{2.8}$$

with the bar indicating time average. Therefore $\overline{I}$ equals the dc current bias and $\tilde{I} = \hat{I}_s \sin\omega t$ is the ac current bias. In general, a constant phase has to be considered in the ac current, but in the present case, which will give an approximate solution for $\Omega = \omega_s / \omega_c \neq \overline{u}$ this is of no importance. This constant phase is therefore omitted. The differential equation analogous to (2.2) is now

$$\frac{d\varphi}{d\tau} + \sin\varphi = \overline{i} + \tilde{i} \ . \tag{2.9}$$

No strict analytical solution to this equation is known.

However, it can be solved using analogue [3] or digital computers [4]. A result of [3] is shown as curve R in Fig. 2 for the rather large amplitude $\hat{i}_s = \hat{I}_s/I_c = 0.3$. At this amplitude a visible deformation of the $\overline{u}(i_0)$ characteristic occurs, but the amplitude is much larger than that which is present in usual detector applications. In practice, noise rounding also occurs, see curve B.

Approximately, Eq. (2.9) can also be solved analytically. We adopt here a method described by [5]. In a perturbation of the autonomous case ($\hat{I}_s = 0$) we assume that with $0 < \hat{I}_s << I_c$ the dc current bias I_o is changed by ΔI_o, keeping $\bar{u}$ constant:

$$\bar{i} = i_o + \Delta i_o . \tag{2.10}$$

The phase $\varphi(\tau)$ will then deviate only a small amount $\Delta\varphi$ from $\varphi_o(\tau)$, the solution in the autonomous case, see Sect. 2.1:

$$\varphi = \varphi_o + \Delta\varphi . \tag{2.11}$$

Δi and $\Delta\varphi$ are now developed into power series of $\hat{i}_s$

$$\Delta i = i_1 + i_2 + i_3 \ldots \tag{2.12}$$

$$\Delta\varphi = \varphi_1 + \varphi_2 + \varphi_3 \ldots \tag{2.13}$$

where i_k and φ_k are proportional to $\hat{i}_s^k$.

Now, the nonlinear term in (2.9) with φ from (2.11) and (2.13) is developed into a Taylor series:

$$\sin\varphi = \sin\varphi_o + \varphi_1 \cos\varphi_o + \left[\varphi_2 \cos\varphi_o - \frac{\varphi_1^2}{2} \sin\varphi_o\right] \ldots \tag{2.14}$$

Inserting Eqs. (2.10) to (2.14) into (2.9) and comparing equal powers of $\hat{i}_s$ we obtain

$$\dot{\varphi}_o + \sin\varphi_o = i_o \qquad \text{for } k = 0 \tag{2.15}$$

$$\dot{\varphi}_1 + \varphi_1 \cos\varphi_o = i_1 + \tilde{i} \qquad \text{for } k = 1 \tag{2.16}$$

$$\dot{\varphi}_2 + \varphi_2 \cos\varphi_o = i_2 + \frac{\varphi_1^2}{2} \cdot \sin\varphi_o \qquad \text{for } k = 2. \tag{2.17}$$

These equations can be solved iteratively, keeping in mind that $\bar{u}$ is unchanged, i. e. $\overline{\dot{\varphi}_k} = 0$ for $k > 0$.

The solution of (2.15) was performed in section 2.1, giving Eq. (2.7a) for $\varphi_o(\tau)$.

To solve Eqs. (2.16) and (2.17) for φ_k, $k = 1,2$, we state that they are of the form

$$\frac{d\varphi_k}{d\tau} + \varphi_k \cos\varphi_o = f_k(\tau) . \tag{2.18}$$

The general solution of (2.18) is

$$\varphi_k = \exp\left(-\int^{\tau} \cos\varphi_o d\tau'\right) \cdot \int^{\tau} \exp\left(\int^{\tau'} \varphi_o d\tau''\right) \cdot f_k(\tau')\, d\tau' \tag{2.19}$$

as can be verified by inserting into (2.18). $\varphi_o(\tau)$ is the phase of the autonomous, i. e. dc-current-biased Josephson junction, for which (2.7b) holds. Inserting the latter equation into (2.19) yields

$$\varphi_k = \dot{\varphi}_o \cdot \int^{\tau} \frac{f_k(\tau')}{\dot{\varphi}_o(\tau')} d\tau' . \tag{2.20}$$

To apply the solution (2.20) to (2.16) we have to put $k = 1$ and replace $f_1(\tau')$ by $i_1 + \tilde{i}$. The condition $\dot{\varphi}_1 = 0$ can only be fulfilled if the term under the integral sign in Eq. (2.20) contains no dc component. We therefore have

$$\overline{(i_1 + \tilde{i})\, \dot{\varphi}_o^{-1}} = 0 \tag{2.21}$$

or with (2.4) for $|i_o| > 1$

$$\overline{[i_1 + \tilde{i}]\, [i_o + \cos(\overline{u}\tau)]} = 0 . \tag{2.22}$$

With $\tilde{i}$ having no dc component per definition, we obtain

$$i_1 = \frac{1}{i_o} \overline{[\tilde{i} \cos(\overline{u}\tau)]} = \frac{\hat{i}_s}{i_o} \overline{[\cos(\Omega\tau) \cdot \sin(\overline{u}\tau)]} \tag{2.23}$$

with $\Omega = \omega_s / \omega_c$. From (2.23) follows that

$$i_1 = 0 \qquad \text{for} \qquad \overline{u} \neq \Omega . \tag{2.24}$$

This means there is no linear change in dc current introduced by a small ac current.

Still holding $k = 1$, the phase difference φ_1 now follows from (2.20) with $k = 1$ and $f_1(\tau) = \tilde{i} = \hat{i}_s \sin\Omega\tau$ and (2.4) as

$$\varphi_1 = \dot{\varphi}_o \int^{\tau} \frac{\hat{i}_s \sin\Omega\tau'}{\dot{\varphi}_o} d\tau' \tag{2.25}$$

$$= \frac{\hat{i}_s}{i_o + \cos(\overline{u}\tau)} \int^{\tau} [i_o + \cos(\overline{u}\tau')] \cdot \sin(\Omega\tau')\, d\tau'$$

Using standard integral tables we arrive at

$$\varphi_1 = \frac{-\hat{i}_s}{i_o - \cos \overline{u}\tau} \left\{ \frac{i_o \cos \Omega\tau}{\Omega} - \frac{\cos (\Omega - \overline{u})\tau}{2(\Omega - \overline{u})} - \frac{\cos (\Omega + \overline{u})\tau}{2(\Omega + \overline{u})} \right\}. \tag{2.26}$$

Next, we come to the solution of (2.17), which is the equation of second-order differences. The solution is obtained from (2.20) with $k = 2$ and $f_2(\tau) = i_2 + \frac{1}{2}\varphi_1^2 \sin\varphi_o$. The argument is similar in the case where $k = 1$: Because of an unchanged $\overline{u}$ we must state $\overline{\dot{\varphi}_2} = 0$. This can be fulfilled if the term under the integral in (2.20) has no dc component. With $\dot{\varphi}_o$ from (2.4), we therefore have

$$\overline{(i_2 + \frac{1}{2}\varphi_1^2 \sin\varphi_o)(i_o - \cos \overline{u}\tau)} = 0. \tag{2.27}$$

Using (2.7a), from (2.27) we obtain

$$i_2 = -\frac{1}{i_o} \overline{(1 - i_o \cos \overline{u}\tau)\, \varphi_1^2}. \tag{2.28}$$

φ_1 from (2.26) must now be inserted. When taking the time average, the identity

$$\frac{1 - i_o \cos \overline{u}\tau}{(i_o - \cos \overline{u}\tau)^2} = -2 \sum_{k=1}^{\infty} \frac{k \cos (k\overline{u}\tau)}{(i_o + \overline{u})^k} \tag{2.29}$$

can be used. The result for the second order dc current term is

$$i_2 = \Delta i = -\frac{\hat{i}_s^2}{4\, i_o (\Omega^2 - \overline{u}^2)} \quad \text{for } \overline{u} \neq \Omega. \tag{2.30}$$

The right-hand side of Eq. (2.30) diverges at $\overline{u} = \Omega$, where it is not valid. Far enough away from this voltage, however, the profile of the dc characteristic is only slightly changed. In these regions the voltage response at constant current bias can be approximated from (2.30) by use of the differential resistance r_d in the autonomous case, see Eq. (2.6):

$$\Delta\overline{u} = -r_d\, \Delta i = \frac{\hat{i}_s^2}{4\, \overline{u} (\Omega^2 - \overline{u}^2)} \quad \text{for } \overline{u} \neq \Omega. \tag{2.31}$$

These important relations for the change of the $\overline{u}(i_o)$ characteristic caused by a small sinusoidal ac current were first found by [6] using a slightly different type of solution.

Equation (2.30) and (2.31) show that the Josephson junction can be used as a detector with quadratic behaviour, as can a semiconductor pn junction and a Schottky junction, too.

Like these standard quadratic detectors the Josephson junction acts as a wide-band detector when biased at $\overline{u} << \Omega$ or at $\overline{u} >> \Omega$. In the latter case, the detection mechanism is classical, i. e. it is directly related to the non-zero curvature of the $\overline{u}(i_o)$ characteristic. The wide-band Josephson detector in the narrower sense, however, uses a bias at $\overline{u} << \Omega$ where the voltage response in the noise-free case considered can increase infinitely with $\overline{u} \rightarrow 0$, according to Eq. (2.31).

In contrast to standard detectors, in addition to amplitude information the Josephson detector offers information on the frequency of the incident signal. It can be directly found from that voltage $\overline{u}_c$, where according to (2.30) and (2.31) the singularity in the measured $\overline{u}(i_o)$ characteristic is expected. Just at this voltage, however, (2.30) and (2.31) are not valid. In order to extend the validity of the equations to all voltages, noise in the junction must be taken into account. This will be done in the next section.

2.3 Oscillation Linewidth

Noise arises from the junction itself, predominantly as thermal noise, and from an external circuit. The noise can be modelled by a noise current source in parallel to the junction in Fig. 1. This noise current source will act on the Josephson junction. The resultant effect on the dc characteristic can be treated as follows. The noise current causes an additional voltage, i. e. voltage fluctuations across the junction. Especially the low frequency noise components can be understood to modulate the dc voltage $\overline{u}$ across the junction. In the autonomous case, this means that the instantaneous oscillating frequency of the junction is changed randomly within a linewidth γ_1 around the centre frequency which is given by the noise-free case. A pronounced change in the $\overline{u}(i_o)$ characteristic only occurs at very small voltages $\overline{u}$. The right angle in the $\overline{u}(i_o)$ characteristic is smoothed.

When there is an impressed sinusoidal ac current, the influence on the dc response (2.30) or (2.31) can be understood from a statistically time modulated voltage $\overline{u}$ [6]. This time-dependent voltage $\overline{u}$ must be inserted e. g. into (2.30) and then time averaging must be performed. With $\sigma(x)$ being the spectral density with a maximum at $x = \overline{u}$ and a half power width $2\gamma_1$, we can also calculate the average Δi from (2.30) by inserting $i_o = \sqrt{1 + \overline{u}^2}$ and replacing

$$\frac{1}{\sqrt{1+\overline{u}^2}\,(\Omega^2 - \overline{u}^2)} \rightarrow \int_0^\infty \frac{\sigma(x)}{\sqrt{1+x^2}\,(\Omega^2 - x^2)}\,dx\,. \tag{2.32}$$

For $\sigma(x)$ as a Gaussian-distributed frequency spectrum, this was performed in [6]. The results concerning the $\Delta i(\overline{u})$ behaviour near $\overline{u} = \Omega$ are similar to those obtained from [7] by a different method:

$$\Delta i = \frac{\hat{i}_s^2}{8\,\overline{u}\,\sqrt{1+\overline{u}^2}} \left[\frac{\overline{u} + \Omega}{(\overline{u} + \Omega)^2 + \gamma_1^2} + \frac{\overline{u} - \Omega}{(\overline{u} - \Omega)^2 + \gamma_1^2} \right]. \tag{2.33}$$

For $\gamma_1 \to 0$, Eq. (2.33) reduces to (2.30). For γ_1 and $|\overline{u} - \Omega| << \Omega$, the second term in the brackets of Eq. (2.33) dominates the first one, and Eq. (2.33) becomes

$$\Delta i \approx \frac{\hat{i}_s^2}{8\Omega\sqrt{1+\Omega^2}} \frac{\overline{u} - \Omega}{\gamma_1^2 + (\overline{u} - \Omega)^2} . \tag{2.34}$$

This describes a crossover of the dc characteristics with and without ac injection at $\overline{u} = \Omega$ instead of the singularity as expected from the noise-free case. Eq. (2.34) shows that the voltage response $|\Delta\overline{u}|$ has maxima

$$|\Delta\overline{u}|_{max} = \frac{\hat{i}_s^2}{16\,\Omega\sqrt{1+\Omega^2}\,\gamma_1} \quad \text{at } \overline{u} = \Omega \pm \gamma_1 \tag{2.35}$$

making possible a frequency selective mode of operation.

Fig. 3 shows the current response according to Eq. (2.33) for $\Omega = \omega_s / \omega_c = 0.3$ and various linewidths γ_1. Corresponding $\Delta u(\overline{u})$ characteristics with similar shapes have also been measured [6].

It should be emphasized that these calculations only hold for amplitudes $\hat{i}_s$ small enough. This characteristic, which is quadratic for all values of $\overline{u}$ at small $\hat{i}_s$, becomes linear around $\overline{u} = \Omega$ when the amplitude $\hat{i}_s$ is high enough to induce the first order step of constant voltage, i. e. a Shapiro step. In a detector application, however, the amplitude is normally sufficiently small.

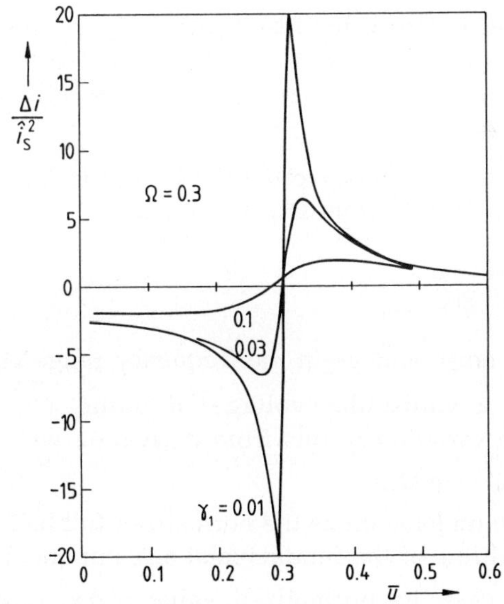

Figure 3
Current response of Josephson detector according to Eq. (2.33) with $\omega_s / \omega_c = \Omega = 0.3$ and γ_1 as parameter.

The question remains, how broad the linewidth γ_1 is in practical situations. It is calculated here for the limit of only intrinsic white Johnson noise. The noise current source of normalized value $i_N(\tau)$ will then have the mean square value

$$\overline{\Delta i_N^2} = 4\, k_B\, T\, \omega_c\, \Delta\Omega_N / (2\pi\, I_c^2\, R) \tag{2.36}$$

within a bandwidth $\Delta\omega_N = \omega_c \Delta\Omega_N$. The resultant voltage fluctuations have two origins. First, noise current components at Ω_N cause voltage fluctuations at the same frequency, and at frequencies low enough, the ratio of both is given by the differential resistance $r_d = i_o/\overline{u}$. Secondly, noise current components at normalized frequencies $\overline{u} + \Omega_N$ and $\overline{u} - \Omega_N$ are converted down to Ω_N and add quadratically to the low frequency noise voltage. In order to calculate the conversion coefficient, we make use of Eq. (2.20) with $k = 1$, $\dot{\varphi}_o$ from (2.4) and

$$f_1(\tau) = i_N \sin(\overline{u} + \Omega_N)\, \tau. \tag{2.37}$$

With this sinusoidal current of amplitude $i_N = \sqrt{\overline{\Delta i_N^2}}$ we have replaced the noise current with the same mean-square value within the frequency band $\overline{u} + \Omega_N$ to $\overline{u} + \Omega_N + \Delta\Omega_N$. Now, only those components of φ_1 are calculated which oscillate with Ω_N, and from these only the dominant component for small Ω_N is considered. From φ_1 the noise voltage $u_N = d\varphi_1/d\tau$ of frequency Ω_N follows. The mean square value results in

$$\overline{u_N^2} = \frac{1}{4} \frac{\overline{\Delta i_N^2}}{\overline{u}^2}. \tag{2.38}$$

Also taking into account down-converted noise from frequencies $\overline{u} - \Omega_N$ and the noise components directly from Ω_N, we obtain

$$\overline{u_N^2} = r_d^2\, \overline{\Delta i_N^2}\, \left(1 + \frac{1}{2 i_o^2}\right) \tag{2.39}$$

as the square of the noise voltage in the frequency range Ω_N to $\Omega_N + \Delta\Omega_N$. This is treated as a sinusoidal voltage of value $\sqrt{\overline{u_N^2}} \cdot \cos\Omega_N\tau$. The corresponding phase variation results from derivation with respect to time τ, giving $-(\sqrt{\overline{u_N^2}} / \Omega_N) \cdot \sin \Omega_N \tau$.

Now, $2\gamma_1$ will be understood as the normalized full half-power width of the broadend spectral line of the fundamental ac component in the junction. In the autonomous case, its normalized value $i_1(\tau) = \hat{i}_1 \cos(\overline{u}\tau)$ equals the corresponding voltage component from (2.4) for $|i_o| > 1$. In the argument of the cosine function, the phase noise calculated above must be added:

$$i_1(\tau) = \hat{i}_1 \cdot \cos\left(\overline{u}\tau - \frac{\sqrt{\overline{u_N^2}}}{\Omega_N} \cdot \sin \Omega_N \tau\right) \tag{2.40}$$

which gives

$$i_1(\tau) \approx \hat{i}_1 \left\{\cos \overline{u}\tau + \frac{\sqrt{\overline{u_N^2}}}{2\Omega_N}\left[\cos\left(\overline{u} + \Omega_N\right)\tau - \cos\left(\overline{u} - \Omega_N\right)\tau\right]\right\} \tag{2.41}$$

for

$$\frac{\sqrt{\overline{u_N^2}}}{\Omega_N} << 1 . \tag{2.42}$$

Equation (2.41) describes a frequency carrier at normalized frequency $\overline{u}$ with sidebands at $\overline{u} + \Omega_N$ and $\overline{u} - \Omega_N$. The power of these sidebands represents the noise power in the frequency regions $\overline{u} + \Omega_N$ to $\overline{u} + \Omega_N + \Delta\Omega_N$ and $\overline{u} - \Omega_N - \Delta\Omega_N$ to $\overline{u} - \Omega_N$, respectively. We thus obtain the spectral current density

$$s_i(\Omega) = \frac{\hat{i}_1^2\, \overline{u_N^2}}{8\, \Omega_N^2} . \tag{2.43}$$

With $\Omega_N^2 = (\Omega - \overline{u})^2$ and large Ω_N, Eq. (2.43) represents slopes according to a Lorentz curve. Because of (2.42), Eq. (2.43) is not valid very close to the carrier. We now assume that also the spectral distribution near the carrier obeys a Lorentzian distribution

$$s_{iL}(\Omega) = p_i \frac{\gamma_1/\pi}{\Omega_N^2 + \gamma_1^2} \tag{2.44}$$

with

$$p_i = \int_0^\infty s_{iL}(\Omega)\, d\Omega \tag{2.45}$$

Furthermore, we assume that the power p_i, which is normalized to some load resistance, in the case with noise equals the power $p_i = \hat{i}_1^2/2$ in the noise-free case. Adjusting (2.44) to (2.43) at $\Omega_N^2 >> \gamma_1^2$, and using (2.36) and (2.38) we then obtain

$$\Delta f = \frac{2\,\gamma_1 \omega_c}{2\pi} = \frac{1}{\pi}\left(\frac{2e}{\hbar}\right)^2 k_B T\, R\, r_d^2 \left(1 + \frac{1}{2\, i_0^2}\right) \tag{2.46}$$

for the full half-power frequency width of the spectral line.

Equation (2.46) shows, that the linewidth becomes smaller, the smaller the normal resistance R is. Accordingly, the frequency resolution of the Josephson junction as a spectral detector increases with decreasing R. According to (2.46), at $r_d \approx 1$ and $1/(2i_0^2) << 1$, the linewidth is equal to ≈ 40

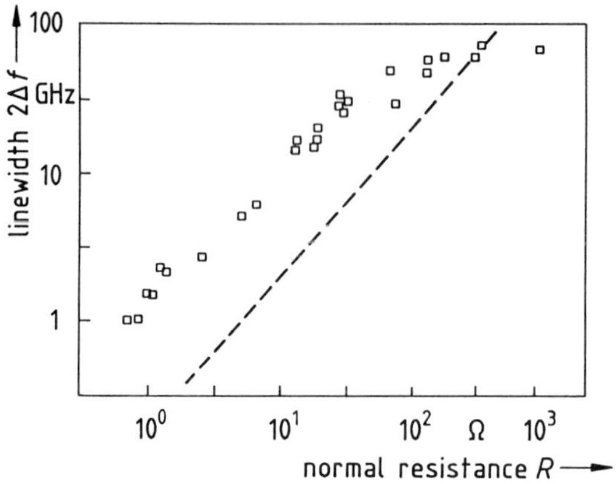

Figure 4
Oscillation linewidth Δf of Nb-Nb Josephson point-contact junction at T = 4.2 K. experimental, – – – calculation. After [9].

MHz per Ohm of the junction normal resistance and per Kelvin of the absolute temperature.

Spectral linewidths of the point contacts and microbridges have been measured. Fig. 4 shows a result from a Nb-Nb point contact [8], [9], oscillating at f = 760 GHz for various normal resistances R. The dashed line shows calculated results from (2.46) with T = 4,2 K and the term $1/(2i_o^2)$ neglected. Comparison shows that for 3 Ω ≤ R ≤ 30 Ω a fit can only be obtained by assuming an effective noise temperature T^* of about 15 to 20 K, i. e. 4 to 5 times the temperature of the liquid helium bath used. This can be explained by additional shot noise ($e\overline{U} > k_BT$) and quantum noise ($hf > k_BT$). The nonlinear dependence beyond 100 Ω may be caused by the external circuit which shunts the junction. The behaviour below 3 Ω may be influenced by heating effects. The minimum linewidth of Δf = 960 MHz in the experiments at 760 GHz was obtained for R = 0.7 Ω. At $R \approx 1$ Ω, the frequency dependence of the linewidth for 240 GHz ≤ f ≤ 1.8 THz was weak.

A reduction of the linewidth can be expected by special means of noise shunts, such as were recently used in parametric amplifiers using Josephson junctions [10].

2.4 External Circuit

The current sensitivity and the voltage sensitivity of a quadratic detector are defined as the ratios of Δi or $\Delta\overline{u}$, respectively, to the rf power P which is dissipated in the detector. The rf signal has up to now only been characterized by its current $\hat{I}_s \sin\omega_s t$. Its relation to the power P is given by

$$P = \frac{\hat{I}_S^2}{2} \cdot R_s \tag{2.47}$$

where R_s is the real part of the input impedance $Z_s = R_s + jX_s$ at signal

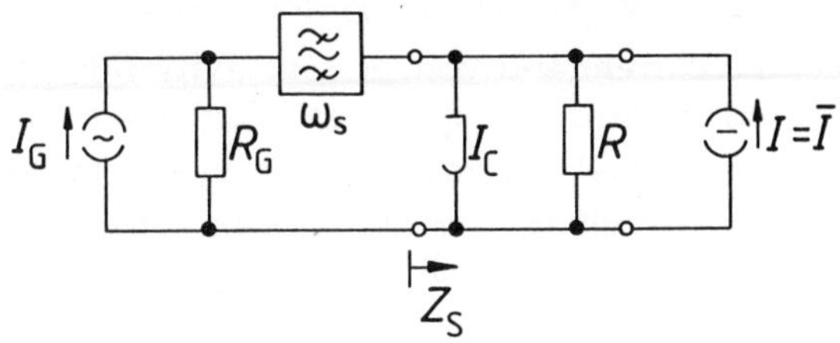

Figure 5
Equivalent circuit of a Josephson junction shunted by the external circuit through a band-pass filter.

frequency ω_S. This applies to a situation which is sketched in Fig. 5: The external circuit has one path for a dc current $I = \bar{I}$. The second path is connected through a band pass filter allowing for currents with frequency ω_S, only. At all other frequencies, this part of the circuit should act as an open circuit in order to fit to the model used in the calculations.

The input impedance Z_S has been calculated approximately by analytical methods and strictly by numerical computation. Fig. 6 shows some numerical results from [4] in the noise-free case for $\Omega = 1$ and $\hat{i} = 0.2$ as an example. It can be seen that there is a pronounced variation in the input resistance in the vicinity of the first step. Moreover, there is a region of negative input resistance immediately below the first step. (2.47) should therefore be restricted to positive values of R_S . In fact, such a negative input resistance can be observed experimentally, although noise rounding occurs at the peaks of Fig. 6.

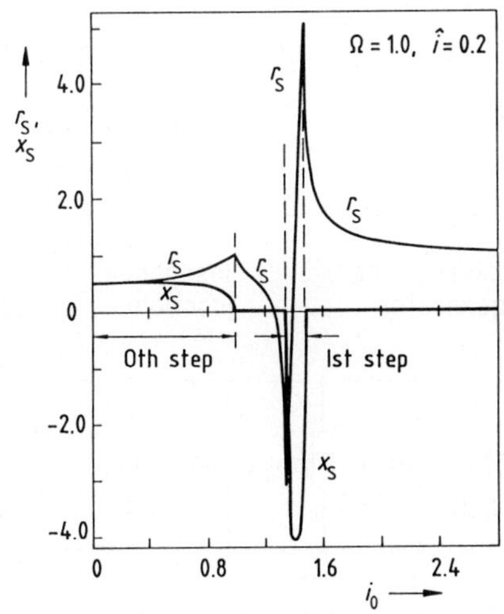

Figure 6
Real part $r_S = R_S / R$ and imaginary part $x_S = X_S / R$ of the input impedance $Z_S = R_S + jX_S$ at $\Omega = 1$ and $\hat{i} = 0.2$. After [4].

In principle, the current and voltage sensitivity can be calculated from this model, but it would require very special microwave circuits to realize a more or less effective impedance matching at ω_S and at the same time the above-mentioned open circuit at all frequencies except at ω_S.

A different model of the external circuit is a transmission line shunting the junction and having a matched load and a generator at its end [11], [5], see Fig. 7a. Typical microwave transmission lines are blocking dc. This property is modelled by a high pass filter with a characteristic frequency ω_g << ω_S. Noise sources $I_{GN}(t)$ and $I_N(t)$ are included. The equivalent circuit Fig. 7a serves to find an effective junction with modified data compared with the original one. This effective junction is shown in Fig. 7b. The load conductance $1/R_G$ is added to the junction conductance. This must be cancelled, however, for frequencies smaller than ω_g, by connecting $-R_G$ in parallel. The noise source $I^*_{GN}(t)$ equals the noise source $I_{GN}(t)$ but without the frequency components below ω_g. The modified junction's characteristic data (index "M") can be shown to be

$$I_c^{(M)} = I_c \tag{2.48a}$$

$$R^{(M)} = \frac{R}{1+g} \tag{2.48b}$$

$$\Omega^{(M)} = \frac{\Omega}{1+g} \tag{2.48c}$$

$$I_N^{(M)} = I_N + I^*_{GN} \tag{2.48d}$$

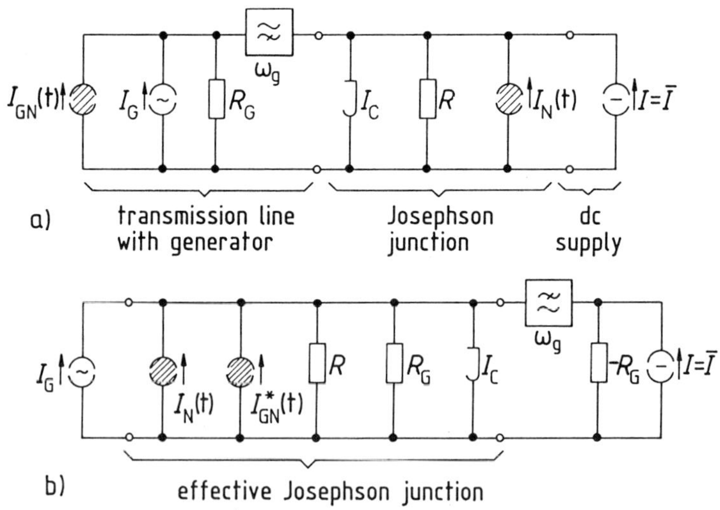

Figure 7
Equivalent circuit of a Josephson junction connected to an ideal transmission line, a) original equivalent circuit, b) reduced equivalent circuit with an effective Josephson junction, after [11].

with $g = R/R_G$.. The rf source is now described by an ideal current source whose inner conductance is zero. This is just the case dealt with in sections 2.2 and 2.3. The results can be transferred to the present case with the modified junction [11]. They are presented here for a case of interest in practice of the voltage $\Delta\overline{U}$ divided by the available rf power $P_{av} = \hat{I}_G^2 R_G/8$. In the noise-free case one obtains

$$\frac{\Delta\overline{U}}{P_{av}} = \frac{8\, r_d^*}{I_c} \frac{g}{4\,(1+g)^2\,(i_o + g\,\overline{u})\,(\Omega^2 - \overline{u}^2)} \quad \text{for } \overline{u} \neq \Omega \tag{2.49}$$

(For r_d^* see below). $\Delta\overline{U}/P_{av}$ vanishes for $g \to 0$ and $g \to \infty$ as well, representing impedance mismatch in both cases. But Eq. (2.49) makes it possible to maximize $\Delta\overline{U}/P_{av}$ by an appropriate choice of the transmission line impedance $R_G = R/g$.

For $\overline{u} \approx \Omega$, one obtains similar to (2.34)

$$\frac{\Delta\overline{U}}{P_{av}} = \frac{r_d^{*2}}{I_c} \frac{g\,(\Omega - \overline{u})}{(i_o + g\,\overline{u})\left[(\Omega - \overline{u})^2 + \gamma_1^{(M)2}\right]} \tag{2.50}$$

where $\gamma_1^{(M)}$ is the modified oscillation linewidth which can also be calculated from [11]. In the case of a small linewidth compared with ω_g, i.e. case b) of [11], this modified linewidth is

$$\gamma_1^{(M)} = \gamma_1 \frac{1}{(1+g)^2 - g\,r_d} \frac{\left[1 + \dfrac{1 + g\,t_G}{2\,(i_o + g\,\overline{u})^2}\right]}{\left[1 + \dfrac{1}{2\,i_o^2}\right]} \tag{2.51}$$

with $t_G = T_G/T$ being the ratio of noise temperatures T_G of R_G and T of R.

With respect to variation of Ω, $|\Delta\overline{U}/P_{av}|$ according to (2.49) is maximum at $\Omega = \overline{u} \pm \gamma_1^{(M)}$.

On the condition which leads to Eq. (2.51), namely a small linewidth compared with ω_g., the differential resistance r_d^*, occurring in (2.49) and (2.50) is given by

$$r_d^* = \frac{r_d}{(1+g)^2 - g\,r_d}\,. \tag{2.52}$$

3. Noise Equivalent Power

Detectors are used to give information on very weak signals. Their minimum detectable power P_{min} is therefore an important characteristic value. It is defined as that rf power which causes a change in dc voltage $|\Delta\overline{U}|$ of just such magnitude which equals the r.m.s. noise voltage

$$\sqrt{\overline{U_N^2}}$$

at the detector output. In general, this noise voltage may include contributions from the following and usually amplifying electronic circuit, but this will not be dealt with at present. We will also omit frequency-dependent noise, e.g. $1/f$-noise, whose influence is often reduced by using a chopped signal instead of a continuous wave and using a band-pass filter at the detector output. Then, strictly the definition of P_{min} must consider the fundamental harmonic of $\Delta U(t)$ instead of $|\Delta \overline{U}|$. The results concerning P_{min} are then changed by a constant factor of the order of 1, only, therefore only the simplified definition will be used here.

Under these circumstances, and with P_{min} being treated as the power which is available from the signal generator, we have

$$P_{\text{min}} = \frac{\sqrt{\overline{U_N^2}}}{\left|\Delta \overline{U}/P_{av}\right|}. \tag{3.1}$$

For the autonomous junction, we obtain the mean-square noise voltage within a post-detection bandwith B from (2.39) and (2.36) as

$$\overline{U_N^{2(A)}} = 4\, k_B\, T\, B\, R\, r_d^2 \left(1 + \frac{1}{2\, i_0^2}\right). \tag{3.2}$$

In the case of the junction being shunted by the transmission line which is connected to the signal generator and for the noise conditions as stated in section 2.4, the mean-square noise voltage is changed just like the linewidths [11], see Eq. (2.51)

$$\overline{U_N^2} = \overline{U_N^{2(A)}} \cdot \frac{\gamma_1^{(M)}}{\gamma_1}. \tag{3.3}$$

Under the aforementioned condition of white noise, P_{min} turns out to be proportional to $\sqrt{B}$. The noise equivalent power NEP can therefore be calculated as

$$\text{NEP} = \frac{P_{\text{min}}}{\sqrt{B}} = \frac{\sqrt{\overline{U_N^2}}}{\left|\Delta \overline{U}/P_{av}\right| \sqrt{B}} \tag{3.4}$$

With $\Delta \overline{U}/P_{av}$ from (2.49) we then have a general expression for all biasing points $\overline{u}$ and signal frequencies Ω.

In the frequency selective mode operation with $\overline{u} \approx \Omega$, we use (2.50) instead of (2.49) to obtain P_{min} and the NEP. At $\overline{u} \approx \Omega$, the voltage response is resonant and the rf input bandwidth is approximately equal to $2\Delta f^{(M)} = 2\omega_c^{(M)} \gamma_1^{(M)}/2\pi$. For such a situation with a pre-detection bandwith of

$2\Delta f^{(M)}$ and a post-detection bandwidth of B, a system noise temperature T_s can be calculated according to [12] as

$$P_{\min} = k_B\, T_s \sqrt{2\ \Delta f^{(M)} \cdot B}\ . \tag{3.5}$$

In [13], T_s is calculated and minimized with respect to variations in Ω - $\overline{u}$. The minimum T_s is reached for $\Omega - \overline{u} = \pm\ \gamma_1^{(M)}$ and is equal to

$$T_{s\,\min} = 2\sqrt{\pi}\left\{T_G + \frac{T}{g}\left[3 + 2\,(1+g)^2\ \overline{u}^2\right]\right\}. \tag{3.6}$$

Besides the factor $2\sqrt{\pi}$, the noise from the generator impedance, or background noise, enters additively into T_s, and below, it can be assumed to be equal to zero. Furthermore, g may then be varied to minimize T_s resulting in

$$g = g_{\mathrm{opt}} = \sqrt{1 + \frac{3}{2\,\overline{u}^2}}\ . \tag{3.7}$$

This gives a rule for designing the impedance of the transmission line in order to obtain minimum system temperature. It is then given by

$$T_{s\,\min\,\min} = T \cdot \left\{4\,\overline{u}^2\left(1 + g_{\mathrm{opt}}^{-1}\right) + 6\ g_{\mathrm{opt}}^{-1}\right\}. \tag{3.8}$$

Experimental data for the system noise temperature of detectors in the selective mode of operation are at least a factor of 20 higher than expected from (3.8). The latter are in the order of 20 to 40 K. The reasons for these discrepancies are primarily insufficient impedance matching and deviations of the point contacts used from the RSJ model.

4. Spectrometer with Wide Frequency Span

So far, we have assumed only a sinusoidal signal to be detected, which causes a characteristic singularity in the $\overline{u}(i_o)$ characteristic. Several sinusoidal signals at different frequencies will cause several such singularities displaced on the $\overline{u}(i_o)$ characteristic. This makes it possible to measure line spectra, see e.g. [14]. The Josephson detector is, however, also able to measure continuous spectra of incoherent radiation. This was first proposed by [15]. In the following sections we will describe the related theory and some experimental set-ups and results.

4.1 Theory

From Eq. (2.33) we can establish a formula for $\Delta i(\overline{u})$ in the case where the incident power is not monochromatic at $\omega_s = \Omega/\omega_c$, but has a spectral distribution $S_i(\omega) = I_c^2 s_i(\Omega)$ of the current. We assume that $s_i(\Omega)$ varies only slowly over normalized frequency intervals of the order of γ_1 and vanishes for $\Omega < \gamma_1$. In (2.33), $\hat{i}_s^2/2$ can then be replaced by $s_i(\Omega)d\Omega$. An integration of the right-hand side of (2.33) over $0 \leq \Omega \leq \infty$ gives the total deviation $\Delta i(\overline{u})$ caused by all spectral components of $s_i(\Omega)$. In the case of the junction being shunted by the matched transmission line, this can be done starting from an equation similar to (2.49). But in order to show the basic principle, we will here only consider the simplest case, i.e. without external circuit and with $\gamma_1 = 0$. We can then use (2.30) and obtain

$$\Delta i(\overline{u}) = -\frac{1}{4\, i_o} \int_o^{\infty} \frac{s_i(\Omega)}{\Omega^2 - \overline{u}^2}\, d\Omega . \tag{4.1}$$

With the extension of $s_i(\Omega)$ to negative frequencies according to $s_i(-\Omega) = s_i(\Omega)$, (4.1) can be written as

$$\Delta i(\overline{u}) = -\frac{1}{8\, i_o \overline{u}} \int_{-\infty}^{\infty} \frac{s_i(\Omega)}{\Omega - \overline{u}}\, d\Omega . \tag{4.2}$$

The integral in (4.2) is a Hilbert transform $H[s_i(\Omega)]$. With the usual normalization

$$H[s_i(\Omega)] = \frac{1}{\pi} \int_{-\infty}^{\infty} \frac{s_i(\Omega)}{\Omega - \overline{u}}\, d\Omega \tag{4.3}$$

and with the inverse Hilbert transform $H^{-1}[H[s_i(\Omega)]] = s_i(\Omega)$

$$H^{-1}[g(\overline{u})] = -\frac{1}{\pi} \int_{-\infty}^{\infty} \frac{g(\overline{u})}{\overline{u} - \Omega}\, d\overline{u} \tag{4.4}$$

Eq. (4.2) becomes

$$\Delta i(\overline{u}) = -\frac{\pi}{8\, i_o\, \overline{u}} \cdot H[s_i(\Omega)] . \tag{4.5}$$

The spectrum can then be calculated from

$$s_i(\Omega) = H^{-1}[g(\overline{u})] \tag{4.6}$$

using the measurable function

$$g(\overline{u}) = -\frac{8}{\pi}\, i_o \cdot \overline{u} \cdot \Delta i(\overline{u}) = -\frac{8}{\pi} \sqrt{\overline{u}^2 - 1} \cdot \overline{u} \cdot \Delta i(\overline{u}). \tag{4.7}$$

4.2 Experiments

First experiments with an evaluation based on Eq. (4.6) were performed by [15], [16] measuring the instrumental function of a far infrared grating monochromator at f = 600 GHz by passing the radiation of a high-pressure mercury lamp through this monochromator. This radiation was then directed to a Nb-Nb Josephson point contact at T = 4.2 K. The measured linewidths of the monochromator at slit values of 10 mm and 20 mm were 18 GHz and 36 GHz, respectively, in good agreement with theoretical expectations. The linewidth of the Josephson oscillation, and thus the frequency resolution of the spectrometer, was estimated as 3 GHz.

In another experiment [17] the frequency resolution of a Josephson spectrometer has been measured in the 70 GHz range, also using Nb-Nb point contact junctions and two monochromatic signals which were generated by klystrons. The half-power linewidth displayed of each of the signals was about 600 MHz. Because the linewidths of the klystron-generated signals are assumed to be much smaller, this probably reflects the linewidth of the internal Josephson oscillation. The two signals were about 2 GHz apart and produced two clearly discernible peaks in the displayed spectrum $s_i(\Omega)$.

Further experiments with a planar Josephson junction, integrated into a planar circuit on a chip and with a computer-controlled measuring and evaluation process, will be described in the following [18]. Fig. 8 shows the principle of the measuring set-up, consisting mainly of the planar integrated front end with the Josephson junction, the liquid-helium cryostat, the dc measurement facilities, and the computer. The incident rf radiation is directed to the planar antenna by a plane mirror through the two cryostat windows made of fused quartz. The dc voltage-current characteristic of the Josephson junction is sampled by setting the dc current and measuring the dc voltage. This procedure is controlled by the computer. After the characteristics with and without rf radiation have been sampled, the spectral density of the rf radiation is obtained from a numerical Hilbert transform.

The broadband superconducting planar front end of the spectrometer, see Fig. 9, consists of the receiving antenna A, the impedance transformer T, and the planar integrated Josephson junction J. Two dc pads P are connected by high-impedance lines L and serve as dc current and voltage contacts.

To reach a spectral resolution of 200 MHz, the normal resistance R of the Josephson junction should be about 1 Ω. For matching the impedance, the junction is integrated at the end of a parallel plate line with a wave impedance of 1 Ω. From (3.6) we know, that a lower generator impedance would be preferable to obtain a lower system temperature, but this did not seem realizable in practice.

The antenna is intended to be a broadband, i.e. frequency-independent-gain antenna, according to the nominal operating band of the spectrometer, extending from 30 GHz to about 1 THz. The antenna used is a planar tapered V-antenna with a continuously varying strip width. The shape of the

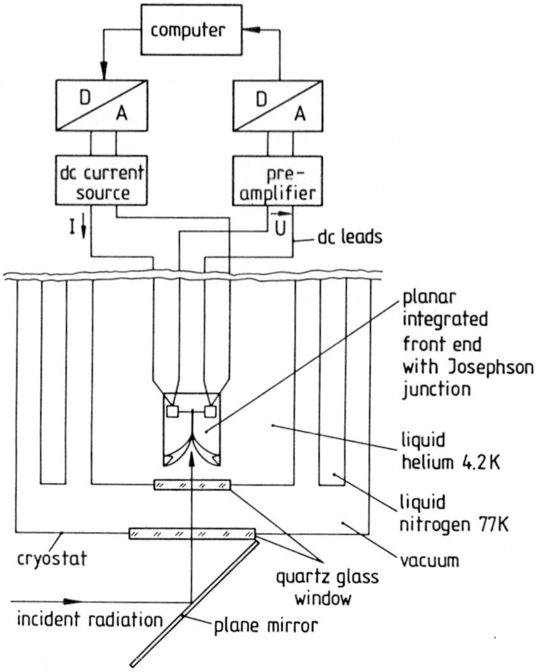

Figure 8

Measuring set-up of the Josephson spectrometer.

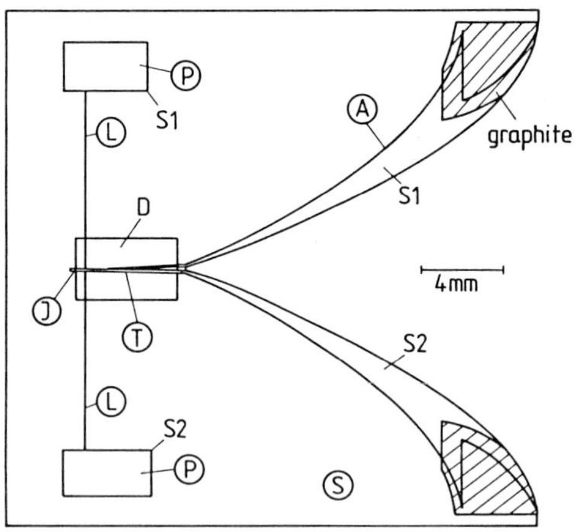

Figure 9
Layout of the planar integrated front end. A - antenna, J - Josephson junction, L - high impedance line, P - dc pads, S - substrate, T - taper; S_1, S_2 - superconducting layers 1 and 2, respectively.

antenna is synthesized according to [19]. The antenna is resistively loaded by graphite films at the ends of the strips. Antenna gain measurements on scaled models provide an average gain of 2 dBi, at least between 1 GHz and 18 GHz.

The rather high feedpoint impedance of the antenna is transformed to 1 Ω by a continuous broadband taper which has the cross section of a quasi-coplanar strip line at the high-impedance end and a parallel-plate line at the low-impedance end. The geometry of the taper is found from a synthesis method for TEM tapers [20]. Antenna and taper are designed for the use of RT/duroid, i.e. glass fiber reinforced teflon, as substrate.

The planar Josephson junctions used had an Nb/Nb-oxide/PbAu sandwich structure. The contact area is so small, only a few μm^2, and the critical current so high, that no hysteresis occurs in the $\overline{u}(i_o)$ characteristic and no geometrical resonances are expected at frequencies below 1 THz. Electrostatic discharge can improve the $\overline{u}(i_o)$ characteristic. The junctions then show approximately the characteristic square-law behaviour $\overline{u} = \sqrt{i_o^2 - 1}$ for $|i_o| > 1$, and they are likely to obey, at least approximately, the RSJ model.

A planar version of the front end of the spectrometer was chosen because of the following principal advantages over a three-dimensional circuit:

- impedance match possible and therefore operation over wide frequency bands
- better reproducibility
- integration of a planar Josephson junction which is mechanically more stable and needs no adjustment
- reliable contacts between the components of the front end.

The computer in Fig. 8 controls the dc current source via a D/A converter and reads the dc voltages via an A/D converter, both converters using 14-bit quantization. To increase the evaluation speed, the Hilbert transform is replaced by Hilbert filtering, using a finite impulse response algorithm [21]. With this method, it took 1 s to sample the $\overline{u}(i_o)$ characteristic, 2.5 s to the end of the computation, and 5 s for the spectrum to be displayed on the screen.

The first front ends were realized by thin-film technique on 0.3 mm thick glass substrates and tested at 4.2 K inside the cryostat. The incident rf radiation is generated by klystrons (line spectrum).

Fig. 10 shows a typical dc voltage-current characteristic of a junction used, without incident microwave power. The characteristic is free of hysteresis and indicates $R = 0.5\ \Omega$ and $I_c = 0.65$ mA. The line spectrum of a klystron at 70 GHz is measured with the spectrometer, and the plotted result is shown in Fig. 11. The dynamic range of the measurement set-up was about 25 dB in the version which is presented in [18].

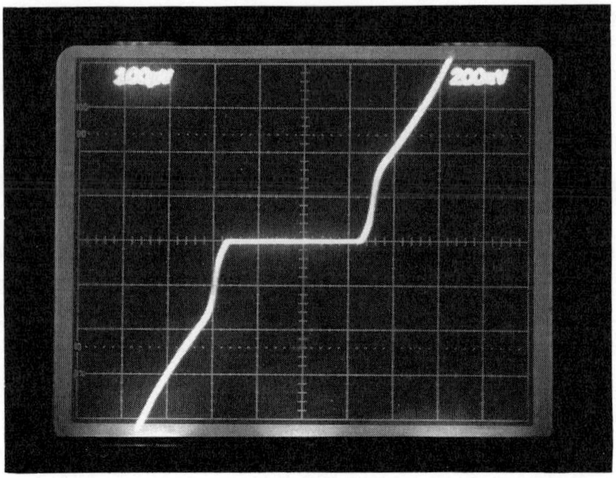

Figure 10
Measured dc voltage-current characteristic without rf radiation. Vert.: voltage 20 μV/div.; hor.: current 0.2 mA/div.

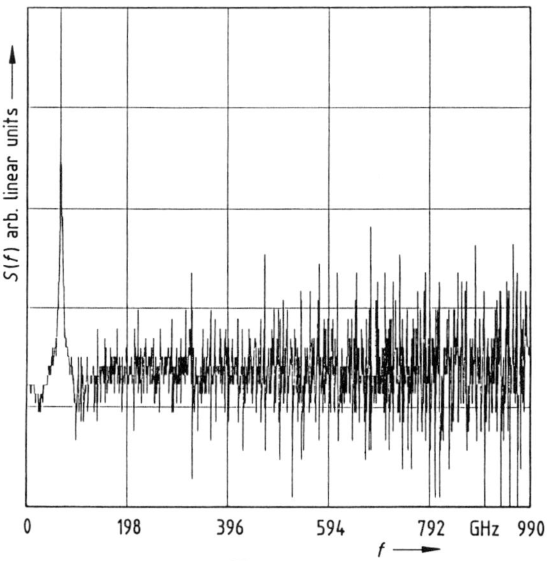

Figure 11
Line spectrum of a 70 GHz klystron, measured with the broadband spectrometer.

The minimum detectable power measured, referred to the antenna feeding point, was about -73 dBm at 70 GHz with an audio-amplifier bandwidth of about 100 kHz. This reflects an NEP of 1.6×10^{-13} W/$\sqrt{\text{Hz}}$. At a first glance, this seems to be very high when compared with the theoretical limit of 1.8×10^{-17} W/$\sqrt{\text{Hz}}$, taking into account only intrinsic thermal noise of the junction [22]. However, a careful noise analysis of the whole system in

Fig. 8 was performed, which considered noise from the D/A converter, the dc source, the approximation of limiting the Hilbert transform to a Hilbert filter, the audio-amplifier, the quantization, and the A/D converter, among other sources, [22]. This analysis showed that the noise of the voltage audio-amplifier used was much too high for this purpose. With the noise of this amplifier taken into account, the theoretical NEP was 6.8×10^{-14} W $/\sqrt{\mathrm{Hz}}$. The residual discrepancy with the measured value can easily be explained by non-ideal impedance match. Future use of a less noisy voltage amplifier will considerably decrease the NEP and will also increase the dynamic range. According to [15], a value as high as 60 dB should be possible. A chopper in the signal path, together with the method of synchronized detection, will work in the same direction.

An investigation of the full half-power width of a displayed spectral line of a klystron showed that it was about 700 MHz. The spectral resolution of the spectrometer, however, can be said to be about 200 MHz.

5. Outlook

A detailed theoretical investigation into the Josephson junction as a spectral detector began around the year 1972 [5], [23], [24]. Experiments had already started earlier, see e.g. [14]. However, to the author's knowledge, up to now there is no regular use of Josephson junctions in this function, or in other microwave-receiver functions.

One reason for this is that the infrared and millimeter waves were of almost no commercial interest in the past. Regarding the millimeter waves, there is some evidence of change in the future.

A second reason is possibly the difficulties a potential user experiences in cooling the device down to liquid helium temperatures. Perhaps the effort can be considerably reduced if junctions with Josephson-like behaviour from high-T_c material, prove suitable for these applications. First tests with granular bridges from $Y_1Ba_2Cu_3O_7$ as a spectrometer device operating at liquid-nitrogen temperature have been successfully performed [25].

A third reason is certainly that existing measuring set-ups must be improved from the engineering point of view. This includes the development and the use of a chopper which should work electronically at a high chopping frequency and the aforementioned reduction of noise in the connected electronic circuits.

References

[1] W. Buckel, Supraleitung, 2. Aufl., Weinheim, Physik Verlag 1977

[2] J. Hinken, Supraleiter-Elektronik, Grundlagen und Anwendungen in der Mikrowellen technik, Berlin, Springer 1988

[3] P. Russer, J. Appl. Phys. **43** (1972) pp. 2008-2010

[4] F. Auracher, T. van Duzer, J. Appl. Phys. **44** (1973) pp. 848-851

[5] K.K. Likharev, Dynamics of Josephson Junctions and Circuits, New York, Gordon and Breach 1986

[6] H. Kanter, F.L. Vernon, J. Appl. Phys. **43** (1972) pp. 3174-3183

[7] K.K. Likharev, B.T. Ulrich, Systems with Josephson Junctions, Moscow Univ. Publ. 1978

[8] Yu.Ya. Divin, N.A. Mordovets, Sov. Techn. Phys. Lett. **9** (1983) pp. 108-110. Early translation of the original article from Russian by S.S. Tinchev is gratefully acknowledged

[9] Yu.Ya. Divin, N.A. Mordovets, On limiting resolution of far infrared spectroscopic techniques based on Josephson effect, Proc. 1984 Conf. IR and MM waves, Takarazuka, Th-9-5, 427

[10] L.S. Kuzmin, K.K. Likharev, V.V. Migulin, E.A. Polunim, N.A. Simonov, X-band parametric amplifier and microwave SQUID using single-tunnel-junctions superconducting interferometer, Superconducting Quantum Interference Devices and their Applications, SQUID´85, Hrsg. H.D. Hahlbohm und H. Lübbig, Berlin, New York, de Gruyter 1985, p. 1029

[11] V.P. Zavaleyev, K.K. Likharev, Radio Eng. Electr. Phys. **23** (1978) pp. 113-121

[12] J.D. Kraus, Radioastronomy, New York, McGraw-Hill 1966

[13] V.P. Zavaleyev, K.K. Likharev, Radio Eng. Electr. Phys. **23** (1978) pp. 120-130

[14] D.G. McDonald, V.E. Kose, K.M. Evenson, J.S. Wells, J.D. Cupp, Appl. Phys. Lett. **15** (1969) pp. 121-122

[15] Yu. Ya. Divin, O.Yu. Polyanski, A.Ya. Shul´man, Sov. Techn. Phys. Lett. **6** (1980) pp. 454-455

[16] Yu.Ya. Divin, O.Yu. Polyanski, IEEE Trans. Mag. **MAG-19** (1983) pp. 613-615

[17] U. Stumper, J.H. Hinken, W. Richter, D. Schiel, L. Grimm, Electr. Lett. **20** (1984), p. 540

[18] J.H. Hinken, J. Niemeyer, R. Fromknecht, M. Klinger, R. Pöpel, W. Richter, D. Schiel, U. Stumper, E. Vollmer, W. Engelland, L. Grimm, Proc. 18th Europ. Microw. Conf. (1988) pp. 177-182

[19] E. Vollmer, J.H. Hinken, Proc. 16th Europ. Microw. Conf. (1986) p. 673

[20] F. Sporleder, H.-G. Unger, Waveguide tapers, transitions and coupler, Stevenage, P. Peregrinus 1979

[21] L.R. Rabiner, R.W. Schofer, Bell Syst. Techn. J **53** (1974) pp. 363-390

[22] R. Beyer, Ansprechempfindlichkeit eines Josephson-Spektrometers, Studienarbeit, Institute of High-frequency Engineering, Technical University of Braunschweig, 1988 (unpublished)

[23] K.K. Likharev, V.K. Semenov, Radio Eng. Electr. Phys. **18** (1975) pp. 1734-1741

[24] K.K. Likharev, V.K. Semenov, Radio Eng. Electr. **18** (1975) pp. 1892-1899

[25] M. Daginnus, M. Klinger, M. Schumann, B. Mordas, W. Richter, J.H. Hinken, personal communication

Superconducting Tunnel Junctions for Radioastronomical Receivers

K. H. Gundlach
Institut de Radio Astronomie Millimétrique (IRAM)
Domaine Universitaire, F-38406 St. Martin d'Hères, France

1. Millimeter and Submillimeter Radiation from the Interstellar Medium

The extension of astronomical observations into the millimeter and submillimeter wavelength region has yielded exciting discoveries; numerous molecules have been detected in comets, in stellar atmospheres, circumstellar shells, proto-planetary nebulae and interstellar clouds. Most molecules are observed in rotational emission lines. The rotational excitation usually occurs through collision with H_2 molecules and He atoms with kinetic temperatures of 10 to 200 K. Fig. 1 shows the (1-0) and (2-1) rotation lines of the isotope $C^{18}0$ from the center of a spiral galaxy, 13.5 million light years away from us [1]. The brightness temperature T_B (vertical scale) is the Rayleigh-Jeans temperature of an equivalent black body that would give the same signal

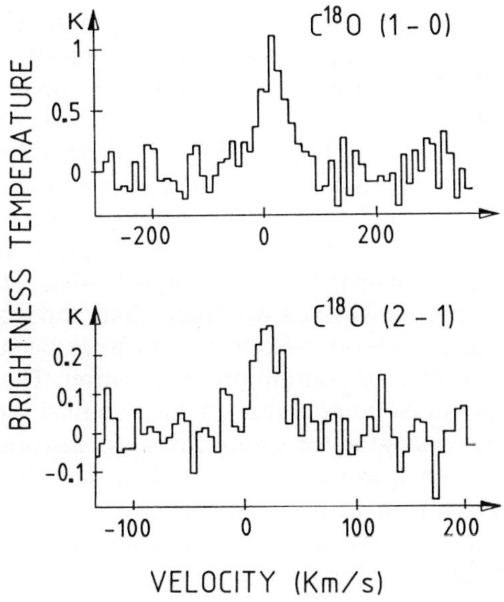

Figure 1
Record of $C^{18}O$ rotational emission lines from the IC 342 spiral galaxy with an SIS receiver on the IRAM 30 m telescope in southern Spain. Observation time τ about 20 minutes [1].

power as the source. The horizontal frequency scale is converted into a velocity scale. In the laboratory, the maxima are located at 109.8 GHz for the (1-0) and at 219.6 GHz for the (2-1) transitions of this isotope. The offset of the maxima from zero velocity indicates that the galaxy moves away from us at about 30 km/s.

At present, more than 70 molecules are known in interstellar space, including complex organic molecules such as ethyl alcohol C_2H_5OH, acetic acid CH_3OOH [2], acetone $(CH_3)_2CO$ [3]; exotic species like isohydrocyanic acid HNC [2] or heavy radicals such as C_5H and C_6H [4]. Other examples are the ring molecules C_3H_2 and SiC_2 [5]. Water vapour H_2O and silicon monoxide SiO occur as maser lines [6].

Certain physical conditions in the interstellar medium favour molecule formation processes which are quite different from those in the terrestrial environment, including a different type of organic chemistry. Thus a number of new organic molecules have been discovered for the first time in the interstellar medium; some of these non-terrestrial molecules have since been synthesized in the laboratory and their spectra measured.

The broadening and Doppler-shifts of molecular lines can give information about the temperature distribution and motions inside interstellar clouds and star-forming regions. One of the important, although poorly understood discoveries of molecular line astronomy, is the outflow of oppositely directed jets from young stellar objects.

Another branch of radio astronomy is the study of continuum radiation. It requires, contrary to the line observations, detectors with large instantaneous bandwidth. Continuum observations include the 3 K cosmic background radiation, thermal radiation from dust in interstellar clouds, synchrotron radiation and free-free emission from ionized regions.

Millimeter and submillimeter astronomy has recently undergone a phase of expansion with large investments in new radio telescopes, such as the 45 m antenna in Nobeyama, Japan, or the 30 m IRAM antenna in Spain, shown in Fig. 2. Other radiotelescopes use, instead of a single dish, an array of several antennas to obtain higher angular resolution. For example, the array in Nobeyama consists of five 10 m telescopes. An IRAM array of four 15 m telescopes is under construction in the French Alps. The new telescopes are larger and/or of higher surface accuracy than those of the preceding generation. So far, cooled Schottky-diode mixers have been mainly used for line and in some cases also for continuum observation. The observations are limited by the receiver noise temperature, which should ideally be less than 100 K at millimeter wavelengths for ground based telescopes. The search for heterodyne receivers with lower noise levels than have been obtained from Schottky-diode receivers was an important stimulus for the development of receivers using the Josephson effect, the quasiparticle current in superconductor-insulator-superconductor (SIS) junctions or in superconductor-insulator-normal conductor (SIN) junctions. Up to now, only the quasiparticle SIS mixer is replacing millimeter-wave Schottky mixers. Current developments are concentrated on improving millimeter-wave quasiparticle

Figure 2
The 30 m IRAM telescope located in the Sierra Nevada, Spain, at an altitude of 2850 m.

SIS mixers, extending their use into the submillimeter range and exploring the application of Josephson and SIN mixers for the submillimeter band.

2. Description of Receivers for Radio Astronomy

2.1 Direct Detectors

The direct detector, also known as the video detector, is generally of interest for continuum observations. So far, bolometers are mostly used. The incoming electromagnetic radiation is converted into heat which changes the temperature and thus the resistance of the device. The information on the phase of the incoming radiation is lost at the output of the bolometer. It is an incoherent detector. The sensitivity of a direct detector can be characterised by its noise-equivalent power NEP (cf. Sect. 5.1). The best bolometers use doped germanium, cooled to about 0.3 K by a He-3 cryostat. At the operation frequency around 230 GHz an NEP of 3 x 10^{-16} $W/\sqrt{Hz}$ has been achieved. The typical bandwidth is 50 GHz [7, 8]. Results obtained for superconducting tunnel junctions as direct detectors will be discussed in Sect. 5.

Direct detectors can also be used for line observation when combined, e.g. with a Fourier transform spectrometer [9] or an absorption spectrometer [10].

2.2 Heterodyne Detection

For the observations of molecular lines below about 50 GHz, masers are used as input stage amplifiers followed by a heterodyne system. Maser amplifiers have noise temperatures as low as 10 K in the K-band, but require large pump power. The instantaneous bandwidth can be about 200 MHz. Above 50 GHz, amplifiers with sufficiently low noise temperature are not available. The signal is therefore coupled to a low-noise heterodyne system. Fig. 3 shows a block diagram of a heterodyne receiver followed by a back-end spectrometer. The essential component for us is the mixer diode. Its non-linear current voltage (I_{dc} -V) curve is used to mix the signal frequency f_S with the frequency f_{LO} of a local oscillator to generate the intermediate frequency $f_{IF} = |f_S - f_{LO}|$. The latter is generally at about 1.5 or 4 GHz where amplifiers with noise temperatures less than 10 K are available. For a double sideband (DSB) mixer the output at the intermediate frequency is equally sensitive to input signals at either side of f_{LO}, that is at $f_S = f_{LO} + f_{IF}$ and $f_{-S} = f_{LO} - f_{IF}$. The output of a single sideband (SSB) mixer is only sensitive to the signal frequency f_S or its image frequency f_{-S}. The IF amplifier typically has a bandwidth of 400 to 600 MHz. Its output signal is passed to a back-end spectrometer. Each of its channels has a much smaller bandwidth $\Delta\nu$, for instance, 1 MHz or even 100 kHz.

The signal-to-noise ratio achieved in the observation time τ in a channel of bandwidth $\Delta\nu$ of the back-end spectrometer is

$$\frac{S}{N} \approx \frac{1}{2} \frac{T_B \cdot \sqrt{\tau \cdot \Delta\nu}}{T_{Rec} + T_{Atm} + T_{Gr}} \tag{2.1}$$

where T_{Rec} is the noise temperature of the entire receiver, T_{Atm} the noise from the atmosphere and T_{Gr} the noise from the ground. Under good conditions $T_{Atm} + T_{Gr}$ is 40 to 50 K at about 100 GHz. If $T_{Rec} >> T_{Atm} + T_{Gr}$ an improvement of T_{Rec} by a factor of two reduces the observation time by a factor of four or corresponds to a doubling of the surface of the dish. Reduc-

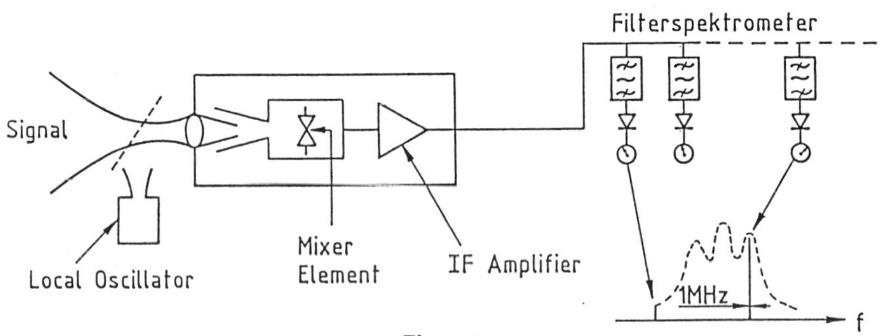

Figure 3
Block diagram of a heterodyne receiver with a back-end spectrometer

tion of T_{Rec} below $T_{Atm} + T_{Gr}$ is less rewarding. Ground-based observations are restricted to atmospheric frequency windows. One therefore also uses airborne and balloon telescopes; space telescopes are in the planning stage. In these cases, it would be interesting to reduce T_{Rec} to about 10 K.

Referring to the block diagram in Fig. 3, the receiver noise temperature may be written as

$$T_{Rec} = T_{in} + L_{in}\, T_M + L_{in}\, L_M\, T_{IF} \tag{2.2}$$

Here L_{in} and T_{in} are the loss and the noise temperature, respectively, associated with the receiver input; T_M and T_{IF} are the noise temperatures of the mixer and IF amplifier. The mixer conversion loss $L_M = P_S / P_{IF}$ is the ratio of the signal power available at the mixer input to the power P_{IF} coupled to the IF amplifier. It can be seen from Eq. (2.2) that for a mixer with conversion loss ($L_M > 1$) IF amplifier noise will be important, but for a mixer with conversion gain ($L_M < 1$) the contribution of the IF amplifier noise to the receiver noise temperature will be reduced. We shall discuss this later.

3. The Quasiparticle and the Josephson Current in SIS Tunnel Junction

The SIS tunnel junction consists of two superconductors separated by an insulator, generally less than 5 nm thick. In the superconducting state we are no longer dealing with the electrons of the Bloch-Sommerfeld model but with electrons interacting with each other via the lattice; these electrons are called quasiparticles [12]. The quasiparticles near the Fermi level form Cooper pairs with a binding energy 2Δ up to a few meV. Josephson effects occur when Cooper pairs tunnel. The d.c. Josephson super current flows without voltage drop across the SIS junction. A finite voltage generates the a.c. Josephson super current. Finite voltages also give rise to tunneling of quasiparticles. This effect can be described in the semiconductor model of the quasiparticle states in the superconductors comprising the SIS junction (Fig. 4). There is an energy gap of a width of 2Δ for the quasiparticles in the superconducting electrodes. The density of quasiparticle states $D(\varepsilon)$ has a square-root singularity above and below the energy gap. When a d.c. voltage V is applied across the ideal SIS junction, current cannot flow at zero temperature for $|V| < 2\Delta/e$. At $|V| = 2\Delta/e$ quasiparticles tunnel from a region of high density of states on one side into a region of high density of states on the other side. Thus the current rises very sharply at $|V| = 2\Delta/e$; it jumps from $I_{dc} = 0$ to $I_{dc} \approx (1.65\,\Delta)/(eR_N)$. By a further increase of V, the I_{dc}-V curve approaches a straight line, the slope of which represents the so-called normal state resistance R_N of the junction. Fig. 4b shows a set of theoretical I_{dc}-V curves from Ref. [13] for various temperatures between T = 0 and the transition temperature $T = T_c$. It should be noted that the theoretical I_{dc}-V curve has

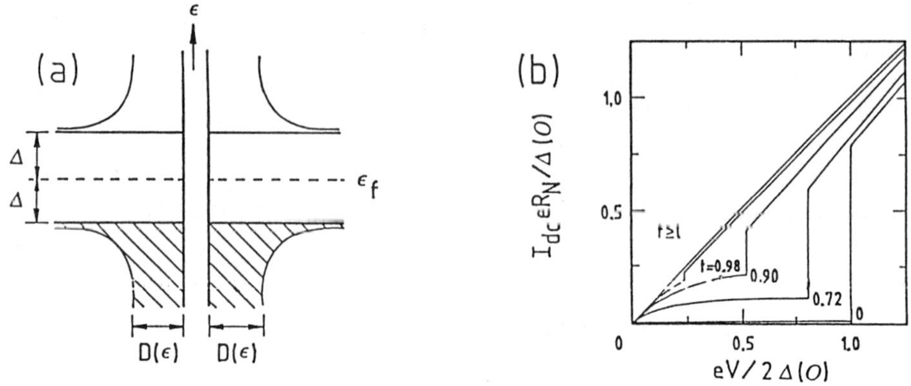

Figure 4
(a) Energy band model of an SIS junction and (b) theoretical I_{dc} -V curves for various reduced temperatures $t = T/T_c$ [13].

finite regions of infinite slope even at finite temperature $0 < T < T_c$. Although tunnel junctions of very good quality can be made, nobody has yet succeeded in reaching the theoretical curves shown in Fig. 4b. Imperfections in the tunnel barrier and at the interface between the electrodes and the barrier cause leakage current and broadening of the sharp current onset at the gap voltage.

A shortcoming in the semiconductor model is that Cooper pair states are not displayed. This is different in the model of ADKINS [14] where the Cooper pairs are located at the Fermi level which is separated from a continuum of quasiparticle states by the energy gap Δ, as shown in Fig. 5. In this model, quasiparticle tunneling occurs by breaking a Cooper pair, leaving one quasiparticle on the left-hand side in the quasiparticle continuum whereas the other quasiparticle of the Cooper pair tunnels into the quasiparticle continuum at the right-hand side. As energy must be conserved, the process starts at $|V| = 2\Delta/e$.

Microwave irradiation of a frequency $\omega/2\pi$ induces steps of width $\hbar\omega/e$ in the I_{dc} -V *characteristic, as exemplified in* Fig. 6. This effect occurs when the width δV of the quasiparticle current onset at the gap voltage is small compared to $\hbar\omega/e$. The steps have been explained in terms of photon-assisted tunneling [15] illustrated in Fig. 5b; the absorption of m photons per quasiparticle provides the energy to open a new tunneling channel at the d.c. bias voltages

$$V = (2\Delta - m\hbar\omega)/e \qquad m = 0, 1, 2 \ldots \tag{3.1}$$

Negative integers of m correspond to radiation-induced transition of quasiparticles into the level of high density of states. Classical theory cannot explain the photon structure in the I_{dc} -V characteristic.

Tien and Gordon [15] and later TUCKER [17] have evaluated the effect of microwave radiation on SIS junctions using quantum theory. Writing the voltage across the junction in the form

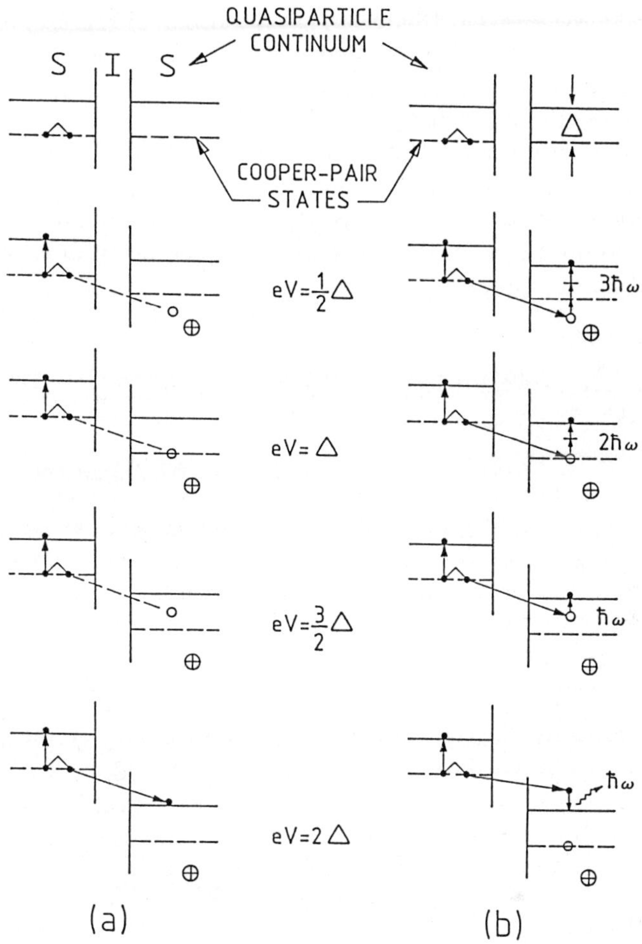

Figure 5
(a) Quasiparticle tunneling in the model of ADKINS [14] (b) Photon-assisted tunneling.

$$V(t) = V_0 + v \cdot \cos(\omega t) \tag{3.2}$$

where V_0 is time independent and the cosine term is generated by the microwave field, TUCKER derived for the time-dependent quasiparticle tunnel current the expression [17]

$$I(t) = \sum_{m,\ell=-\infty}^{+\infty} J_m(\alpha) J_{m+\ell}(\alpha) \Big[I_{dc}(V_0 + m\hbar\omega/e) \cdot \cos(\ell\omega t) + I_{KK}(V_0 + m\hbar\omega/e) \sin(\ell\omega t) \Big] \tag{3.3}$$

Here $J_m(\alpha)$ and $J_{m+\ell}(\alpha)$ are the Bessel functions with indices m and $m+\ell$ and the argument $\alpha = ev/\hbar\omega$. The term $I_{dc}(V_O + m\hbar\omega/e)$ represents the d.c. *I-V* curve without irradiation and shifted in voltage by integral multiples of the photon voltage $h\omega/e$. Finally, $I_{KK}(V)$ is the Kramers-Kronig transform of the d.c. *I-V* curve without irradiation.

An interesting result from Eq. (3.3) is that the applied a.c. voltage $v{\cdot}\cos(\omega t)$ not only generates a dissipative, that is in-phase current but also a reactive, out-of- phase, current. This implies that the SIS junction reacts with time delay to an a.c. voltage. The current at frequency ω is obtained from Eq. (3.3) for $\ell = \pm 1$ with the result

$$I_\omega = \sum_{m=-\infty}^{+\infty} J_m(\alpha)\{J_{m+1}(\alpha) + I_{m-1}(\alpha)\}\cdot I_{dc}(V_o + m\hbar\omega/e)\cdot\cos(\omega t)$$
$$+ J_m(\alpha)\{J_{m+1}(\alpha) - J_{m-1}(\alpha)\}\, I_{KK}(V_o + m\hbar\omega/e)\cdot\sin(\omega t) \qquad (3.4)$$

The d.c. *I-V* curve under irradiation follows from Eq. (3.3) setting $\ell = 0$. The resulting expression

$$I_{dc} = \sum_{m=-\infty}^{+\infty} J_m^2(\alpha)\, I_{dc}(V_o + m\hbar\omega/e) \qquad (3.5)$$

already derived by Tien and GORDON [15], accounts for the photon steps observed in the quasiparticle characteristic shown in Fig. 6.

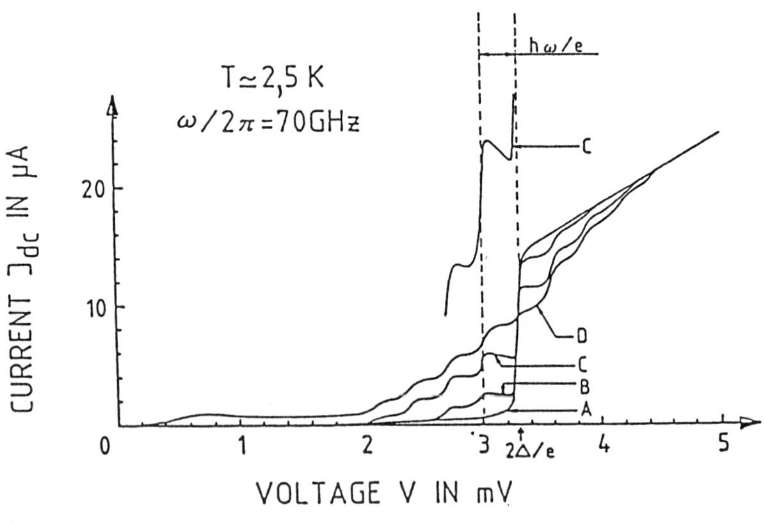

Figure 6
Current-voltage curves of a Pb/Bi/In-oxide-Pb/Bi SIS junction (A) without and (B) to (D) with microwave irradiation of increasing power [16].

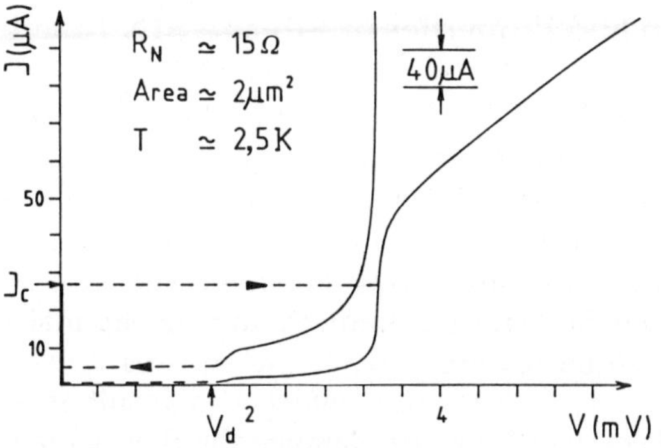

Figure 7

Maximum zero-voltage current I_c and drop-back voltage V_d of an SIS junction

So far, we have only discussed quasiparticle tunneling. Tunneling of Cooper pairs affects the quasiparticle current-voltage curve of the SIS junction. The d.c. Josephson effect causes a zero-voltage current. According to the BCS theory, its maximum value at $T = 0$ may be written as

$$I_c = \frac{\pi \cdot \Delta(o)}{2eR_N} \tag{3.6}$$

Using a current source, the operating point jumps from zero voltage to $V \approx 2\Delta/e$ when the impressed current exceeds I_c. This is illustrated in Fig. 7. The finite voltage generates an a.c. Josephson current. If the junction capacitance C is sufficiently large, the Josephson oscillation is almost sinusoidal, particularly at larger voltages and thus has only little effect on the quasiparticle I_{dc} -V curve around the gap voltage. There is, however, the so-called drop-back voltage V_d. If one reaches V_d by reducing the impressed d.c. current, the junction switches back - like an RCL sweep oscillator [18] - to the zero-frequency state, that is, to zero voltage. In Fig. 7 $V_d \approx 1.6$ mV. Evaluation of V_d yields [19]

$$V_d \approx \sqrt{\frac{\hbar I_c}{eC}} \tag{3.7}$$

The drop-back voltage can conflict with the application of quasiparticle direct detectors and mixers at submillimeter wavelengths. This will be discussed in Sect. 5 and 7.

4. Fabrication and Properties of SIS and SIN Junctions

4.1 Lead Alloy Junctions

At present, junctions for radio astronomy applications are usually based on lead alloy. Such junctions can be fabricated in simple evaporation systems operating at about 5 x 10^{-4} Pa. Pure lead has a transition temperature $T_c = 7.3$ K and an energy gap $\Delta(o) = 1.3$ meV. The Pb-PbO-Pb junction is vulnerable to mechanical stresses induced during thermal cycling between room and liquid helium temperature. These stresses cause dislocation flow and/or grain-boundary deformation which can result in the formation of hillocks and whiskers, which in turn damage the thin oxide barrier. Thus Pb-PbO-Pb junctions become short-circuits upon thermal cycling between room and liquid helium temperature. Even junctions kept at room temperature can become short circuits after some hours. The growth of hillocks in this type of junction can be observed with an optical microscope.

LAHIRI [20, 21] reported that the addition of a few percent of indium to the Pb base electrode drastically improves the junction resistance to thermal cycling. It is believed that the presence of In in the Pb film, particularly in the film grain boundaries, can cause significant grain boundary strengthening, thus reducing grain-boundary deformation due to mechanical stresses.

To further improve the thermal stability of lead alloy junctions, IBM incorporated small amounts of gold in both electrodes in addition to the indium in the base electrode [21]. The addition of gold produces intermetallic compounds such as $AuPb_3$ and $AuIn_2$ which enhance the resistance to stress relaxations. Unfortunately, the intermetallic compounds have lower transition temperatures than lead and thus degrade the sharpness of the I_{dc} -V characteristic. The Au/Pb/In-oxide-Pb/Au junctions have a gap voltage of 2.45 mV.

Later, LAHIRI et al. proposed using ε-phase Pb/Bi counter electrodes with 29 wt% bismuth [22]. It is interesting to note that bismuth is not normally a superconductor. However, it becomes superconducting if it is deposited by quenched condensation at 4 K and not warmed up above 20 K [23]. The Pb/Bi alloy with 29 wt% bismuth becomes superconducting at 8.4 K and has an energy gap of 1.7 meV. The Au/Pb/In-oxide-Pb/Bi junctions have a gap voltage of 2.9 mV.

The work on Pb/Bi counter electrodes pioneered by IBM and our aim to fabricate junctions with a large gap voltage and a sharp I_{dc} -V curve led us to try Pb/Bi alloy for both electrodes. It was found that Pb/Bi/In-oxide-Pb/Bi junctions are reasonably stable to thermal cycling, have a gap voltage of 3.4 mV and can display an I_{dc}-V characteristic of good quality; see for example Figs. 6 and 7. The addition of about 10 % indium to the Pb/Bi base electrodes is necessary to make the junction resistive to thermal cycling and to achieve current densities of the order of 1000 A/cm^2 [24].

For SIN junctions, one generally uses a Pb/In [25] or Pb/Bi/In [26] base electrode and a counter electrode of a normal metal. It was found that the evaporation of silver, for example, does not damage the tunnel barrier on Pb/In or Pb/Bi/In; bismuth or aluminium can also be used, but gold creates short circuits [27]. SIN junctions with a Pb/In base electrode have a gap voltage of 1.3 mV and with a Pb/Bi/In base electrode a gap voltage of 1.7 mV.

From the experiments reported by BAKER and MAGERLEIN [28, 29] it may be concluded that the tunnel barrier of junctions with Pb/In, Au/Pb/In and Pb/Bi/In base electrodes consists predominantly of indium oxide. Different methods can be used for the oxidation:

(i) The base electrode is simply exposed to room air for some minutes.

(ii) The base electrode is kept in the vacuum system and exposed to dry oxygen at 0.1 to 5 Pa for up to 30 minutes. The sample is either at room temperature or heated up to 70 °C.

(iii) The base electrode, again kept in the vacuum system, is exposed to a d.c. plasma discharge using pure oxygen at a pressure of about 0.1 Pa.

(iv) The tunnel barrier may also be formed by the r.f. oxidation and sput ter process developed by Greiner [30]. The rate of oxide growth is initially large, but decreases with time. The sputter rate is almost time independent. After typically 20 to 40 minutes, these two mechanisms are in equilibrium. The junction normal state resistance R_N can be controlled by the oxygen pressure and the r.f. voltage.

Advantages of lead alloy junctions:

a) The fabrication is relatively easy, expensive equipment is not necessary to obtain junctions areas below 1 μm^2.

b) The capacitance per unit area is 30 to 40 fF/μm^2 [31], compared to 60 fF/μm^2 for Nb-Al_2O_3-Nb and 80 fF/μm^2 for NbN-MgO-NbN junctions discussed in the next sub-section.

c) The normal state resistance R_N can be changed when the junction is mounted in the cryostat at its operating temperature [32]. This is very useful for the optimization of SIS receivers [33, 34].

d) The gap voltage of 3.4 mV of the Pb/Bi/In-oxide-Pb/Bi junction is the highest used so far in radio astronomy and is particularly of in-terest for sub-millimeter quasiparticle receivers. The gap voltage of Nb-Al_2O_3-Nb junctions is 2.8 to 3 mV and of NbN-MgO-NbN junctions up to 5.2 mV.

Disadvantages of lead alloy junctions:

a) It is desirable to evaporate the SIS junctions together with an integrated low-pass filter for the IF output signal and/or with an antenna structure. As lead alloy is soft, the evaporation layers can easily be scratched.

b) Storage at room temperature for a period of more than about a week can be a problem. In room-ambient air, at room temperature, the resistance R_N generally increases with time. The rate at which R_N increases varies from junction to junction; in some cases, R_N even decreases with time. The only reliable way to keep R_N constant is to store the junction at sufficiently low temperature. In practice, one uses a container with liquid nitrogen (77 K) or a deep freezer (-20 ^oC). This makes the use of lead alloy junctions difficult when receivers have to be shipped over long distances and precludes their use in space-based applications at present under discussion.

4.2 Refractory Metal Junctions

In the future, lead alloy will probably be replaced by refractory metals such as Nb, NbN or Nb_3 (Ge, Si). These superconductors are mechanically very hard, have high tensile strength and SIS junctions made of them (particularly for Josephson computer circuits) are reported to be very stable with respect to storage at room temperature and repeated thermal cyclings between room and liquid He temperature [35 - 37]. Interesting for high-frequency mixers is the large gap voltage of about 5 mV of NbN junctions [36, 37, 41, 45].

The preparation of Nb-based junctions is more difficult, and the equipment required is considerably more expensive compared with the preparation of lead alloy junctions. The melting point of Nb is 2468 ^oC, its transition temperature of 9.2 K is sensitive to impurities, particularly to O_2 and N_2 [38]. The sharpness of the d.c. I-V curve is strongly influenced by the quality of the superconductor within a distance from the tunnel barrier of the order of the Ginzburg-Landau coherence length ξ. For lead alloy $\xi \approx 80$ nm, but for Nb, ξ is only about 10 nm. Thus for Nb junctions, particularly good metal-barrier interfaces are required. Concerning the native oxide barrier, the only insulating oxide of niobium, Nb_2O_5, is unstable and easily forms conducting channels of suboxides leading to a poor d.c. *I-V* curve. However, Nb junctions of very high quality can be obtained when appropriate artificial tunnel barriers such as oxides of Al [35], Mg, Y, Er [39] or Zr [40] are used.

We briefly describe the fabrication of Nb-Al_2O_3-Nb junctions. First, an Nb layer is deposited by d.c. magnetron sputtering at a rate of 100 nm/min in a vacuum system with a background pressure below 5 x 10^{-5} Pa. This is followed by sputter depositon of Al at a rate of 10 nm/min. The Al film which is about 5 nm thick, is then oxidized by exposing it to pure oxygen. Finally, the second Nb layer is deposited in the same way as the first one. The resulting Nb-Al_2O_3-Nb sandwich, prepared without breaking vacuum, covers the whole substrate [35, 40, 42, 43].

To illustrate how individual junctions with small areas can be made, we take the work of IMAMURA et al. [44]. Fig. 8a shows the tri-layer composed of the lower Nb film, the Al_2O_3 barrier and the upper Nb film. A photoresist pad is first laid over the area where the junction will be formed. Then the upper Nb film is completely anodized around the photoresist pad as shown in Fig. 8b. The anodized tri-layer is then patterned and passivated by sputtered SiO_2. This is illustrated in Fig. 8c. Subsequently, a contact hole in the SiO_2 is formed above the junction with reactive CHF_3 gas to give the structure in Fig. 8d. A cross section of the junction after the deposition of an Nb contact strip is shown in Fig. 8e.

With this method, junctions with areas down to 1.5 μm^2 could be fabricated. Because the Nb underneath the resist pad is also partially oxidized, the process described had to be modified to obtain smaller areas. In the modified version, the photoresist anodization mask in Fig. 8b is replaced by a double layer of sputtered SiO_2 and electron beam resist as illustrated in Fig. 8f. High quality junctions with areas below 1 μm^2 could then be obtained. One example is shown in Fig. 9a.

As with Nb, junctions of NbN were first made with native oxide tunnel barrier. It has poor tunneling characteristics and a large dielectric constant. Several types of artificial barriers such as Ge, a-SiO and Al_2O_3 have been tried. The best all-NbN junctions were obtained by SHOJI et al. [36, 37], using amorphous MgO as tunnel barrier.

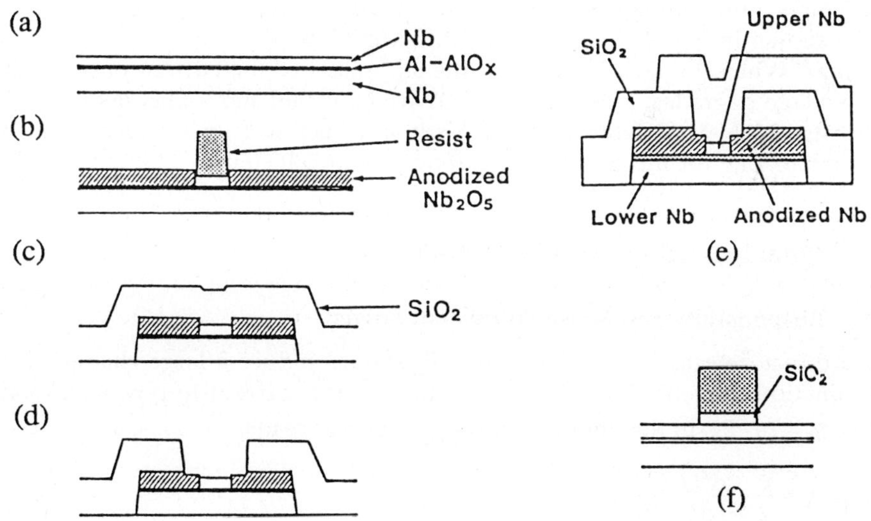

Figure 8
Fabrication sequence (a) - (d) and cross section (e) of an Nb-Al_2O_3-Nb junction. The modified version of (b) to obtain junction areas below 1.5 μm^2 is shown in (f). According to IMAMURA et al. [44].

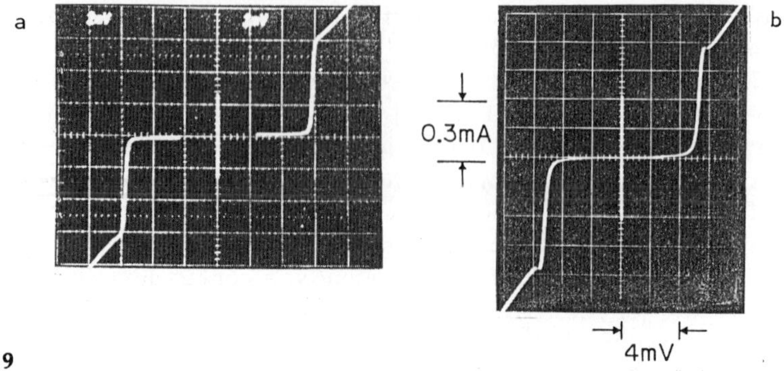

Figure 9
I_{dc} -V characteristics of (a) an Nb-Al_2O_3-Nb junction with an area of 0,7 μm^2 [44] and (b) of an NbN-MgO-NbN junction with an area of 400 μm^2 [45].

In the preparation of NbN-MgO-NbN junctions, the "whole wafer" route is also used without breaking vacuum. NbN films are deposited by r.f. reactive sputtering in 6 % N_2-Ar mixture at about 1 Pa with a deposition rate of about 100 nm/min. The MgO films are prepared by r.f. sputtering using an MgO target in Ar at 1.3 Pa and a deposition rate of 0.5 nm/min. To precisely control the thickness of the MgO film, the substrate is rotated during the deposition of MgO. During the deposition of the NbN counter electrode the substrate is heated up to 150 °C.

The preparation of individual junctions requires a number of process steps similar to those for all-Nb junctions [36, 37, 41, 45]. Up to now, NbN-MgO-NbN junctions are far inferior in quality to Nb-Al_2O_3-Nb. The I_{dc} -V curve shown in Fig. 9b belongs to a NbN-MgO-NbN junction with an area of 400 μm^2. When the area is reduced, the sharpness of the current onset at the gap voltage degrades. Nevertheless, AOYAGI et al. [46] and VILLEGIER et al [47] fabricated NbN-MgO-NbN junctions with areas below 1 μm^2, which could be very interesting for mixers operating above about 500 GHz.

5. Quasiparticle Direct Detectors

5.1 Responsivity and Noise Equivalent Power

For a direct detector, the signal power P_S should induce a large change ΔI of the junction current. The classical result for the current-to-power responsivity $\eta_c = \Delta I / P_S$ in the limit of small signal power reads

$$\eta_c = \frac{K}{2} \cdot \frac{d^2 I_{dc}/dV^2}{dI_{dc}/dV} \tag{5.1}$$

The coupling coefficient K is the ratio of absorbed to incident signal power. In the operating point V_o just below the gap voltage of a high-quality SIS junction, the curvature $d^2 I_{dc} / dV^2$ and thus η_c can become arbitrarily large.

Here the classical theory is no longer valid and Eq. (5.1) has to be replaced by the corresponding quantum mechanical expression. Starting from Eq. (3.5), one obtains in the limit of small radiation power ($\alpha << 1$) the result

$$\eta_c = K \frac{e}{\hbar\omega} \frac{I_{dc}(V_o + \hbar\omega/e) - 2\, I_{dc}(V_o) + I_{dc}(V_o - \hbar\omega/e)}{I_{dc}(V_o + \hbar\omega/e) - I_{dc}(V_o - \hbar\omega/e)} \tag{5.2}$$

It is seen that the derivatives in the classical expression (5.1) are replaced by finite differences. For a high quality I_{dc} -V characteristic, the current responsivity from Eq. (5.2) approaches the quantum limit $\eta_c = e/\hbar\omega$ for a matched junction: one additional quasiparticle per incident signal photon traverses the tunnel barrier. This gives e.g. $e/\hbar\omega$ = 3450 A/W for $\omega/2\pi$ = 70 GHz. It should be noted that η_c decreases with increasing frequency.

The signal power resulting in a signal-to-noise ratio of one for a post-detection bandwidth Δf = 1 Hz is called the noise equivalent power, henceforth NEP. At $T \approx 4$ K and the operating point V_O just below the gap voltage, $V_O > 2\,kT/e$. In this case, shot noise dominates and for a matched junction [48, 49] one obtains.

$$\mathrm{NEP} = \hbar\omega\Delta f \sqrt{\frac{2}{e\Delta f}(I_{dc}(V_o) + I_S(V_o))} \tag{5.3}$$

Here I_{dc} (V_O) is the current without radiation and I_S (V_O) the signal current. Each term under the square root can be understood as the number N of electrons passing the barrier in the observation time $\Delta t = 1/\Delta f$. For an ideal junction at $T = 0$, I_{dc} $(V_o) = 0$. The remaining noise is then due to the signal itself. The ultimate noise limit NEP = $\sqrt{2}\,\hbar\omega \cdot \Delta f$ is reached when one signal electron passes the barrier in the observation time Δt. At 70 GHz, one obtains NEP $\approx 5 \times 10^{-23}$ W. In a real junction, I_{dc} (V_o) may be about 1 μA yielding NEP $\approx 2 \times 10^{-16}$ W at 70 GHz; about 4×10^6 signal photons are needed to obtain a signal-to-noise ratio of one [49].

The first direct detection experiments with SIS junctions have been made by Richards et al. at 36 GHz [50]. The best current responsivity was about half the quantum limit $e/\hbar\omega$ and the lowest NEP = 2.6×10^{-16} W. HARTFUSS and GUNDLACH [48, 49] performed similar experiments at 70 GHz. The best current responsivity was also about half the quantum limit, but the NEP $\approx 10^{-15}$ W.

Although these NEP values are admittedly very low, the SIS video detector can hardly compete with ^{3}He-cooled bolometers [7, 8]. For the SIS detector, NEP increases with ω (Eq. (5.3)) and bolometers have very large bandwidth. For space application and others for which ^{3}He-cooling poses problems, the SIS video detector could be of interest.

5.2 Frequency Limitation

The lower frequency limit is roughly given by $\hbar\omega_\ell = e\delta V$ where δV is the width of the current rise at the gap voltage. For frequencies $\omega < \omega_\ell$ the junction will operate as a classical rather than a quantum detector. Of greater interest is the upper frequency limit. We first assume that the Josephson effect can be completely suppressed by a magnetic field. For the ideal I_{dc} -V characteristic and the operating point V_o slightly below $2\Delta/e$, as shown in Fig. 10.

$$I_{dc}(V_o) = I_{dc}(V_o - \hbar\omega/e) = 0 \quad \text{for} \quad \hbar\omega < 4\Delta \tag{5.4}$$

Thus from Eq. (5.2) the current responsivity of a matched junction becomes

$$\eta_c = e/\hbar\omega \quad \text{for} \quad \hbar\omega < 4\Delta \tag{5.5}$$

It is seen that the quantum limit $e/\hbar\omega$ can be attained up to twice the gap frequency $4\Delta/\hbar$. *When* ω reaches the value $4\Delta/\hbar$ the current responsivity drops to $(0.5\,e)/\hbar\omega$ and then decreases with $1/\omega^2$.

If the Josephson d.c. current cannot be completely suppressed by a magnetic field, the dropback voltage V_d, discussed in Sect. 3, will influence the upper frequency limit. The photon point $V_1 = V_o - \hbar\omega/e$ should not pass into the unstable region below V_d. Hence the upper frequency limit can be expected to be given by the relation $\hbar\omega_{max} = (2\Delta - eV_d)$.

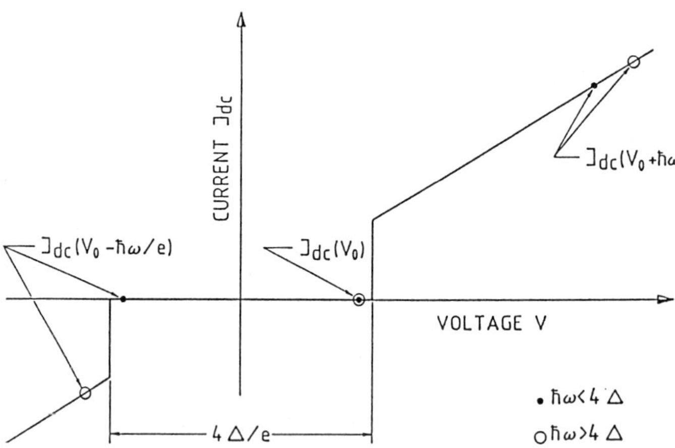

Figure 10
Idealized d.c. quasiparticle characteristic illustrating that the current responsivity $\eta_c = e/\hbar\omega$ can be obtained up to twice the gap frequency $4\Delta/h$.

5.3 Possible Gain Mechanism

Smith et al. [51] have suggested that SIS junctions made from different superconductors with unequal gaps Δ_1 and Δ_2 will function as direct detectors with gain. The dynamic resistance of S_1IS_2 junctions can be negative in the voltage range $|\Delta_1 - \Delta_2| \leq eV \leq (\Delta_1 + \Delta_2)$ as shown in Fig. (11).

Thus for various bias voltages V_0 the current responsivity evaluated by Eq. (5.2) can be higher than $e/\hbar\omega$. Two examples are illustrated in Fig. (11). Gain is expected even if V_0 is slightly below the cusp at $eV = |\Delta_1 - \Delta_2|$, so that a small signal voltage cannot sample the region of negative conductance. The gain mechanism is obviously a quantum effect; the photon point $(V_0 + \hbar\omega/e)$ reaches the region of negative resistance.

To our knowledge, a current responsivity larger than $e/\hbar\omega$ has not so far been experimentally verified. It is of interest for high frequency (> 100 GHz) direct detectors. Thus junctions of considerably larger energy gap than in Fig. (11) are required. We also note that the actual cusp shape of an S_1IS_2 junction is a strong function of temperature [52], an effect which is not advantageous for practical application.

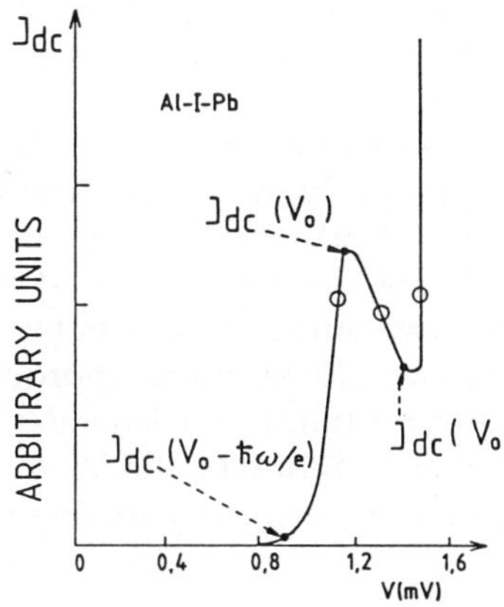

Figure 11
I_{dc} -V characteristic of an Al-Al_2O_3-Pb junction displaying a region of negative conductance [53].

6. Classical Mixing with the Schottky Diode

For a classical mixer, the diode is considered a non-linear resistor. It is a passive element which responds without time delay to external r.f. signals. Quantum effects of the photon field are negligible. The d.c. *I-V* characteristic of the diode is sensitive to the signal power, but not to the signal frequency. A diode operating as a non-linear resistor cannot produce conversion gain unless there is a region of negative dynamic resistance in the d.c. I-V curve.

The theoretical lower limit of the conversion loss is $L_M = 2$ for the SSB and $L_M = 1$ for the DSB mixer [19, 54].

The d.c. I-V characteristic of a Schottky diode cooled to about 15 K may be written as

$$I_{dc} = I_0 \cdot \left\{ exp\left(\frac{eV}{E_0}\right) - 1 \right\} \tag{6.1}$$

To obtain a strong non-linear current-voltage characteristic for mixing, the tunneling parameter E_0 should be as small as possible. So far, $E_0 \approx 8$ meV has been achieved. The current from Eq. (6.1) gives rise to shot noise

$$\langle i^2 \rangle = 2e\, I_{dc} \cdot \Delta f \tag{6.2}$$

where Δf is the bandwidth. Assuming that this noise current arises from Nyquist noise, one can derive an equivalent diode noise temperature T_D with the result

$$T_D = E_0 / 2k_B \tag{6.3}$$

We note that T_D decreases with decreasing E_0. One obtains $T_D \approx 40$ K for $E_0 = 8$ meV [54]. A further noise source is due to the spreading resistance of the Schottky diode.

It has been shown that the ideal classical mixer is a switch. The local oscillator switches the diode between high and low conductance states. The exponential I_{dc} -V curve requires unrealistically large local oscillator voltage to approach the ideal behaviour of the switch [55]. After the discovery of quasiparticle tunneling between two superconductors by GIAEVER [53], it was suggested that the SIS junction would be more appropriate to realize the switch for frequency mixing, because it has a considerably stronger non-linear current-voltage characteristic than a Schottky diode. It was, however, also recognised that the classical picture of frequency mixing is too simple for the SIS junction.

7. Quantum Mixing with the SIS Junction

7.1 Theoretical and Experimental Results of Quasiparticle Mixing

TUCKER [17, 56], SHEN [57], HARTFUSS and TUTTER [58, 59], TUCKER and FELDMAN [19], FELDMAN [60, 61], WINKLER [62] and others have analyzed the performance of quasiparticle mixing with SIS junctions starting from Eq. (3.3). The results may be summarized as follows:

(i) In contrast to the classical resistive mixer, the quasiparticle mixer can have conversion gain: $L_M < 2$ for a SSB and $L_M < 1$ for a DSB mixer. We have already mentioned that according to Eq. (2.2),the conversion gain reduces the contribution of IF amplifier noise to the receiver noise temperature. The conversion gain, first observed by SHEN et al. [63] and KERR et al. [64], is proportional to the dynamic resistance R_d at the photon steps shown in Fig. 6. This dynamic resistance depends on the signal source impedance, on R_N,and he local oscillator power. Even regions with negative values of R_d can be obtained, as seen in Fig. 6. Negative dynamic resistance implies that arbitrarily large conversion gain is available [64, 65].

Large values of R_d involve the problem of impedance matching to the IF amplifier which usually requires a load resistance of about 50 Ω. A transformer between the junction and the IF amplifier can be used for impedance matching. So far, such transformers considerably reduce the desired IF bandwidth of 500 MHz or more to about 50 MHz [66].

Until now, conversion gain has not been utilized in receivers being used in radio telescopes. In practice, however, SIS mixers have less conversion loss than cooled Schottky mixers.

(ii) The quantum theory of mixing has revealed that the noise temperature T_M of the quasiparticlemixer can reach the quantum limit $\hbar\omega/2k_B$ set by the uncertainty relationship [19, 54, 59, 60]. Other mixers can in principle also operate at the quantum limit of noise. It appears, however, that this limit can be more easily reached with the SIS junction. Obvious reasons are the low operation temperature and the small energy gap of the superconductor. This allows operation in the mV range, therefore the current level necessary for impedance matching remains low compared to semiconducting devices. As a consequence, the resulting shot noise can be very low.

(iii) As a consequence of the strong non-linear current-voltage characteristic, the required local oscillator power is two to three orders of magnitude lower than for cooled Schottky mixers. This is of great practical importance as high frequency oscillators producing sufficient power to pump the Schottky mixer, are expensive and have short life times. At frequencies above about 150 GHz, Schottky mixers have sometimes been operated with degraded performance because of in-sufficient local oscillator power [54, 67]. For SIS mixers frequency multipliers driven by a lower frequency source, such a Gunn oscillator or klystron, can provide sufficient local oscillator power.

(iv) SIS junctions have neither d.c. series resistance nor spreading resistance. This is an advantage when tuning out their capacitance. Series and spreading resistance also can generate additional noise and give rise to loss of local oscillator and signal power.

7.2 Realization of Quasiparticle Heterodyne Receivers for Radioastronomical Observations

The SIS junction must be cooled to about 4 K or less. Thus quasiparticle mixers are cryogenically more complicated than cooled Schottky mixers which operate at 15 K. A block diagram of a cryostat for SIS receivers is shown in Fig. 12 [68]. A commercially available closed cycle He refrigerator cools the radiation shields at 70 K and 15 K. When the reservoir of liquid He is pumped, the boil off gas can cool the mixer with the junction down to 2.5 K.

The signal from the telescope and the local oscillator are coupled to the feed horn in front of the mixer via a teflon lens at room temperature, styrofoam windows on the 70 K and the 15 K radiations shields and a helium-tight quartz window on the mixer chamber. A single backshort serves as tuning element for input power matching and compensation of the junction capacitance. The IF output signal of the SIS junction is coupled to a low-noise amplifier on the 15 K stage; for example, an HEMT (high electron mobility transistor) of center frequency 3.95 GHz, bandwidth 650 MHz, and noise temperature 12 K [34].

The noise temperature referred to the receiver input is obtained from the expression

$$T_{Rec} = (T_{hot} - YT_{cold})/(Y-1) \tag{7.1}$$

where T_{hot} and T_{cold} are the temperatures of hot and cold black-body absorbers placed successively in front of the cryostat input window and Y is the ratio of the corresponding IF output powers P_{hot}/P_{cold}. Curve (a) and (b) in Fig. 13 display output powers as a function of voltage for T_{cold} = 77 K and T_{hot} = 295 K, respectively. These curves called gain curves, conversion peaks, or photon peaks, clearly reflect the quantum nature of the SIS mixer. The

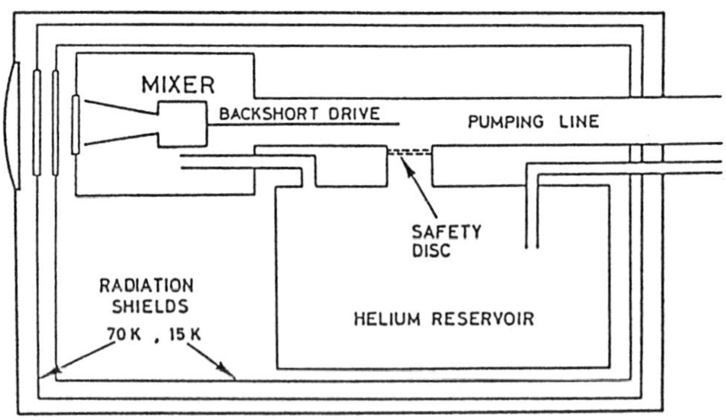

Figure 12

Schematic diagram of the IRAM hybrid cryostat for SIS receivers [68].

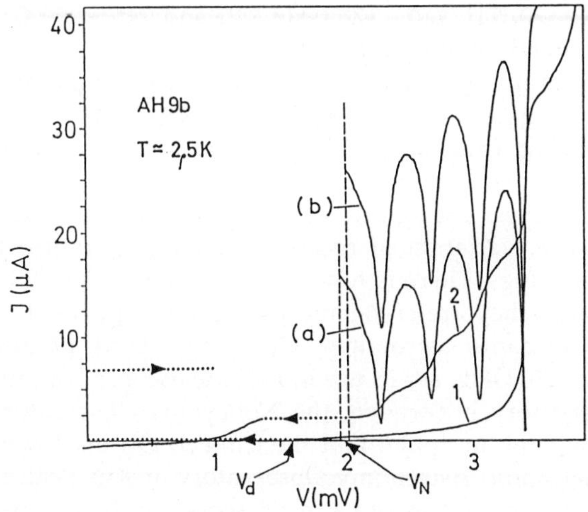

Figure 13
I-V curves (1) without and (2) with local oscillator power. Also shown are the IF output noise power (a) with cold and (b) with hot load, the drop-back voltage V_d and the threshold voltage V_N (defined later). Local oscillator frequency is 94 GHz [69].

maxima occur almost in the middle of the photon-assisted tunneling steps. For a three-port mixer L_M , T_M and thus T_{Rec} should be lowest at the first conversion peak below the gap voltage. In Fig. 13 the DSB receiver noise temperatures are, however, 70 K on the first and third conversion peaks and 65 K on the second peak.

It is required that mm-wave heterodyne receivers have IF bandwidth of 500 MHz or more and can operate over a signal frequency band as large as possible. In Fig. 14 we quote DSB receiver noise temperatures as a function of the local oscillator frequency of IRAM SIS receivers covering the frequency

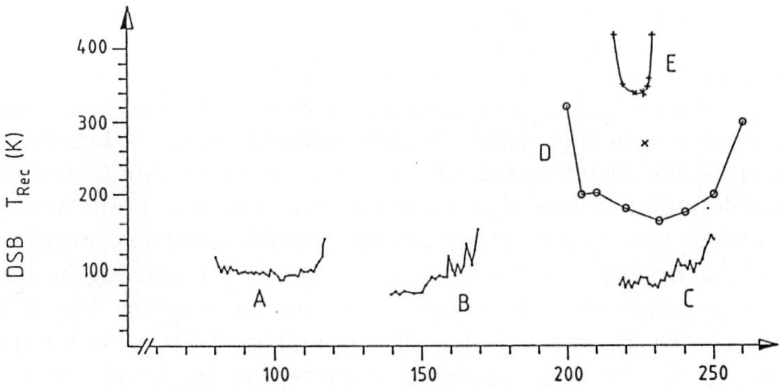

Figure 14
Curves (A), (B), and (C): noise temperatures of IRAM SIS receivers operating from 80 - 115 GHz [70, 71], from 140 - 170 GHz [71, 72], and from 215 - 250 GHz [34]. Curves (D) and (E) noise temperatures of cooled Schottky receivers from Ref. [76] and [77], respectively.

ranges 80 - 115 GHz [70, 71], 140 - 170 GHz [72] and 215 - 250 GHz [34]. All receivers have an IF bandwidth of at least 500 MHz.

Up to the present, the lowest receiver noise temperatures in the 100 GHz frequency range are about 80 K SSB [73, 74] and 100 K SSB [75]. The IRAM receiver operating from 215 to 250 GHz has probably the lowest noise temperature reported so far for this frequency range. It is of course difficult to compare different receivers taking into account input and output bandwidth, operation temperature, variation of receiver noise temperature with signal frequency and other things. There is however, little doubt that SIS heterodyne receivers, even though they are cryogenically more complicated, will replace Schottky receivers in radio astronomy. Currently about 12 SIS receivers, operating from 45 to 250 GHz, are in use in radio telescopes [74], including the Onsala Space Observatory in Sweden, the Nobeyama Observatory in Japan, the IRAM 30 m telescope in Spain, and the Caltech Owens Valley Observatory, or the National Radio Astronomy Observatory at Kitt Peak in the USA. Other radio observatories such as the 3 m telescope on the Gornergrat of the University of Cologne, the 4 m telescope of the University of Nagoya, or the IRAM array of four 15 m telescopes under construction in the French Alps are beginning to use SIS receivers. Up to now, lead alloy SIS junctions have been used in almost all SIS receivers. Nb-Al_2O_3-Nb junctions are used by the Nobeyama Observatory in Japan, and the National Radio Astronomy Observatory in the USA has recently begun to use them.

7.3 Upper Frequency Limit

There is strong interest in heterodyne receivers for submillimeter astronomy. Theory predicts low noise quasiparticle mixing with SIS junctions up to $f_\ell = 4\Delta/h$ if there were no interference from Josephson effects [62, 78, 79]. This result is similar to that for quasiparticle direct detectors discussed in Sect. 5.2.

Noise due to the "drop-back" voltage

The SIS mixer produces extremely large noise below a certain threshold voltage V_N. In Fig. 13 $V_N \approx 2$ mV. The existence of V_N is attributed to the drop-back voltage V_d shown in Figs. 7 and 13, and discussed in Sect. 3. In a classical model the sum of the instantaneous d.c. and a.c. voltages should not enter into the unstable region below V_d; otherwise the operating point jumps in each r.f. cycle to the zero-voltage state, a process which generates strong noise [25]. In Fig. 13 the first three conversion curves are fully resolved, the fourth conversion curve passes into the region of strong noise below V_N. In this experiment $f_{LO} \approx 94$ GHz and the junction gap voltage 3.4 mV. As the mixer is operated on the first or second conversion curve, the drop-back noise does not pose problems. Increasing f_{LO} to 220 - 250 GHz, and using a junction with the same gap voltage, only the first coversion curve could fully be resolved [34, 80]. The second conversion curve had already passed into the noisy region below V_N. Thus the upper frequency limit for low-noise mixing is $f_\ell \approx 250$

GHz in this case. The result is in reasonable agreement with Eq. (6.9) of TUCKER and FELDMAN [19] which yields $f_\ell \approx 250$ GHz by taking $eV_{LO} = 2\hbar\omega_{LO}$, $\omega R_N C = 5$ and $2\Delta = 3.4$ meV. Increasing the frequency above about 250 GHz, it is no longer possible to fully resolve the first quasiparticle conversion curve. In addition, Josephson steps of lower order and therefore larger amplitude reach the region of the first photon step. Switching between Josephson steps can occur, a process which also generates strong noise.

Other phenomena

If Josephson steps enter the region of the first photon step used for quasiparticle mixing, Josephson oscillation and mixing will have a strong influence on quasiparticle mixing. It is still an open question if and how seriously this interferes with low-noise mixing. In point contacts, low-noise Josephson mixing was never really successful. The situation for SIS junctions is assumed to be different as their relatively large capacitance should reduce the down conversion of noise from higher sidebands [19]. In any case, Josephson mixers are probably more difficult to optimize than quasiparticle mixers. A mixer theory including quasiparticle and pair tunneling in SIS junctions is necessary. But such a theory is very complex and it takes some time to provide practical results for designing SIS mixers for frequencies approaching or exceeding the gap frequency $2\Delta/h$ [81].

A disadvantage of the quasiparticle mixer is that it saturates at rather low signal power level, typically at a few pW at 100 GHz. This is calculated and experimentally verified for smooth conversion curves as in Fig. 13 [19, 82, 83]. The Josephson steps passing into the region of the first photon-assisted tunneling step with increasing frequency can more or less strongly modulate the quasiparticle conversion curves [83 - 85]. FELDMAN et al. [82] pointed out that this effect considerably reduces the saturation power obtained for smooth conversion curves.

The Josephson effect can in principle be suppressed by applying a magnetic field so that one flux quantum is within the junction cross-sectional area. For submicron tunnel junctions the required magnetic field can become so large that it reduces the gap voltage and degrades the sharpness of the current onset at the gap voltage. In addition, experience shows that Josephson steps and the drop-back voltage can often not be completely quenched by a magnetic field in high current density junctions with areas less than 1 μm^2. This is, clearly seen in the experiment of WENGLER et al. at 466 GHz for example, [84]. The conversion curve is so strongly modulated by Josephson structure that the mixer probably worked using a mixture of Josephson and quasiparticle modes.

On the other hand, WINKLER et al. [86] studied mixing with Al-Al_2O_3-(In)Al junctions around 70 GHz, which is close to the gap frequency $2\Delta/h = 90$ GHz of Al. They could avoid drop-back noise by applying a magnetic field. If such a result could in future also be achieved for junctions with an energy gap $2\Delta > 3.5$ meV, correspondingly smaller area, and higher current density, low-noise quasiparticle mixing approaching the THz frequency range would be possible. However, it is not clear yet whether Joule heating and/or nonequilibrium effects due to large quasiparticle injection into the electrodes in such junctions will pose problems.

8. Mixing with SIN Junctions

A way to completely avoid Josephson effects is to use superconductor-insulator-normal metal (SIN) junctions. In this case, only one electrode has the energy gap 2Δ and the singularities in the density of states as shown Fig. 15a. This has two disadvantages. First, the gap voltage is only $V_g = \Delta/e$ and therefore the theoretical upper frequency limit only $f_\ell = 2\Delta/h$. Second, the theoretical d.c. I-V characteristic is less sharp than for SIS junctions, as can be seen by comparing Figs. 15b and 4b. Consequently, the lower frequency limit for quantum mixing with an SIN junction is higher than for the corresponding SIS junction.

RUDNER et al. [25] made the first SIN mixer experiments. They used a six junction Pb/In-oxide-Ag junction series array and found a minimum mixer conversion loss of 9.1 db, at 75 GHz. Receiver noise temperatures were not reported. The conversion loss of 9.1 db must be regarded as excellent considering the relatively low frequency for which the photon voltage $\hbar\omega/e$ is much smaller than the range of the current rise at the gap voltage. Theoretically, such an SIN mixer is expected to improve with increasing frequency. A subharmonically pumped SIN mixer is planned for 375 GHz local oscillator frequency [87].

WINKLER and CLAESON [78] simulated high frequency mixer experiments using Al-oxide-Ag junction around 75 GHz. The gap frequency of this

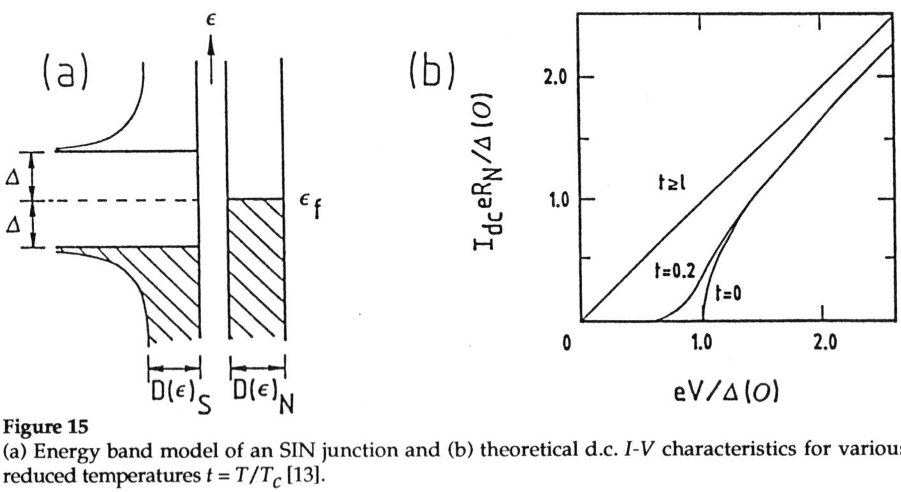

Figure 15
(a) Energy band model of an SIN junction and (b) theoretical d.c. I-V characteristics for various reduced temperatures $t = T/T_c$ [13].

SIN junction is $\Delta/h = 41$ GHz, so 75 GHz almost corresponds to the theoretical upper frequency limit $2\Delta/h$ of SIN junctions.

BLUNDELL and GUNDLACH [26] did experiments in the frequency range 220 to 230 GHz with Pb/Bi/In-oxide-Ag junctions. They have a gap frequency $\Delta/h = 410$ GHz and therefore theoretically an upper frequency limit of 820 GHz. A series array of three junctions, each with an area of about 1.2 μm^2, was used. We recall from Sect. 3 that for a single junction, photon-assisted tunneling is observed when the width δV of the current onset at the gap voltage is sharp on the scale of the photon voltage $\hbar\omega/e$. For a series array of three SIN junctions, the gap voltage is $3\Delta/e$ and the width of the current onset at the gap voltage should be small on the voltage scale $3\hbar\omega/e$. For 230 GHz one obtains $3\hbar\omega/e$ = 2,85 mV. Therefore the d.c. *I-V* characteristics shown in Fig. 16, curve (2) is not sharp enough to observe photon-assisted tunneling in the pumped characteristic, curve (1), and quantum structure in the conversion curve (B) and (C). Consequently, quantum mixing around 230 GHz cannot be expected. Nevertheless, the experiment in this frequency range gave encouraging results. The conversion curve (A) and (B) correspond to curves (a) and (b) in Fig. 13. From Eq. (7.1) the height of the maxima in Fig. 16 results in a receiver noise temperature of approximately 230 K. This has to be regarded as a good result considering that, as a consequence of the relatively round current-voltage characteristic, the SIN array appears to operate as a classical rather than as a quantum mixer. At a higher frequency where the d.c. *I-V* characteristic becomes sharper on the voltage scale $\hbar\omega/e$, the SIN mixer should be able to compete with the SIS mixer without having problems arising from Josephson effects.

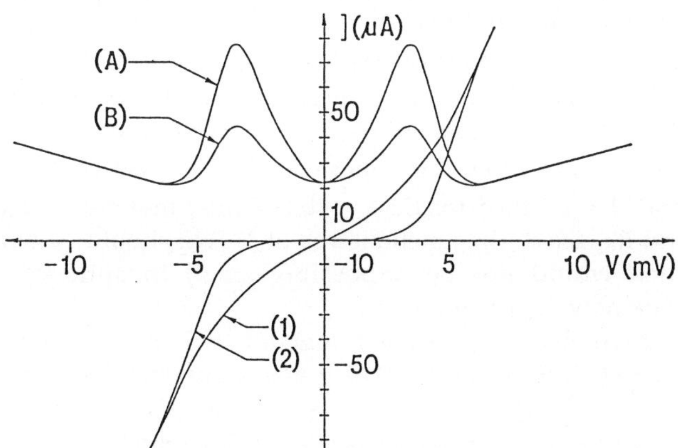

Figure 16
D.C. *V* curves of a Pb/Bi/In-oxide-Ag three-junction array at a temperature of 2.3 K with (1) and without (2) local oscillator irradiation at 218 GHz. Curves (A) and (B) are the IF output response to hot (295 K) and cold (77 K) loads placed at the receiver input [26].

9. Outlook for Quasiparticle Tunnel Junctions of the High-T_c Superconductor

The arrival of the high-T_c ceramic-like superconducting oxides has raised a number of questions concerning the fabrication and the application of SIS junctions made from these new materials. We will first discuss the problems of the fabrication of SIS junctions and then look at their interesting application as high-frequency quasiparticle mixers.

The Ginzburg-Landau coherence length ξ of the new superconductor is small; a typical value is $\xi \approx 2$ nm [88]. The coherence length is the shortest distance over which the Cooper-pair density and the energy gap Δ can change. Quasiparticles taking part in the tunneling process probe the superconductor within a distance ξ from the superconductor-tunnel barrier interface. Consequently, imperfections extending into the superconductor at a length ℓ from the tunnel barrier, have only little effect on the I_{dc}-V characteristic for $\xi >> \ell$. We have already mentioned in Sect. 4.2 that $\xi \approx 80$ nm for lead. It is therefore relatively easy to fabricate high quality SIS junctions from lead and certain lead alloys. A coherence length of 2 nm requires an almost ideal interface between the tunnel barrier and the superconductor. This makes the fabrication of good quality SIS junctions extremely difficult. The choice of which insulator or semiconductor should be used as the tunnel barrier and whether the resulting junction would have the stability required for practical devices, is still unclear.

The new superconductor is of great interest because its large energy gap Δ would permit quasiparticle mixer operation in the THz frequency range without interference from Josephson effects. This, however, requires junctions of very high current densities and small areas. An important design parameter for the quasiparticle mixer is the relaxation parameter $\omega C R_N$. According to TUCKER and FELDMAN [19], one should take $\omega C R_N \geq 4$. The normal state resistance R_N is set for impedance matching. The specific capacitance C/A of the junctions is also given. The junction area A would therefore scale as $1/\omega$. From our experience [34] a good choice at 230 GHz is $R_N \approx 50\ \Omega$ and $A \approx 1.5\ \mu m^2$ for $C/A \approx 40$ fF/μm^2, making $\omega C R_N = 4.3$. If we keep C/A and R_N fixed, the junction area at 1 THz should be about 0.35 μm^2; a value which would not be impossibly small in spite of the granular structure of the new superconductor.

The current density of such a junction will be very large. Assuming $\Delta = 15$ meV, the BCS equation (3.6) gives a current density of more than 10^5 A/cm^2.

Until now it has not been easy to achieve such large values in thin films. In addition, experience with the classical superconductor shows that the quality of the tunnel barrier degrades more and more with increasing current density. Another problem may be posed by Joule heating at such high current densities. If the single junction of resistance R_N is replaced by a series array of n junctions, each with the resistance R_N/n, the area of each junction in the array will be n times larger, but the current density is the same for the single junction and for the array junction; so this does not reduce Joule heating.

In spite of all this, we will assume for the moment that high-quality SIS junctions of the new high-T_c material can one day be fabricated. Let us assume that our junction will have a transition temperature $T_c = 90$ K and a gap voltage $V_g = 30$ mV. From experience with the classical superconductor, it is known that a sharp quasiparticle I_{dc}-V characteristic requires an operation temperature $T < T_c/2$ or even $T \leq T_c/3$. So $T \approx 30$ K will probably be necessary. However, current low noise IF amplifiers require an operation temperature of 15 K or less. It would therefore be reasonable to operate the SIS junction also at $T \approx 15$ K.

It should be known whether the quasiparticle mixer, operated at a physical temperature of 15 K, would not be too noisy and therefore whether the mixer can reach its absolute minimum noise temperature i.e. the quantum limit, $\hbar\omega/2k_B$. The discussion will be restricted to an SSB mixer which is generally used in molecular line observation. For an SSB mixer all sidebands but that of the signal are rejected, so that radiation can only enter the signal port.

According to Feldman, conditions for attaining the quantum limit of the mixer noise are [60]:

(i) In expression (3.5) for I_{dc}, taking an I_{dc}-V characteristic, a bias point V_o and local oscillator voltage V_{LO}, so that only the term with the index m = 1 is significant. This can be fulfilled if the I_{dc}-V characteristic is of sufficiently high quality, if the local oscillator frequency $\omega_{LO} < 4\,\Delta/\hbar$ (cf. Fig. 10) and $\alpha = eV_{LO}/\hbar\omega_{LO} < 0{,}7$.

(ii) A sufficiently low operation temperature, so that shot noise dominates. All current correlation matrix elements entering in the calculation of the mixer noise have the factor [17, 19, 60]

$$I(V_o + m\hbar\omega_{LO}/e) \cdot \coth \frac{V_o + m\hbar\omega_{LO}/e}{2k_BT} \tag{9.1}$$

Shot noise dominates if

$$eV_O + m\hbar\omega_{LO} > 2k_BT \tag{9.2}$$

As a cautious estimate we assume that the relevant value of m is not only m = 1, but also includes m = 0 and m = -1. The usual bias point is in the middle of the first photon step:

$$V_O = (2\Delta - \hbar\omega_{LO}/2)/e \tag{9.3}$$

Thus for m = -1 the condition (9.2) reads

$$2\Delta - \frac{3}{2}\hbar\omega_{LO} > 2k_BT \quad (9.4)$$

For T = 15 K the term $2k_BT$ = 2.6 meV. Hence the large energy gap guarantees domination of shot noise up to $\omega_{LO}/2\pi \approx$ 4 THz at a junction temperature of 15 K.

In conclusion, a high-quality SIS junction with the large gap voltage of the high-T_c ceramic-like superconductor would permit low-noise quasiparticle mixing at the physical temperature of 15 K presently required for low-noise IF amplifiers. Operation at 15 K would offer a considerable practical advantage over traditional quasiparticle mixers which require cooling with liquid helium to 4 K or less. Finally, the large gap voltage would greatly reduce perturbation by Josephson-related phenomena and thus be very helpful in extending the frequency of correct quasiparticle mixing into the THz range.

References

[1] A. Eckart, D. Downes, R. Genzel, A.I. Harris, W. Wild and D.T. Jaffe, to appear in Astron. and Astrophysics 1988

[2] M. and G. Winnewisser, Phys. Bl. **39** (1983) p. 9 and 289

[3] F. Combes, M. Gerin, A. Wootten, G. Wlodarczak, F. Clausset and P. Encrenaz, Astron. Astrophys. **180**, L 13 (1987)

[4] M.Guélin, J. Cernicharo, C. Kahane, J. Gomes-Gonzales and C.M. Wlmsley, Detection of a Heavy Radical in IRC + 10216, Astron. Astrophys. **175** (1987) p. 25

[5] M. Guélin, Organic and Exotic Molecules in Space, in Molecules in Physics, Chemistry and Biology, ed. D. Reidel, in press

[6] D. Downes, H2= Masers in Star-Forming Regions, in Birth and Infancy of Stars, Editors R. Lucas, A. Omont and R. Stora, North-Holland, Amsterdam, p. 557

[7] E. Kreysa, Int. Symp. on Millimeter and Submillimeter Wave Radio Astronomy U.R.S.I., Granada, Spain (1984) p. 153

[8] E. Kreysa, private communication

[9] J. Chamberlain, The Principle of Interfrometric Spectroscopy, Wiley, New York (1979)

[10] H.J. Hartfuss and K.H. Gundlach, Int. J. of Infrared and Millimeter Waves **2** (1981) p. 809

[11] H. Mattes, Technischer Bericht Nr. 66 des Max-Planck-Institutes für Radioastronomie (1986)

[12] E.A. Lynton, Superconductivity, Chapman and Hall Ltd. and Science Paperbacks (1974)

[13] P.L. Richards and T.M. Shen, IEEE Trans. Electron Devices **ED-27** (1980) p. 1909

[14] C.J. Adkins, Reviews of Modern Physics (1964) p. 211

[15] P.K. Tien and J.P. Gordon, Phys. Rev. **129** (1963) p. 647

[16] K.H. Gundlach, R. Blundell and E.J. Blum, Mikrowellen Magazin **11** (1985) p. 32

[17] J. Tucker, IEEE J. Quantum Electron. **QE-15** (1979) p. 1234

[18] K.H. Gundlach, J. Kadlec and W. Heller, J. Appl. Phys. **48** (1977) p. 1688

[19] J. R. Tucker and M.J. Feldman, Reviews of Modern Physics **57** (1985) p. 1055

[20] S.K. Lahiri, J. Appl. Phys. **41** (1970) p. 3172

[21] S.K. Lahiri, J. Vac. Sci. Techn. **13** (1976) p. 148

[22] S.K. Lahiri, S. Basavaiah and C.J. Kircher, Appl. Phys. Lett. **36** (1980) p. 334

[23] W.H.-G. Müller and F. Baumann, Thin Solid Films **48** (1978) p. 193

[24] K.H. Gundlach, S. Takada, M. Zahn and H.J. Hartfuss, Appl. Phys. Lett. **41** (1982) p. 294

[25] S. Rudner, M.J. Feldman, E. Kollberg and T. Claeson, J. Appl. Phys. **52** (1981) p. 6372

[26] R. Blundell and K.H. Gundlach, Int. J. of Infrared and Millimeter Waves **8** (1987) p. 1573

[27] K.H. Gundlach, unpublished results

[28] J.M. Baker and H.J. Magerlein, J. Appl. Phys. **54** (1983) p. 2556

[29] H.J. Magerlein, J. Appl. Phys. **54** (1983) p. 2569

[30] J.H. Greiner, J. Appl. Phys. **45** (1974) p. 32

[31] J.H. Magerlein, Trans. Magn. **MAG-17** (1981) p. 286

[32] K.H. Gundlach and H. Konishi, Appl. Phys. Lett. **46** (1985) p. 441

[33] R. Blundell and K.H. Gundlach, Extended Abstracts of the "1987 International Superconductivity Electronics Conference (ISEC 87)", August 28-29, 1987, Tokyo, Japan, p. 388

[34] R. Blundell, M. Carter and K.H. Gundlach, Int. J. Infrared and Millimeter Waves **9** (1988) p. 361

[35] H.A. Huggins and M. Gurvitch, J. Appl. Phys. **57** (1985) p. 2103

[36] A. Shoji, M. Aoyagi. S. Kosaka, F. Shinoki and H. Hayakawa, Appl. Phys. Lett. **46** (1985) p. 1098

[37] A. Shoji, SQUID´85-Superconducting Quantum Interference Devices and their Appl., 1985 Walter de Gruyter & Co., Berlin - New York, p. 631

[38] C.C. Koch, J.O. Scarborough and D.M. Kroeger, Phys. Rev. **B 9** (1974) p. 888

[39] J. Kwo, G.K. Wertheim, M. Gurvitch and D.N.E. Buchanan, IEEE Transactions on Magnetic **MAG-19** (1983) p. 795

[40] O. Michikami, K. Tanabe, H. Asano and Y. Katoh, Superconductivity Electronics, Editor KO HARA, Ohmasha, Kanda Tokyo 101 (1987) p. 186

[41] H.G. Le Duc, S.K. Khanna and J.A. Stern, International Journal of Infrared and Millimeter Waves **8** (1987) p. 1243

[42] M. Gurvitch, M.A. Washington and H.A. Huggins, Appl. Phys. Lett. **42** (1983) p. 472

[43] S. Morohashi, F. Shinoki, A. Shoji, M. Aoyagi and H. Hayakawa, Appl.Phys. Lett. **46** (1985) 1179

[44] T. Imamura, H. Hoko and S. Hasuo, Extended Abstracts of the "1987 International Superconductivity Electronics Conference (ISEC´87) August 28-29, 1987, Tokyo, Japan, p. 57

[45] A. Shoji, M. Aoyagi, S. Kosaka and F. Shinoki, Editor KO HARA, Ohmsha, Konola Tokyo 101, (1987) p. 182

[46] M. Aoyagi, A. Shoji, S. Kosaka and F. Shinoki, SCE 86-28 (1986) p. 43

[47] J.C. Villegier and R. Radparvar to appear in the Proceedings of the Applied Superconductivity Conference, San Fransisco, August 1988

[48] H.J. Hartfuss and K.H. Gundlach, SQUID´80 - Superconducting Quantum Interference Devices and their Applications, 1980, Walter de Gruyter and Co., Berlin - New York, p. 851

[49] H.J. Hartfuss and K.H. Gundlach, Intern. Journal of Infrared and Millimeter Waves **2** (1981) p. 809

[50] P.L. Richards, T.M. Shen, R.E. Harris and F.L. Lloyd, Appl. Phys. Lett. **36** (1980) p. 480

[51] A.D. Smith, W.R. McGrath, P.L. Richards, R.E. Harris, F.L. Lloyd, D.E. Prober and P. Santhanan, IEEE Transactions on Magnetics **MAG-19** (1983) p. 490

[52] L. Solymar, Superconducting Tunneling and Applications, Chapman and Hall Ltd. London, 1972

[53] I. Giaever, Phys. Rev. Lett. **5** (1960) 464

[54] T.G. Phillips and D.P. Woody, Ann. Rev. Astron. Astrophys. **20** (1982) p. 285

[55] M.R. Barber, **MTT-15** (1969) p. 629

[56] J.R. Tucker, Appl. Phys. Lett. **36** (1980) p. 477

[57] T.M. Shen, IEEE J. Quantum Electron. **QE-17** (1981) p. 1151

[58] H.J. Hartfuss and M. Tutter, Int. J. Infrared and Millimeter Waves **2** (1983) p. 809

[59] H.J. Hartfuss and M. Tutter, Int. J. Infrared and Millimeter Waves **5** (1984) p. 717

[60] M.J. Feldman, IEEE Transactions on Magnetics **MAG-23** (1987) p. 1054

[61] M.J. Feldman, Int. J. of Infrared and Millimeter Waves **8** (1987) p. 1287
[62] D. Winkler, Department of Physics, Göteborg 1987, ISBN 91-7032-307-0, Chalmers Bibliotheks Tryckeri
[63] T.M. Shen, P.L. Richards, R.E. Harris and F.L. Lloyd, Appl. Phys. Lett. **36** (1980) p. 777
[64] A.R. Kerr, S.-K. Pan, M.J. Feldman and A. Davidson, Physica **1088** (1981) p. 1369
[65] W.R. McGrath, P.L. Richards, A.D. Smith, H. van Kempen, D.E. Prober, R.A. Batchelor and P. Santhanam, Appl. Phys. Lett. **39** (1981) p. 655
[66] A.V. Räisänen, D.G. Crété, P.L. Richards and F.J. Lloyd, Int. J. of Infrared and Millimeter Waves **7** (1986) p. 1835
[67] E.J. Blum, Advances in Electronics and Electron-Physics **36** (1981) p. 97, Academic Press Inc.
[68] E.J. Blum, Advances in Cryogenic Engineering **31** (1986) p. 551
[69] K.H. Gundlach, R. Blundell, J. Ibrügger and E.J. Blum, Walter de Gruyter and Co., Berlin - New York, (1985) p. 987
[70] R. Blundell, K.H. Gundlach and E.J. Blum, Electron. Lett. **19** (1983) p. 498
[71] R. Blundell, K.H. Gundlach, E.J. Blum, J. Ibrügger and H. Hein, U.R.S.I. (International Union of Radio Science), Int. Symp. on Millimeter and Submillimeter Wave Astronomy, Granada, Spain, Sept. 1984
[72] J. Ibrügger, M. Carter and R. Blundell, Int. J. of Infrared and Millimeter Waves **8** (1987) p. 595
[73] S.K. Pan, M.J. Feldman, A.R. Kerr and P. Timble , Appl. Phys. Lett. (1983) p. 786
[74] S.K. Pan, A.R. Kerr, J.W. Lamb and M.J. Feldman, National Radio Astronomy Observatory, Charlottesville, Virginia, Electronics Division Intern. Report N^{O} 268 (1987)
[75] Inatani et al. private communication (1987)
[76] N.R. Erickson, IEEE Transactions on Microwave Theory and Technique **MTT-33** (1985) p. 1179
[77] N.J. Keen, K.D. Mischerikow and G.A. Ediss, 16th European Microwave Conference, Dublin, Ireland, published by Microwave Exhibitions and Publishers Ltd. (1986) p. 267
[78] D. Winkler and T. Claeson, J. Appl. Phys. **62** (1987) p. 4482
[79] M.J. Feldman, Int. J. of Infrared and Millimeter Waves **8** (1987) p. 1287
[80] R. Blundell and K.H. Gundlach, SPIE **598** Instrumentation for Submillimeter Spectroscopy (1985) p. 16
[81] H.H. Gröbke, to appear in Infrared Physics, special Issue for CIRP **4**, International Conference on Infrared Physics, ETH Zürich, August 1988
[82] M.J. Feldman, S.-K. Pan and A.R. Kerr, Extended Abstracts of the "1987 International Superconductivity Electronics Conference (ISEC´87)", August 28-29, 1987, Tokyo, Japan, p. 290
[83] E.C. Sutton, IEEE Trans. Magn. **MAG-17** (1983) p. 684
[84] M.J. Wengler, D.P. Woody, R.E. Miller and T.G. Phillips, Int. J. of Infrared and Millimeter Waves **6** (1985) p. 697
[85] B. Ellison and R.E. Miller, Int. J. of Infrared and Millimeter Waves **8** (1987) p. 609
[86] D. Winkler, T. Claeson and S. Rudner, SQUID´85, edited by H.D. Hahlbohm and H. Lübbig, Walter de Gruyter and Co., Berlin (1985) p. 1005
[87] D. Winkler, W.R. McGrath, B. Nilsson, T. Claeson, J. Johansson and E. Kollberg, SPIE, **598** Instrumentation for Submillimeter Spectroscopy (1985) p. 33
[88] P. Chaudhari, Proc. of the 18th Int. Conf. on Low Temperature Physics, Kyoto 1987, Jap. J. Appl. Phys. **26** (1987) p. 2023

Low-Temperature Scanning Electron Microscopy of Superconducting Thin Films and Tunnel Junctions

R. P. Huebener
Physikalisches Institut, Lehrstuhl Experimentalphysik II
Universität Tübingen, D-7400 Tübingen

1. Introduction

Recently Low-Temperature Scanning Electron Microscopy (LTSEM) has been shown to represent a powerful tool for investigating deviations from spatial homogeneity in superconducting thin films and thin-film devices. The particular importance of LTSEM lies in the fact that it probes directly the functional electronic behavior of the superconducting thin-film configuration with high spatial resolution. Such a spatially resolving functional test becomes imperative for the understanding of the physical behavior of superconducting thin-film devices and for identifying the possible origins of abnormal device behavior. In the future LTSEM can be expected to assume a growing role in the quality control of cryoelectronic devices.

The principle of LTSEM is simple. The surface of the specimen to be investigated is scanned by an electron beam while the specimen is cooled to the temperature range of liquid helium (LHe). During the scanning process a proper sample response signal is recorded as a function of the coordinate point (x,y) of the electron beam focus on the sample surface. For this purpose one needs a scanning electron microscope equipped with a well-functioning low-temperature stage. During this process the electron beam can be visualized as a localized heat source thermally perturbing the specimen film [1]. For LTSEM studies of superconducting thin-film configurations two kinds of response signals have been utilized: namely the electrical resistance change $\delta R(x,y)$ and the change $\delta I_{max}(x,y)$ in the maximum pair current of a superconducting thin films or of a superconducting tunnel junction induced by the electron beam. Of course, the change $\delta R(x,y)$ is conveniently recorded in a current-biased (or voltage biased) sample as a voltage signal $\delta V(x,y)$ (or current signal $\delta I(x,y)$).

The idea of thermally perturbing a system in order to learn more about its behavior has been utilized in the past in a different context. Sometime ago, Eigen [2] introduced the principle of *global thermal perturbation* for studying the physico-chemical kinetics of a system from its *temporal response*. Here we apply the principle of *local thermal perturbation* in order to investigate the *spatial structures* in the electronic behavior of the system.

In the following we summarize the results obtained so far on the application of LTSEM to superconducting thin films and tunnel junctions. Previous summaries of these developments have been given elsewhere [3-8], and additional references can be found in these papers. Further, the extension of LTSEM to other areas such as semiconductors and the ballistic phonon signal is discussed in some of the previous summaries [3,4,8].

This paper is organized as follows. In Section 2 we treat the electron beam as a local heat source and discuss the spatial resolution obtained by LTSEM. Section 3 deals with self-generated spatial structures in current-carrying thin-film superconductors such as phase-slip centers and hot spots. Sections 4 to 7 are concerned with superconducting tunnel junctions. Here we discuss inhomogeneous quasiparticle tunneling (Sect. 4) and inhomogeneous pair tunneling (Sect. 5). The distribution of the pair-current density in the presence of barrier-parallel Josephon vortices and of trapped flux quanta is the subject of Sect. 6. Tunnel junction arrays are treated in Sect. 7. In Sect. 8 we discuss the applications of LTSEM to films of high-T_c superconductors and point out possible future trends.

2. Electron Beam as a Local Heat Source

The principal configuration of LTSEM is shown schematically in Fig. 1. The specimen film is attached to the top side of a substrate and is scanned by the electron beam of a scanning electron microscope (SEM), while the bottom side of the substrate is in direct contact with the liquid-He bath. The bath temperature can be reduced below 4.2 K by pumping, such that the temperature range between 4.2 and about 1.5 K is available for these experiments. Further details about the cryogenic arrangements can be found elsewhere [3,4,8]. The recent discovery of the ceramic high-T_c superconductors [9] has stimulated the extension of LTSEM experiments to much higher temperatures, namely to the boiling point of liquid nitrogen and above [10]. For these applications the

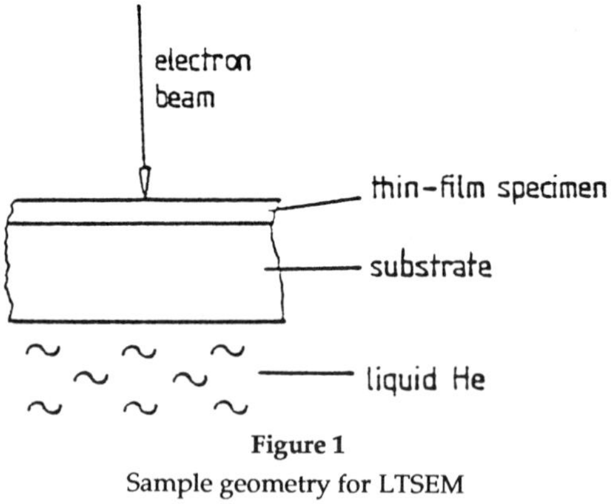

Figure 1
Sample geometry for LTSEM

simple arrangement of Fig. 1 has to be replaced by a sample mounting stage allowing operating temperatures up to, say, 100 K or higher. While our discussion in Sections 3 to 7 essentially deals with the temperature range of 4.2 K and below, in Sect. 8 we return to the extension of LTSEM to much higher operating temperatures in the case of thin-film high-T_c superconductors.

The typical electron beam parameters for LTSEM of thin-film superconductors are 10 - 20 kV and 10 - 100 pA yielding for the beam power 1 μW or less. The typical beam diameter is less than 100 nm. The penetration depth of the beam electrons of 10 - 20 keV energy in the superconducting target materials of interest is typically about 1 μm. The beam energy is dissipated predominantly in the specimen film and not in the substrate material, if the film thickness is not much smaller than this penetration depth. The primary excitation and relaxation processes - inelastic electron-electron scattering, emission of relaxation phonons, etc. - involved in the transfer of the beam energy into the target material take place in a timescale of typically less than 1 ps. As a result, thermalization of the beam energy in the target material is practically accomplished within this timescale. Following this thermalization process, the interaction between the electron beam and the specimen can simply be treated as a local heating effect near the coordinate point (x,y) of the beam focus on the sample surface [1,4,8].

The lateral extension and the response time of the beam-induced thermal perturbation of the specimen film is given by the thermal healing length

$$\eta = \left(\frac{\kappa \cdot d}{\alpha}\right)^{1/2} \tag{2.1}$$

and the thermal relaxation time

$$\tau = \frac{C \cdot d}{\alpha} \tag{2.2}$$

respectively. Here κ, C, and d are the heat conductivity, the heat capacity per volume, and the thickness of the specimen film, respectively. The quantity α is the heat transfer coefficient describing the energy flux from the film into the substrate. In general, the response signal utilized in LTSEM for imaging spatially inhomogeneous sample properties is generated by means of the beam-induced thermal perturbation of the sample. Hence, the spatial resolution of this imaging principle is limited by the healing length η. For increasing the sensitivity of the electronic detection of the response signal it is advantageous to modulate the electron beam at some angular frequency ω and to record the modulated signal by means of phase sensitive detection.

We note that the healing length η of Eq. (2.1) refers to the static case where the temporal variations in the thermal perturbation of the sample are slow compared to the relaxation time τ of Eq. (2.2). Hence, for the operation using a beam modulated at the frequency ω Eq. (2.1) only applies in the limit

$\omega\tau << 1$. The typical frequency range used in the past and satisfying this condition is about 10 - 20 kHz. For higher modulation frequencies not satisfying this condition the frequency-dependent thermal healing length η_ω must be considered. Taking for simplicity the limiting case $\omega\tau >> 1$, one finds for the dynamic healing length

$$\eta_\omega = \left(\frac{2}{\omega \cdot \tau}\right)^{1/2} \cdot \eta = \left(\frac{2D}{\omega}\right)^{1/2} . \tag{2.3}$$

Here D is the thermal diffusivity. The content of Eq. (2.3) is known as the thermal skin effect. According to Eq. (2.3) the healing length η_ω decreases with increasing frequency ω resulting in a corresponding improvement of the spatial resolution obtained by LTSEM. Of course, this improvement of the resolution is obtained only if the modulated signal is recorded. Along with the increase in modulation frequency the amplitude of the beam-induced signal decreases as expected from the corresponding reduction of the size of the thermally modulated sample region. It is this reduction in the signal amplitude which eventually sets an upper limit to the modulation frequency and a corresponding limit to the length η_ω determining the spatial resolution. In the past beam modulation frequencies up to 20 MHz have been used. The frequency dependence of the dynamic healing length η_ω has been found experimentally to be in excellent agreement with the behavior expected from the concept of the beam-induced thermal perturbation [11]. For further details on the thermal model for treating the beam-induced sample perturbation we refer to previous summaries [3,4,8].

3. Spatial Structure in Superconducting Thin Films

Let us consider a superconducting film of width w and thickness d with $w < \xi$ and $d < \xi$, where ξ is the superconducting coherence length. Such a geometry is usually referred to as a one-dimensional superconductor. If the current I through this film is increased monotonically, the current-voltage characteristic (IVC) shows the abrupt appearance of distinct voltage steps at certain current levels indicating the generation of resistive regions at specific sample locations [12]. The voltage steps and the resistive regions are associated with spatially localized phase-slip centers [13]. In a phase-slip center the superconducting order parameter oscillates between zero and its value for zero supercurrent at the Josephson frequency, this oscillation extending over a region of length ξ. In this core of the phase-slip center quasiparticles are generated which diffuse away from the core in both directions reaching a distance given by the quasiparticle diffusion length Λ. Hence, a phase-slip center represents a selfgenerated normal region of length 2Λ in a one-dimensional superconductor.

If the width of the current carrying superconductor becomes larger such that we reach the two-dimensional geometry ($w > \xi$), instead of the phase-slip center one observe the nucleation of trains of flux quanta traversing the film perpendicularly to the current direction at distinct locations of the longitudinal sample coordinate. Again, the IVC shows a step structure, each voltage step being associated with the nucleation of an additional train of flux quanta. Recently the dynamics of this current-induced resistive state has been studied in detail using microbridges of type-I superconductors such as Pb and In [12, 14-16]. In these materials in the dynamic resistive state trains of multiquantum flux tubes traverse the specimen instead of trains of single flux quanta, resulting in correspondingly larger signals for the experimental observations.

If the dynamic current-induced resistive state is established, let us consider what happens if the applied current is increased further and further. With increasing power dissipation the sample temperature eventually becomes large enough locally for the development of a hotspot. Such a hotspot is generated because of a thermal instability resulting from the S-shaped resistance-temperature curve characterizing a superconductor [17]. A selfgenerated hotspot consists of a section of distinct length where the temperature of the superconducting film exceeds the critical temperature T_C. The temperature profile of a hotspot along the longitudinal sample coordinate is shown schematically in Fig. 2. A detailed analysis of the thermal behavior of such a hotspot has been given by Skocpol et al. [18], and further discussions and references can be found elsewhere [4,8,12,19,20].

In principle, the three dissipative structures developed in a current-carrying superconducting film and discussed so far, namely phase-slip centers, moving trains of flux quanta or flux tubes, and hotspots, can all be imaged by LTSEM. If we operate the sample at the proper bias current and scan the sample surface with the electron beam, the beam-induced voltage change $\delta V(x,y)$ recorded as a function of the coordinate point (x,y) of the beam focus is expected to be nonzero only if the beam focus is located near the dissipative spatial structure generated by the applied sample current. Of course, here we assume that the beam irradiation represents only a sufficiently small thermal sample perturbation.

In the past LTSEM has been used highly successfully for imaging hotspots in thin-film superconductors [4,19,20]. These experimental results are in excellent agreement with the beam-induced voltage signal $\delta V(x,y)$ expected theoretically. In particular, one has found that a nonzero signal $\delta V(x,y)$ only arises at the two boundaries of the hotspot where the temperature profile passes through the value T_C. The width of the peak of the signal $\delta V(x,y)$ at each hotspot boundary is given by twice the thermal healing length η or η_ω (Eqs. (2.1) and (2.3)). The two signal peaks are shown schematically at the bottom of Fig. 2.

It is interesting to note that in a similar way LTSEM has served recently for imaging selfgenerated structures in semiconductors [21,22]. The structures

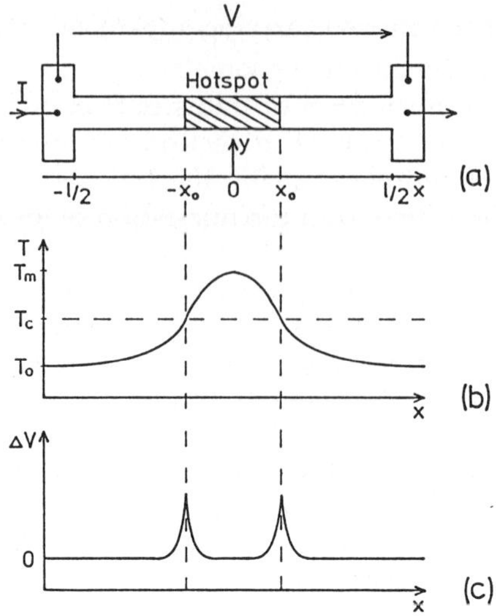

Figure 2
Hotspot in a thin-film superconductor. (a) sample geometry and location of the hotspot (hatched area), (b) temperature profile along the sample, (c) electron beam induced voltage signal δV plotted versus the sample coordinate.

in a superconductor discussed above are a consequence of the applied *current* and appear above a critical current level. On the other hand in a semiconductor at low temperatures the observed spatial structures result from the applied *voltage* and appear in the form of current filaments above a critical voltage level. Again the beam-induced signal reaches its maximum near the two boundaries of a current filament, similar to the situation of a hotspot in a thin-film superconductor.

Taking up this analogy between superconductors and semiconductors one step further, we note that *current-driven* spatial structures are also possible in a semiconductor, namely in form of high-field domains leading to the Gunn effect. These high-field domains correspond more closely to the dissipative structure in thin-film superconductors, we have discussed above. On the other hand, the voltage-driven current filamentation we know from semiconductors does not exist in a superconductor because of the validity of the London equations resulting in the Meissner effect.

Finally, returning to the spatial structures in thin-film superconductors, we emphasize that all examples discussed above refer to self-generated dissipative structures developing above a critical nonequilibrium threshold level. However, LTSEM also finds interesting applications for investigating the spatial inhomogeneities of the local values of the critical current density $j_c(x,y)$ and of the critical temperature $T_c(x,y)$, resulting from the intrinsic microstructure of the sample. This aspect becomes particularly important in thin films of the ceramic high-T_c superconductors, and we come back to this point in Sect. 8.

4. Inhomogeneous Quasiparticle Tunneling

Spatial structures in the quasiparticle tunneling process can arise because of two reasons. First, inhomogeneities in the tunneling barrier and in the superconducting electrodes of the junction lead to spatial variations of the local tunneling current density. Here the inhomogeneous behavior results from the microstructure of the tunnel junction as determined by the particular fabrication process. Secondly, spatial structures in the junction behavior can develop also due to the operational nonequilibrium state of the junction. Here the inhomogeneous energy-gap state under the injection of excess quasiparticles is an example. Both types of spatial structures can be investigated by LTSEM. It appears that inhomogeneities in the tunneling barrier represent perhaps the most important case for this application of LTSEM.

For imaging the spatial structures in the quasiparticle tunneling process it is convenient to current-bias the junction at a specific point on the quasiparticle IVC and to record the beam-induced voltage change $\delta V(x,y)$ across the junction. In this operational mode the dc pair tunneling is suppressed by the application of a small external magnetic field. Assuming that the area thermally perturbed by the electron beam is much smaller than the total area of the tunneling window, the voltage signal $\delta V(x,y)$ is given by

$$\delta V(x,y) = -\left(\frac{\partial V}{\partial I}\right)\delta I(x,y) \qquad (4.1)$$

where $\delta I(x,y)$ is the current increment in the perturbed area and $\partial V/\partial I$ is the differential resistance of the junction at the bias current without irradiation. From Eq. (4.1) the interpretation of the voltage signal $|\delta V(x,y)|$ is straightforward [23, 24]. For a bias current in the thermal tunneling regime the signal images the local tunneling conductance, whereas for a bias current in the gap regime it images the local energy gap of the junction electrodes. In general, the spatial resolution of this imaging technique is given by the healing lengths η or η_ω of Eqs. (2.1) and (2.3), approaching 1 μm in favorable cases.

Inhomogeneities in the tunneling barrier resulting in a local enhancement of the tunneling current density lead to inhomogeneous power dissipation in the junction and, hence, to a nonuniform temperature and superconducting energy gap over the junction area [23]. Clearly, LTSEM probing the local tunneling conductance can provide important clues clarifying the origin of the barrier imperfections such as the fabrication process, the formation of hillocks, local mechanical stresses, etc. Of course, the standard modes for surface analysis of an SEM can be combined with LTSEM in the same instrument.

The inhomogeneous energy-gap structure in the superconducting nonequilibrium state during injection of excess quasiparticles represents another important case which has been clarified by the quasiparticle voltage image [24,25]. In this way the nonequilibrium multiple gap structure has been observed for the first time with high spatial resolution. These experiments were performed using PbIn-PbIn-PbIn double junctions with the standard injector-detector configuration. One of the important points in these studies has been the careful evaluation of the possible role played by barrier inhomogeneities in the nucleation process of the multiple gap state. Regarding further details we refer to [24, 25] and a previous summary [8].

Recently, superconducting tunnel junctions have received increasing interest due to promising applications as cryoelectronic particle detectors [26-29]. Here the low energy gap in a superconductor of typically about 1 meV appears highly favorable compared to a semiconductor with a typical energy gap of about 1 eV. Furthermore, since the superconducting energy gap is smaller than the Debye energy, phonons generated by the particle irradiation can contribute to the detector signal by breaking of Cooper pairs, in contrast to the situation for a semiconducting detector. Hence, superconducting detectors promise a higher energy resolution than a detector based on semiconducting materials. In addition to generating a proper response signal for the two-dimensional imaging of structural aspects in a tunnel junction, the experimental setup for LTSEM can also serve for quantitatively evaluating the junction sensitivity as a particle detector; in this case for electrons. The convenient way of varying the beam energy and current in the SEM represents a particularly attractive feature of such experiments. Furthermore, possible variations of the behavior of the tunnel junction detector as a function of the irradiation dose can easily be investigated. Systematic studies along these lines have been started recently [28,29].

5. Inhomogeneous Pair Tunneling

A spatially inhomogeneous distribution of the pair current in a superconducting tunnel junction can result mainly from three different origins. First, the pair current is concentrated along the outer boundaries of the tunneling window because of the Meissner effect. The characteristic length for the penetration of currents and magnetic fields into a tunnel junction is the Josephson penetration depth λ_J which replaces the London penetration depth λ_L valid for a bulk superconductor. The tunneling barrier represents only a weak link connecting the two superconducting electrodes of the junction. Hence, the maximum density of the screening supercurrent in a tunnel junction is reduced considerably compared to the bulk electrode material, and the penetration depth λ_J is correspondingly larger than λ_L. The detail of the current distribution resulting from the Meissner effect depends upon the geometry of the current feeding lines. Of course, the generation of an inhomogeneous distribution of the pair current due to the Meissner effect is only possible if the side lengths of the tunneling window are not smaller than λ_J. Secondly, a strong spatial dependence of the pair current density is observed in the presence of an external barrier-parallel magnetic field or of magnetic flux quanta with an appreciable magnetic field component parallel to the barrier trapped in the junction. In Sect. 6 we discuss this case in more detail. Third, inhomogeneities in the tunneling barrier lead to an inhomogeneous distribution of the pair current similar to the situation for quasiparticle tunneling.

The signal measured by LTSEM during the scanning process for imaging the pair current distribution is the maximum Josephson current $I_{max}(x,y)$ as a function of the coordinate point (x,y) of the beam focus on the sample surface. From these measurements the beam-induced change $\delta I_{max}(x,y)$ of the maximum pair current can be derived. The details of the experimental procedures can be found elsewhere [30-32].

The beam-induced change $\delta I_{max}(x,y)$ utilized for imaging can be interpreted as follows. We start from the equation for the maximum Josephson current density

$$J(x,y) = J_1(x,y) \cdot \sin\varphi(x,y), \tag{5.1}$$

where $J_1(x,y)$ is the local critical current density and $\varphi(x,y)$ the local difference between the phase of the superconductor wave function in both electrodes. In the following we assume $J_1(x,y)$ to be homogeneous. In this case spatial variations of $J(x,y)$ are only connected with the function $\varphi(x,y)$ and not with inhomogeneities in the barrier or the junction electrodes. The local perturbation of the junction by the electron beam is expected to change the critical current density $J_1(x,y)$ by $\delta J_1(x,y)$ in the irradiated region. Again $\delta J_1(x,y)$ is assumed to be homogeneous. If the phase-difference function remains unaffected, the beam-induced change $\delta I_{max}(x, y)$ is given by

$$\delta I_{max}(x,y) = \delta J_1(x,y) \cdot \delta x \cdot \delta y \cdot \sin\varphi(x,y). \tag{5.2}$$

Here $\delta x\, \delta y$ is the area element around the point (x,y) perturbed by the beam. From Eq. (5.2) we see that the signal $\delta I_{max}(x,y)$ images the spatial distribution of the maximum Josephson current density $J(x,y)$ in the unperturbed junction as determined by the function $\sin\varphi(x, y)$ according to Eq. (5.1).

As we have pointed out, Eq. (5.2) is valid only under the assumption that the phase difference function $\varphi(x,y)$ remains unchanged under the beam irradiation. However, generally the beam-induced change $\delta\varphi(x,y)$ is nonzero and must be taken into account. Such a change in the phase difference function not only occurs in the irradiated part of the junction. Instead it extends over the total junction area, including also the nonirradiated portions. Here we are dealing with a nonlocal contribution to the beam-induced signal which must be added to the local signal contribution of Eq. (5.2), and which is more complicated than the latter because of its nonlocal character. Further details and references regarding this nonlocal effect can be found elsewhere [8,31].

The possibility for two-dimensional imaging of the pair current distribution in a tunnel junction by means of LTSEM has stimulated renewed interest in theoretical calculations of such current distributions for two-dimensional geometries. As we see from Eq. (5.1), the two-dimensional distribution of the maximum Josephson current density $J(x,y)$ is related to the spatial variation of the function $\varphi(x,y)$. The spatial dependence of the function $\varphi(x,y)$ is determined by the stationary Sine-Gordon equation

$$\Delta\varphi(x,y) = \frac{1}{\lambda_J^2(x,y)} \sin\varphi(x,y) \tag{5.3}$$

in combination with the proper boundary conditions; Δ denotes the Laplace operator. Theoretical solutions of Eq. (5.3) for predicting the spatial dependence of the pair current density in a tunnel junction have been discussed since the first paper by Owen and Scalapino [33] about 20 years ago. However, up to now such an analysis based on Eq. (5.3) has been possible only for one-dimensional geometries where only one dimension of the tunneling window can be arbitrarily long, whereas the other dimension is smaller than λ_J. Because of insurmountable mathematical difficulties a similar treatment of the two-dimensional case has not been possible so far. Fortunately, a completely different theoretical approach based upon a simple iteration procedure which has been proposed recently [34] provides the desired answer.

The iteration procedure starts with an assumed spatial distribution $J_a(x,y)$ of the maximum Josephson current, from which the spatial dependence of the magnetic flux density in the junction is calculated. From this distribution of the flux density one determines the spatial variation of the phase difference function $\varphi_a(x,y)$, which in turn leads to a new distribution $J_b(x,y)$ of the maximum Josephson current. The result is then the starting point for the next iteration step. External magnetic fields and trapped flux quanta can easily be incorporated by adding to the field of the Josephson current. This new numerical method has been successfully applied for calculating self-field effects including the influence of the geometry of the current feeding lines and the effect of trapped magnetic flux quanta in the junction [35,36]. In this latter case, in addition to the location, the quantum number and polarity of the trapped flux quanta has been determined accurately. It has become clear, that a correct interpretation of the measured images $\delta I_{max}(x,y)$ obtained from LTSEM is only possible in conjunction with theoretical calculations of the two-dimensional variation of the maximum Josephson current density $J(x,y)$. A few examples demonstrating this point are presented in Sect. 6.

Self-field effects on the spatial distribution of the maximum Josephson current density including the influence of the geometrical arrangement of the current feeding lines have been studied experimentally by means of LTSEM using PbIn-PbIn tunnel junctions [32]. The obtained images $\delta I_{max}(x,y)$ agreed well with the theoretical results calculated with the iteration method outlined above.

In most of our discussion of inhomogeneous pair tunneling we have assumed that the tunneling barrier and the superconducting electrodes of the junction are spatially homogeneous. Hence, we have treated the quantities $J_1(x,y)$ and $\delta J_1(x,y)$ of Eqs. (5.1) and (5.2), respectively, as constants. The accurate evaluation of this homogeneity and the sensitive detection of inhomogeneities in the junction is possible by means of the images based on quasiparticle tunneling as discussed in Sect. 4.

6. Vortex States and Trapped Flux Quanta in Tunnel Junctions

In the following we summarize the observations of the spatial structures in the pair current density resulting from an external barrier-parallel magnetic field and from trapped magnetic flux quanta. The latter case is particularly important for cryoelectronic applications since trapped magnetic flux quanta seriously perturb the functional behavior of a superconducting tunnel junction.

If a small barrier-parallel magnetic field is applied to a tunnel junction the magnetic flux is expelled from the junction due to the Meissner effect. Magnetic flux penetration is limited to an outer region of depth λ_J, the Josephson penetration depth. If the magnetic field is increased sufficiently, flux quanta penetrate into the junction interior establishing the different vortex states of the junction. Each vortex state is characterized by a distinct spatial oscillation of the pair tunneling current density along the junction coordinate perpendicular to the direction of the applied magnetic field. As an example a vortex state with four flux quanta in the junction is shown schematically in Fig. 3. It is the spatial oscillation of the pair tunneling current

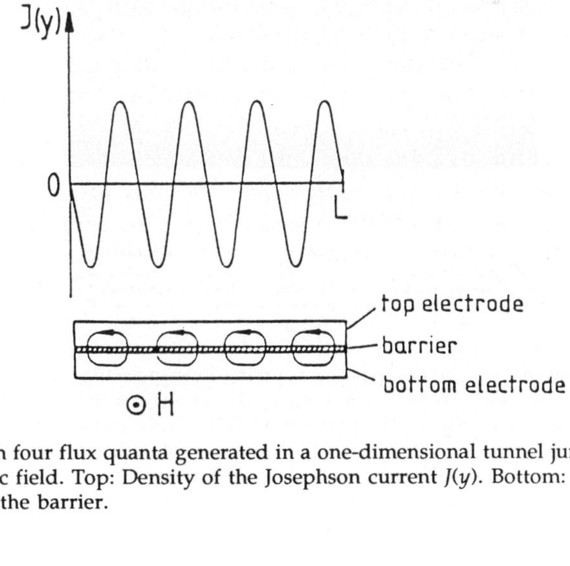

Figure 3
Vortex state with four flux quanta generated in a one-dimensional tunnel junction by a barrier-parallel magnetic field. Top: Density of the Josephson current $J(y)$. Bottom: Josephson vortices seen parallel to the barrier.

density in the various vortex states which leads to the well known modulation of the maximum Josephson current I_{max} as a function of an applied barrier-parallel magnetic field often referred to as the magnetic interference pattern [37].

Imaging of the spatial distribution of the pair current density in the various vortex states of a junction by means of LTSEM is straight-forward [30,31]. Typical results obtained for the 4 - 5 vortex state are shown in Fig. 4. Here the beam-induced signal - $\delta I_{max}(x,y)$ obtained from a series of line scans is plotted vertically as a function of the junction coordinates. The position of the junction and the direction of the line scans are indicated at the bottom. The results shown in Fig. 4 were obtained with a PbIn-PbBi junction of a one-dimensional geometry. The rectangular tunneling window had the length $L = 89$ µm ($L/\lambda_J = 14.4$) and the width $w = 3.5$ µm. The images $\delta I_{max}(x,y)$ clearly show the oscillatory behavior of the pair current density expected from Eq.

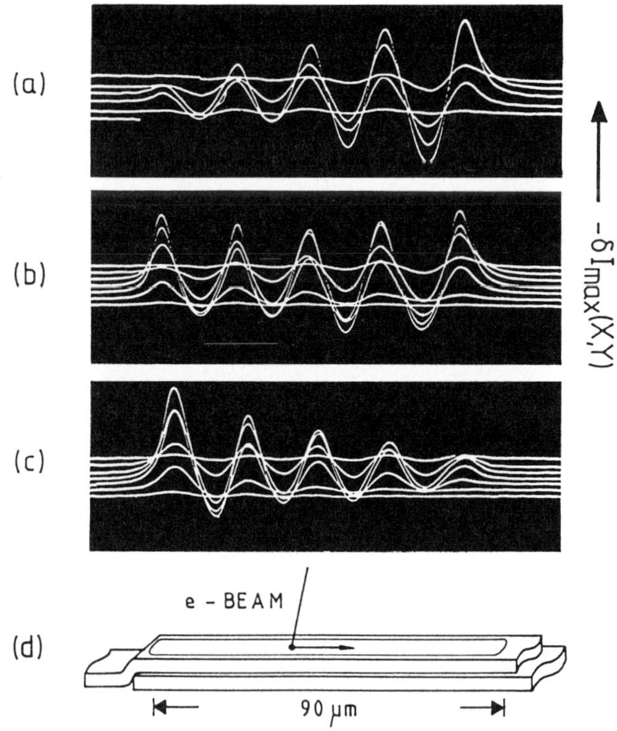

Figure 4
Beam-induced signal $-\delta I_{max}(x,y)$ obtained by scanning longitudinally along a one-dimensional tunnel junction and showing the 4-5 vortex state. The line scans were performed for several values of the transverse coordinate. The position of the junction and the scanning direction are indicated at the bottom (d). Recording (b) was obtained near the local maximum of the 4-5 vortex state in the magnetic interference pattern, whereas recordings (a) and (c) were taken on the low and high-field side of this maximum, respectively (from Ref. [30]).

(5.2) for this vortex state. The recording (b) in Fig. 4 was taken near the local maximum associated with the 4 - 5 vortex state in the magnetic interference pattern. The recordings (a) and (c) were taken on the low- and high-field side of this maximum. The increase and decrease in the maximum amplitude of the signal - $\delta I_{max}(x,y)$ along the longitudinal sample coordinate seen in the recordings (a) and (c), respectively, result from the nonlocal effect associated with the beam-induced change $\delta\varphi(x,y)$ of the phase-difference function and agree with the theoretical predictions. For further details we refer to the earlier papers [8,30,31].

In a one-dimensional PbIn-PbBi junction with a geometry similar to that shown at the bottom of Fig. 4 and having the ratio $L/\lambda_J = 5.2$ the evolution of the different vortex states in an increasing barrier-parallel magnetic field has been observed up to 14 vortices [31]. In addition to these vortex states, the restriction of the flow of the maximum Josephson current to the region near the two junction boundaries at zero magnetic field, as given by the Josephson penetration depth λ_J, could be confirmed by LTSEM imaging. Of course, from these latter experiments the value of λ_J can be obtained directly.

Another important case showing strong spatial variations of the pair current density is associated with magnetic flux quanta trapped in the tunnel junction at zero external magnetic field. Such trapping of flux quanta can occur easily because of flux pinning in the superconducting electrodes of the junction. Because of the thin-film configuration of the junction electrodes, predominantly transverse flux quanta are trapped in these films. A large change in the phase-difference function $\varphi(x,y)$ leading to a correspondingly large change in the distribution of the pair current density is expected only if the trapped flux quanta generate an appreciable magnetic field component parallel to the barrier. Hence, misaligned vortices penetrating both junction electrodes at different locations or vortices which penetrate only a single junction electrode (the flux entering or leaving the junction via the barrier) strongly effect the distribution of the pair current density. On the other hand, a perfectly aligned transverse flux quantum penetrating the junction in a straight line perpendicular to the barrier does not represent such a severe pertubation of the pair current density distribution.

In Fig. 5 is shown the spatial variation of the density of the maximum Josephson current calculated for the case of two flux quanta trapped in the junction. The geometric details are indicated at the top. The flux of the trapped flux quanta enters the junction via the barrier and then transversely penetrates the top electrode. The calculations were performed using the iteration method outlined in Sect. 5. In Fig. 6 we present an image $\delta I_{max}(x,y)$ experimentally observed by LTSEM for a junction having identical characteristic parameters (tunneling window 37 x 85 μm^2; $\lambda_J = 15\ \mu m$) as the case shown in Fig. 5. The perfect agreement between the experimental (Fig. 6) and theoretical (Fig. 5) results assures an interpretation of the experimental

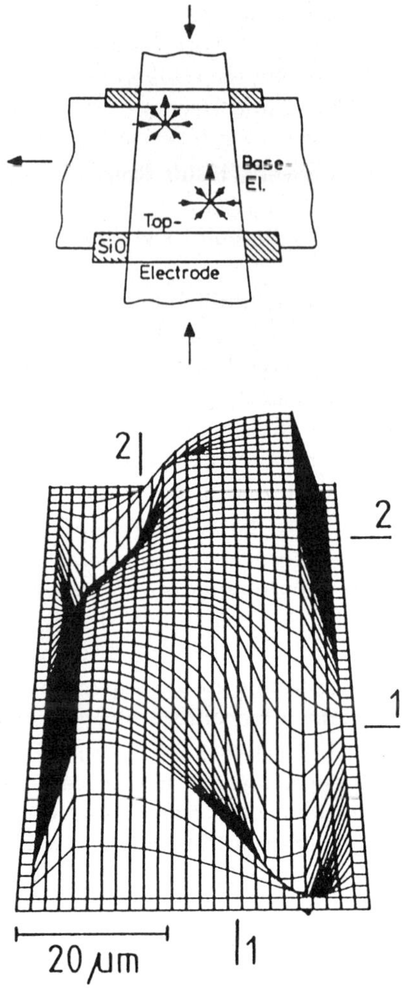

Figure 5
Spatial variation of the density of the maximum Josephson current calculated for the case of two trapped flux quanta in the junction. The position of the centers of the trapped flux quanta is marked by the intersections of lines 1 and 2. The geometric details are also indicated at the top (from Ref. [32]).

image in terms of the trapped vortex configuration shown at the top of Fig. 5. Because of the mutual repulsion it is not surprising that the two trapped flux quanta arrange themselves at the largest possible distance from each other. The results presented in Fig. 5 and 6 strikingly demonstrate the importance of supplementing the two-dimensional images $\delta I_{max}(x,y)$ obtained by LTSEM by detailed theoretical calculations.

Finally, LTSEM also allows to determine quantitatively the magnitude of the pinning forces acting upon the trapped flux quanta. For this purpose one only has to observe the change in position of the flux quantum effected

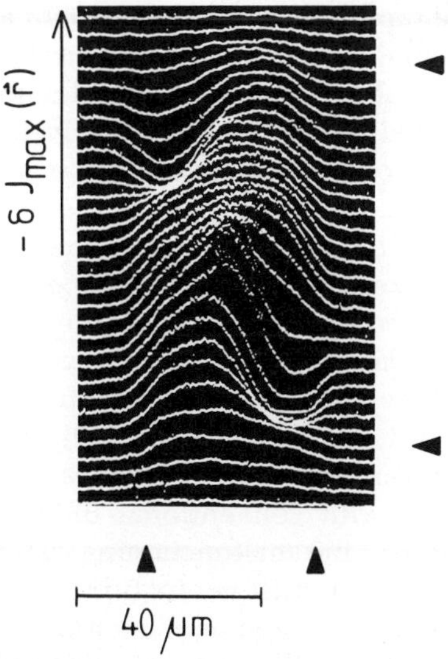

Figure 6
Observed image $\delta I_{max}(x,y)$ for a junction having identical characteristic parameters as the case shown in Fig. 5. The triangles mark the boundary of the tunneling area (from Ref. [32]).

by an external force. Such an external force can easily be applied by means of the Lorentz force of an electric current passing through a junction electrode [38].

A brief summary of the results of this Section can be found in Ref. [7]. The high-resolution observation of trapped flux quanta by means of LTSEM provides an interesting new possiblility for clarifying the nature of the pinning centers involved and for finding ways to eliminate these imperfections which severely perturb the functional junction behavior.

7. Cryoelectronic Circuits and Tunnel Junction Arrays

In addition to the studies of individual tunnel junctions and superconducting thin-film configurations such as microbridges, complex superconducting electronic circuits and large junction arrays can be investigated by LTSEM. Superconducting circuits of increasing complexity are presently being developed for applications as intelligent electromagnetic sensors of extreme sensitivity. Here superconducting interference devices (SQUIDS) often represent important circuit components. The trend is directed towards higher integration including first stage digital data processing based on superconductor technology (Josephson computer) [39,40]. Large arrays of tunnel junctions recently have found increasing interest both due to fundamental aspects [41,42] and due to different applications such as high-frequency devices [43] or frequency-based voltage standards [44,45]. Finally, thin-film superconducting networks are rapidly becoming important subjects of fundamental interest due to their potential for studying fractal structures [46,47].

All examples given above represent complex thin-film configurations, and LTSEM is ideally suited for the spatially resolved study of their physical properties. Clearly, the strong advances made recently in microfabrication technology must be matched by the development of adequate analytical tools. In terms of such a tool LTSEM appears highly promising. The application of LTSEM for spatially resolved functional testing of superconducting circuit configurations, of course, is highly similar to the concept of electron-beam testing already well established in semiconductor technology [48].

Recently, these ideas have been applied for functionally testing the individual junctions in a 166 junction array [49]. Here the shorted tunnel junctions could be picked out immediately by means of LTSEM. Of course, if LTSEM is then combined with conventional SEM or additional standard surface analytical techniques, the malfunctioning junctions can be correlated with anomalies in their microstructure. In turn, this information then can provide hints for improving the microfabrication process.

Beyond the *spatially resolved observation* of many structural aspects, at low temperatures the electron beam eventually may also be utilized for *in-situ repair* of various defects in the circuit. We note again, that similar concepts are well established already in VLSI semiconductor technology.

8. Applications to Thin Films of High-T_C-Superconductors

The recent discovery of the high-T_c copper-oxide based superconductors initiated by the work of Bednorz and Müller [9] has extended the temperature range of superconductivity up to the boiling point of liquid N_2 (77 K) and to higher temperatures [50]. A large part of the current research activities deals with the study of polycrystalline ceramic materials prepared from powders of the different constituent compounds using sintering and solid state reaction techniques at high temperatures. However, it appears that for understanding the underlying physical mechanisms, measurements are needed in particular on well characterized single crystals and films. Here LTSEM with its capability for the spatially resolved observation of the characteristic superconducting properties represents an important and highly promising tool for thin film experiments. Among the many potential applications of the high-T_c copper-oxide based superconductors thin films look particularly attractive for microelectronics. One can envision new hybrid concepts combining superconductors with existing semiconductor technology. As an example, the operation of silicon and gallium arsenide-based transistors at liquid N_2 temperatures looks attractive because of higher mobilities. Superconducting lines could serve for the electrical interconnections between the semiconductor elements on a chip, or between chips.

In the following we summarize the principal idea of using LTSEM for studying the spatial distribution of the critical temperature $T_c(x,y)$ and the critical current density $j_c(x,y)$ in thin-film high-T_c superconductors. Here again (x,y) denotes the coordinate point on the film. A more detailed discussion of the following arguments has been given in Ref. [10].

We start with the two-dimensional distribution $T_c(x,y)$ of the critical temperature. At the critical temperature T_c the electric resistivity ρ of a superconductor shows a transition from the value $\rho = 0$ of the superconducting state to the value $\rho = \rho_n$ of the normal state. In Fig. 7 we show schematically the transition curves for three regions of a specimen film each having a different critical temperature T_c. For simplicity we assume that the normal resistivity ρ_n in the three regions is the same. If the sample film is scanned by a modulated electron beam, the local temperature near the coordinate point (x,y) of the beam focus is modulated. Clearly, the temperature modulation results in a modulation of the resistivity, which is sharply peaked near the temperature corresponding to the steepest slope of the transition curves for the different regions in the sample. We denote the amplitude of the temperature modulation by δT and the total spread of the T_c-values in the sample by ΔT_c. With the further assumption

$$\delta T << \Delta T_c \tag{8.1}$$

the beam-induced resistivity modulation $\delta\rho$ shows the behavior indicated in the lower part of Fig. 7. We see that the response signal $\delta\rho$ strongly differs for the three regions reaching a sharp maximum at the different temperatures corresponding to the steepest slope of the transition curve of each individual region. In principal, for a current-biased (voltage biased) sample the signal $\delta\rho$

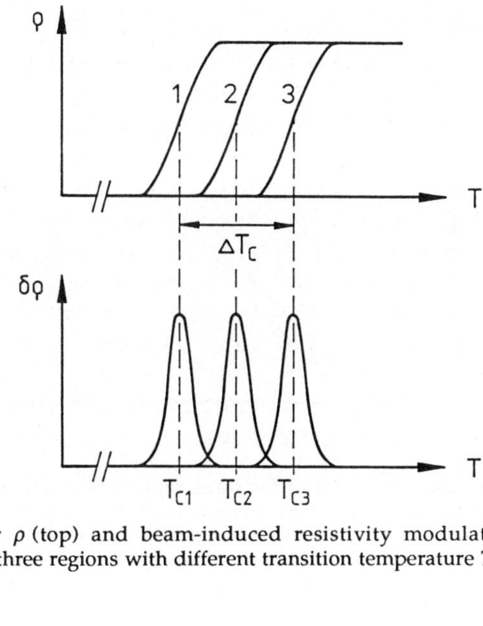

Figure 7
Electric resistivity ρ (top) and beam-induced resistivity modulation $\delta\rho$ (bottom) versus temperature T for three regions with different transition temperature T_c.

can be detected by means of a voltage signal δV (current signal δI). Of course, the translation of the beam-induced local resistivity change $\delta\rho$ into a voltage or current signal requires a careful evaluation of the series/parallel configuration of the separate regions in the (two-dimensional) film. For a detailed discussion of this point we refer to an earlier paper [10].

The experimental procedure for measuring the spatial distribution $T_c(x,y)$ of the critical temperature is then as follows. The specimen temperature is held at a constant value T_S, and the sample is scanned by the modulated electron beam. The maxima of the beam-induced signal $\delta\rho\ (x,y)$ image the regions with $T_c(x,y) = T_s$. By performing this experiment at different values of T_s, the regions with the corresponding values of the critical temperature can be imaged. We note that the superconducting transition can be much sharper locally than the global transition curve of the total sample, which integrates over the spatial distribution of the critical temperature. Therefore, sharp local signals $\delta\rho(x,y)$ can be generated from the small-amplitude temperature modulation during the scanning process, even though the global superconducting transition extends over a wide temperature range.

Next we turn to the two-dimensional distribution $j_c(x,y)$ of the critical current density. In Fig. 8 we show schematically the dependence of the resistive voltage V upon the current density for three regions of a specimen film each having a different critical current density j_c. During the scanning process with a modulated electron beam the local temperature modulation results in a modulation of the corresponding local V-j curve and a corresponding voltage or current signal as indicated schematically in Fig. 8 for the curve on the left. In this experiment the specimen is kept at a constant temperature below T_c upon which the small local beam-induced temperature modulation is superimposed. If the sample film is current-biased, the modulated voltage drop $\delta V(x,y)$ along the film is recorded as a function of the coordinate point (x,y) of the beam focus using phase-sensitive detection. If the bias current is increased in small steps and the sample is scanned after each

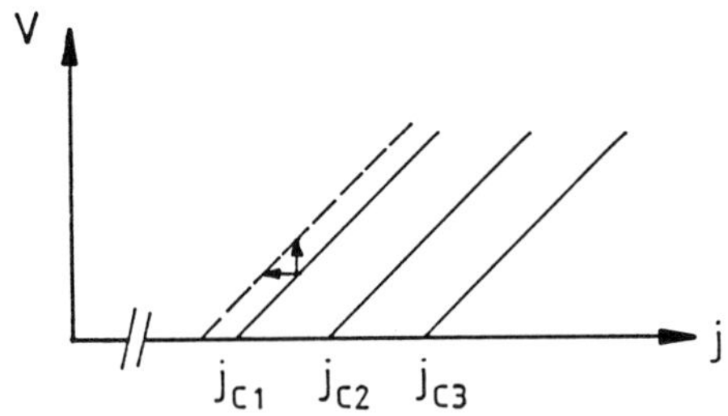

Figure 8
Resistive voltage V versus current density j for three regions with different critical current density j_C. Irradiation with the modulated electron beam results in a modulation of the local V-j curves as shown by the dashed line for the curve on the left. The vertical and horizontal arrow indicates the corresponding beam-induced voltage and current signal, respectively.

step, the first signal $\delta V(x,y)$ appears at the coordinate point with the smallest j_c value. With increasing bias current, regions with different local j_c values in turn generate a signal $\delta V(x,y)$ as soon as the bias current density reaches the corresponding local j_c value. In this way the spatial distribution of the signal $\delta V(x,y)$ as a function of the bias current images the distribution $j_c(x,y)$ of the critical current density [51]. Recently, the concepts outlined above have been experimentally confirmed using polycrystalline superconducting films of Y-Ba-Cu-O [52]. In particular, the spatially resolved imaging of the critical current density distribution $j_c(x,y)$ has shown that the local j_c values can vary by more than a factor of ten in the same film specimen.

It appears clear already, that LTSEM will contribute significantly to the answer of many important questions arising in the materials science of thin-film high-T_c superconductors. A list of such problem areas includes the degree of inhomogeneity of the electric current flow and the possible existence of distinct percolation paths, inhomogeneous flux pinning and flux flow, the role of grain boundaries and of the admixture of foreign phases, etc. Of course, the spatially resolved information on the superconducting properties obtained from LTSEM must be correlated with studies of the microstructure using the standard techniques of surface analysis. At present, it is still uncertain whether superconducting tunnel junctions can ever be fabricated successfully from the copper-oxide based superconductors in the standard form, where the two superconducting electrodes are electrically separated from each other by a thin-film insulating barrier. However, new concepts for the superconducting Josephson electronics based on tunneling across thin-film grain boundaries perhaps become interesting [53]. Again significant contributions can be expected from LTSEM.

Experimentally, extending LTSEM far above the temperature range of liquid He up to the boiling point of liquid N_2 and to higher temperatures is straight-forward. A sample mounting stage with temperature control for the desired temperature range is required. The principal scheme of such an arrangement has been described in Ref. [10].

Acknowledgement

A large part of the work described in this article has been supported financially by a grant of the Stiftung Volkswagenwerk.

References

[1] J. R. Clem and R. P. Huebener, J. Appl. Phys. **51** (1980) p. 2764
[2] M. Eigen, Disc. Faraday Soc. **17** (1954) p. 194
[3] R. P. Huebener and H. Seifert, Scanning Electron Microscopy 1984, III, p. 1053
[4] R.P. Huebener, Rep. Prog. Phys. **47** (1984) p. 175

[5] J. Bosch, R. Gross, and R. P. Huebener, in Josephson Effects - Achievements and Trends, A. Barone, ed. World Scientific Publishing Comp., Singapore (1986) p. 174
[6] R.P. Huebener, J. Bosch, R. Gross, and J. Mannhart, Extended Abstracts of the 1987 International Superconductivity Electronics Conference (ISEC '87), Tokyo, p. 353
[7] R. P. Huebener, J. Mannhart, R. Gross, and J. Bosch, Phys. Bl. **43** (1987) p. 400
[8] R. P. Huebener, Advances in Electronics and Electron Physics, P.W. Hawkes, ed. Academic Press, New York, Vol. **70** (1988) p. 1
[9] J. G. Bednorz and K. A. Müller, Z. Phys. B - Condensed Matter **64** (1986) p. 189
[10] R. P. Huebener, R. Gross, and J. Bosch, Z. Phys. B - Condensed Matter **70** (1988) p. 425
[11] H. Pavlicek, L. Freytag, H. Seifert, and R. P. Huebener, J. Low Temp. Phys. **56** (1984) p. 237
[12] R. P. Huebener, Magnetic Flux Structures in Superconductors, Springer, Berlin, 1979
[13] W. J. Skocpol, M. R. Beasley, and M. Tinkham, J. Low Temp. Phys. **16** (1974) p. 145
[14] K. P. Selig and R. P. Huebener, J. Low Temp. Phys. **43** (1981) p. 37
[15] J. Parisi, B. Mühlemeier, and R. P. Huebener, J. Low Temp. Phys. **60** (1985) p. 45
[16] B. Mühlemeier, J. Parisi, R. P. Huebener, and W. Buck, J. Low Temp. Phys. **64** (1986) p. 131
[17] R. Landauer, Phys. Today **31** (1978) p. 23
[18] W. J. Skocpol, M. R. Beasley, and M. Tinkham, J. Appl. Phys. **45** (1974) p. 4054
[19] R. Eichele, L. Freytag, H. Seifert, R. P. Huebener, and J. R. Clem, J. Low Temp. Phys. **52** (1983) p. 449
[20] L. Freytag, R. P. Huebener, and H. Seifert, J. Low Temp. Phys. **60** (1985) p. 365
[21] K.M. Mayer, J. Peinke, B. Röhricht, J. Parisi, and R.P. Huebener, Physica Scripta **T19** (1987) p. 505
[22] K. M. Mayer, J. Parisi, and R.P. Huebener, Z. Phys. B - Condensed Matter **7** (1988) p. 171
[23] H. Seifert, R. P. Huebener, and P. W. Epperlein, Phys. Lett. **95 A** (1983) p. 326
[24] R. Gross, M. Koyanagi, H. Seifert, and R. P. Huebener, Phys. Lett. **109A** (1985) p. 298
[25] R. Gross, D. B. Schmid, and R. P. Huebener, J. Low Temp. Phys. **62** (1986) p. 245
[26] D. Twerenbold, Phys. Rev. **B34** (1986) p. 7748
[27] D. Twerenbold and H. Zehnder, J. Appl. Phys. **61** (1987) p. 1
[28] R. Gross, U. Klaß, and H. Bender, Extended Abstracts of the 1987 International Superconductivity Electronics Conference (ISEC '87), Tokyo, p. 345
[29] R. Gross, R. P. Huebener, and U. Klaß, Superconductive Particle Detectors, A. Barone, ed., World Scientific, Singapore (1988) p. 144
[30] J. Bosch, R. Gross, M. Koyanagi, and R. P. Huebener, Phys. Rev. Lett. **54** (1985) p. 1448
[31] J. Bosch, R. Gross, M. Koyanagi, and R. P. Huebener, J. Low Temp. Phys. **68** (1987) p. 245
[32] J. Mannhart, J. Bosch, R. Gross, and R. P. Huebener, J. Low Temp. Phys. **70** (1988) p. 459
[33] C. S. Owen and D. J. Scalapino, Phys. Rev. **164** (1967) p. 538
[34] J. Mannhart, J. Bosch, R. Gross, and R. P. Huebener, Phys. Lett. **121A** (1987) p. 241
[35] J. Mannhart, J. Bosch, R. Gross, and R. P. Huebener, Phys. Rev. **B35** (1987) p. 5267
[36] J. Mannhart and R. P. Huebener, Proc. 18th Int. Conf. on Low Temperature Physics, Kyoto, 1987; Japanese J. Appl. Phys. **26** (1987) p. 1565
[37] M. Tinkham, Introduction to Superconductivity, McGraw Hill, New York, 1975
[38] J. Mannhart, J. Bosch, R. Gross, and R. P. Huebener, Phys. Lett. **A122** (1987) p. 439
[39] J. Matisoo, IBM J. Res. Develop. **24** (1980) p. 113
[40] 1987 International Superconductivity Electronics Conference (ISEC '87), Tokyo , Japan, Extended Abstracts
[41] J. Bindslev Hansen and P. E. Lindelof, Rev. Mod. Phys. **56** (1984) p. 431
[42] C. J. Lobb, Physica **126B** (1984) p. 319
[43] J. Bindslev Hansen, T. F. Finnegan, and P. E. Lindelof, IEEE Trans. Magn. **MAG-17** (1981) p. 95
[44] J. Niemeyer, J. H. Hinken, and R. L. Kautz, Appl. Phys. Lett. **45** (1984) p. 478
[45] J. Niemeyer, J. H. Hinken, and W. Meier, IEEE Trans. Instr. Meas. **IM-33** (1984) p. 311
[46] B. Pannetier, J. Chaussy, R. Rammal, and J. C. Villegier, Phys. Rev. Lett. **53** (1984) p. 1845
[47] J. M. Gordon, A. M. Goldman, J. Maps, D. Costello, R. Tiberio, and B. Whitehead, Phys. Rev. Lett. **56** (1986) p. 2280

[48] F. J. Hohn, T. H. P. Chang, D. Kern, W. Bruenger, P. Coane, and M. Lindemann, J. Soc. Photo-Opt. Instrum. Engng **333** (1982) p. 83

[49] J. Bosch, R. Gross, R. P. Huebener, and J. Niemeyer, Appl. Phys. Lett. **47** (1985) p. 1004

[50] T. H. Geballe and J. K. Hulm, Science 239 (1988) p. 367

[51] R. Gross, J. Bosch, and R. P. Huebener, Proc. of the Int. Conf. on High-Temperature Superconductors and Materials and Mechanisms of Superconductivity, J. Müller and J.L. Olsen, ed., North Holland, Amsterdam (1988) p. 1159

[52] R. Gross, J. Bosch, R.P. Huebener, J. Mannhart, C. C.Tsuei, M. Scheuermann, M. M. Oprysko, and C. C. Chi, Nature **332** (1988) p. 818

[53] P. Chaudhari, J. Mannhart, D. Dimos, C.C. Tsuei, J. Chi, M.M. Oprysko, and M. Scheuermann, Phys. Rev. Lett. **60** (1988) p. 1653

Part III

Precision Metrology

Josephson Series Array Potentiometer

J. Niemeyer
Physikalisch-Technische Bundesanstalt, D-3300 Braunschweig

1. Introduction

A Josephson tunnel junction which is operated at nonzero voltages produces an alternating supercurrent the frequency f of which is related to the dc voltage V across the junction by the simple equation:

$$V = (h/2e)\, f \ . \tag{1.1}$$

h is Planck's constant and e the elementary charge. (A detailed description of the Josephson effects is given in [1,2]). The validity of Eq. (1.1) is experimentally proved with extreme accuracy [3]. This makes a Josephson junction to be an ideal frequency to voltage converter.

Within a small frequency range the Josephson oscillator can be phase locked to an external oscillator. In this range the frequency f_e of the external rf bias determines the fundamental frequency of the Josephson rf current and because Eq. (1.1) is valid, also the dc voltage across the junction. A small variation of the dc current bias at this voltage does not therefore cause a voltage variation. This is also the case at the higher harmonics of the external drive frequency so that constant voltage steps appear in the dc characteristic (Fig. 1a) at:

$$V_n = n(h/2e)\, f_e \ . \tag{1.2}$$

The current width ΔI_n of a constant voltage step is dependent on the incident microwave power and the step number n:

$$\Delta I_n = 2I_c \, |J_n(2eV_{rf}/hf_e)| \, . \tag{1.3}$$

I_c is the critical current, V_{rf} the amplitude of the rf voltage across the junction, and J_n the nth order Bessel function [4].

The exactness of Eqs. (1.1) and (1.2) and the possibility of stabilizing the frequency of the rf bias to a very high degree suggest that the microwave-induced constant voltage steps could be used as very reproducible reference voltages in standard instruments. In the case of highly hysteretic tunnel junctions (Fig. 1a) two types of constant voltage steps can be generated,

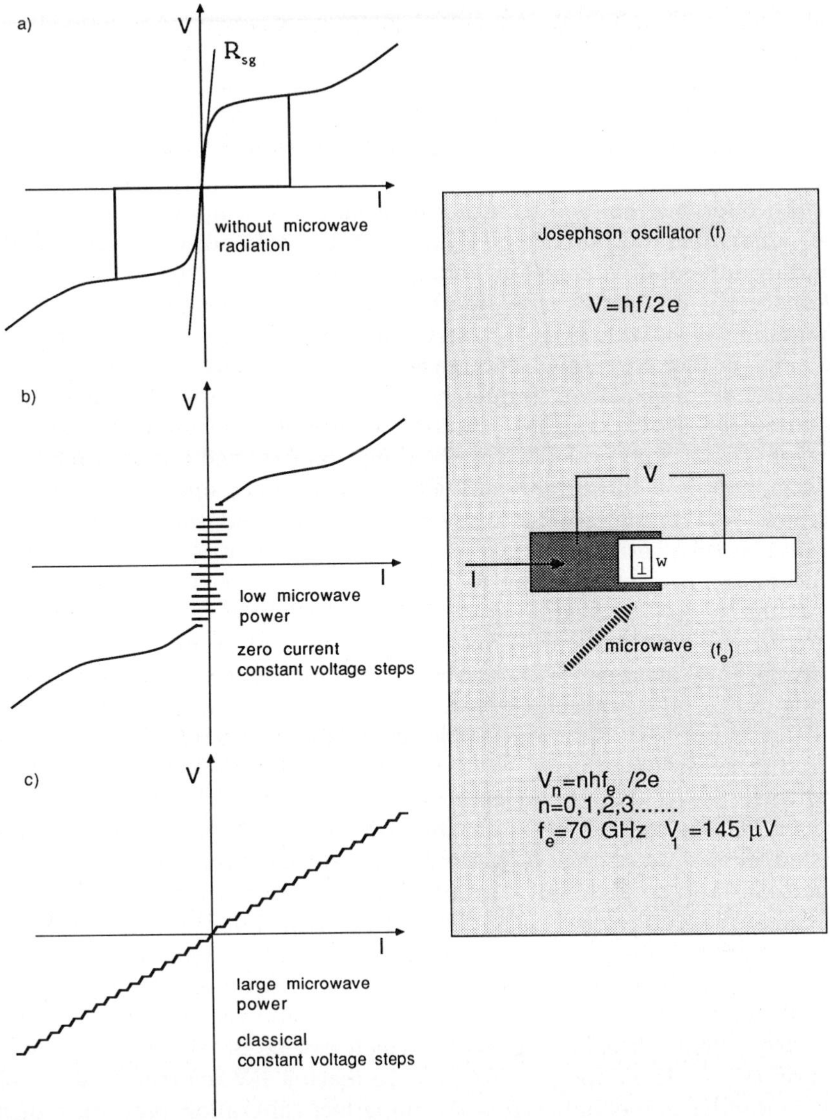

Figure 1
I-V characteristics of a highly hysteretic tunnel junction.
a) Characteristic without microwave radiation. R_{sg} is the subgap resistance of the single particle branch of the curve. b) Zero current constant voltage mode at low incident microwave power. c) Normal constant voltage step mode at high incident microwave power. The insert shows a schematical view of an inline tunnel junction. The junction area $l \times w$ is defined by a window in an insulating layer between the top and base electrode.

depending on the amount of incident microwave power. At relatively low power, constant voltage steps appear in the single particle branch of the *I-V* characteristic (Fig. 1b). As they cross the voltage axis, in principle, no dc current bias is needed to generate a voltage across the junction. These constant voltage steps overlapp, therefore for a certain current bias, several "quantized" voltages are possible in the zero-current step mode. The fact that the reference voltage from a single junction is multivalued with the small fundamental difference of V_1 (Eq. 1.2) between adjacent values allows to induce all possible constant voltage steps (again with the fundamental difference V_1) in a series array, too by combining the single junction output voltages in the suitable way. In a series array the voltage difference between two steps is therefore small enough to be continuously covered by tuning the external microwave frequency. This is important for operating a potentiomeer which should provide a continuous range of reference voltages. There is no possiblity of obtaining unquantized voltages which, on the one hand, is important and advantageous for operating a voltage standard or a potentiometer but on the other hand the stability of the current bias on a certain step is a very delicate problem. A short interruption of the phaselock between the two oscillators by chaos or external noise causes the system to switch to an arbitrary neighbouring step but not to return to the original one. This is especially important because at the relatively low incident microwave power sufficient for generating zero current steps, the phaselock between the Josephson and the external oscillator is not very tight and chaos can be more effective. The size of the desired step can be adjusted to its maximum value by tuning the incident microwave power (Fig. 2). At larger rf power the dc characteristic with the constant voltage steps is driven to the normal state characteristic by photon assisted excess tunneling. In general, the constant voltage steps now no longer overlapp so a chosen current bias is fixed on a certain voltage step and cannot be lost by occasional disturbances (Fig. 1c). This type of constant voltage step has therefore been successfully used as reference voltage in traditional Josephson voltage standards by many institutes (see e.g. [5]). The main disadvantage of the traditional single junction standard is the small reference voltage of a few millivolts which can be obtained from a single junction (V_1 = 145 μV for f_e = 70 GHz). To reach a 1 V reference the use of a voltage divider is inevitable. It is the imperfect calibration procedure of the voltage divider which chiefly limits the precision of the traditional Josephson voltage standards. Also the construction principles of the voltage dividers and the relatively small tuning range of the drive frequency allow the 1 V reference to be only slightly adjusted. This makes it very difficult to use this type of voltage standard as a potentiometer over a wide range of different reference voltages.

For this reason attempts were made in the past to use series arrays of Josephson junctions to obtain larger reference voltages [6,7]. Unfortunately, normal junction fabrication processes produce a junction parameter spread

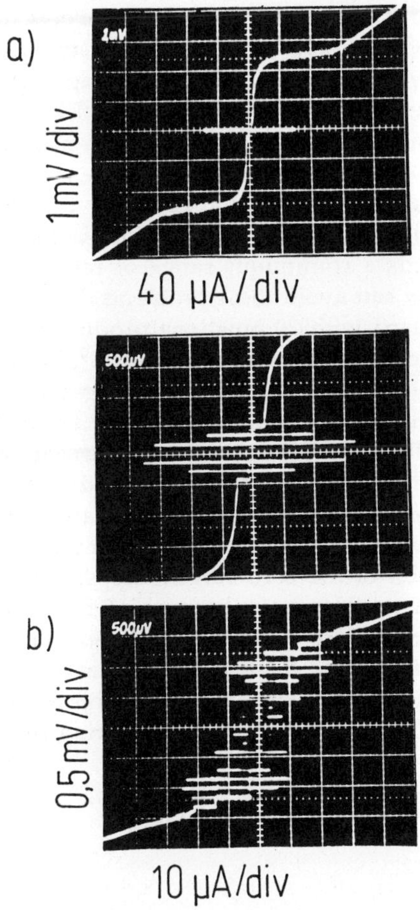

Figure 2
I-V characteristics of a lead alloy tunnel junction.
a) Without microwave radiation. b) With microwave radiation at two different power levels. With increasing microwave power, steps of higher order and decreasing width are induced. $f_e = 70$ GHz.

too large for biasing all the junctions of the array with a single current source on a constant voltage step, especially in the case of high-order steps, the use of which is imperative if only a few junctions are taken to obtain large reference voltages. Besides this, the spread of the various *I-V* characteristics is enlarged by the difficulty in coupling the same amount of microwave power to each junction - the shape of the characteristic depends on the incident rf power because the amount of photon induced excess tunneling current is rf power dependent.

The interesting attempt to use a single biased series array of highly damped, nonhysteretic junctions all locked onto the first order step has not yet been made, although modern photolithography and fabrication processes might allow the production of such complex arrays (cf. Sect. 3). For a drive frequency of 70 GHz, 1000 junctions in a series array would be necessary to reach a 1 volt reference. But the voltage difference between adjacent steps would be increased from V_1 to 1000 V_1 if the *I-V*-characteristics of the single junction were indentical. With slightly different *I-V*-characteristics and sufficiently large constant voltage steps also some intermediate voltage steps might be induced but generally the difference between the steps is arbitrarily increased, too. This is not a useful feature for a potentiometer which needs a continuous range of reference voltages. Only multibiasing a series array can avoid these problems.

A multibiased standard with 20 junctions produced reference voltages up to 100 mV with an uncertainty of a few parts in 10^9, but the complexity of independently biasing each junction of the array suggests that it will be very difficult to extend this design to voltages much above 100 mV [7].

The alternative approach of using zero current steps of highly hysteretic junctions in large series arrays [8] to directly produce large microwave induced reference voltages eliminates the problem of multibiasing, because even a relatively large junction parameter spread only slightly affects the position of the zero current steps in the back switching part of the *I-V* characteristic. However, because of the delicate stability of biasing the array on a certain zero current step, the junction parameters, the microwave distribution circuit and the bias system have to be designed very carefully to guarantee adequate stability of the complete system during operation. This design will be described in the following sections.

2. Circuit Design

a) Single Junction

To reach stable phase lock conditions in an rf driven tunnel junction the rf bias frequency must be larger than the plasma frequency f_p of the junction [9]:

$$f_e >> f_p = (eI_c/\pi hC)^{1/2} = (ej_c/\pi hC_s)^{1/2} \ . \tag{2.1}$$

C is the junction capacitance, C_s the specific junction capacitance, and j_c the critical current density. With $f_e/f_p = 3$, a ratio that satisfies relation (2.1) as has been experimentally proved, a maximum value for the critical current density can be derived:

$$j_{cmax} = (f_e/3)^2 \pi h C_s / e \ . \tag{2.2}$$

Data for different junction electrode and tunneling barrier materials are listed in table 1 (for details see [10,11,12,13]). To avoid disturbing resonance effects, junction width w and length l should be kept smaller than half of the wavelength of the rf bias inside the junction [14]:

$$w, l < (1/2f_e)\,(\mu_o\, C_S\, \{s + \lambda_{L1} + \lambda_{L2}\})^{-1/2} = w_{max}\,. \tag{2.3}$$

μ_o is the magnetic permeability of the vacuum, λ_{L1} and λ_{L1} the London penetration depths of the junction electrodes and s the negligible barrier thickness. This limit is not effective for the junction length, because for this quantity an even more restrictive threshold results from the requirement of a homogenous current distribution over the junction area. If only dc currents are present in the junction, the length of the junction must be smaller than 4 λ_J with

$$\lambda_J = [h/\{4\pi e\, \mu_o\, j_c(s + \lambda_{L1} + \lambda_{L1})\}]^{1/2}\,, \tag{2.4}$$

the Josephson penetration depth. Taking into account that the large displacement currents should be distributed homogenously, too, one obtains the more rigid limit [15]:

$$l < 3\,\lambda_J\,(j_c/2\pi f_e C_s V_n)^{1/2} = (3/2\pi f_e)\,(n\mu_o C_S\{s + \lambda_{L1} + \lambda_{L1}\})^{-1/2} = l_{max}\,. \tag{2.5}$$

From these parameters a maximum critical current I_{cmax} (Tab. 1) can be calculated. Inserting I_{cmax} in Eq. (1.3) and assuming that the rf voltage amplitude across the junction is adjusted to maximize the current step amplitude ΔI_n, one can estimate ΔI_{nmax} for a certain step order (Tab. 1). The

Table 1

Optimum junction design for standard series arrays of Josephson tunnel junctions out of different material combinations. The frequency of the external microwave is assumed to be f_e = 70 GHz. The junction type is characterized by the sequence of base electrode/barrier/top electrode. The maximum step amplitude ΔI_7 is calculated for the seventh constant voltage step with the maximum value of the Besselfunction $J_7(x) = 0.33$. C_S: specific junction capacitance, j_{cmax}: critical current density, l_{L1}, l_{L2}: London penetration depth of the junction electrodes, l_{max}: optimum junction length, w_{max} optimum junction width, I_{cmax}: maximum critical current.

Junction type: / Junction parameter	PbInAu/ PbIn oxide/ PbAu	Nb/ Nb_2O_5/ PbInAu	Nb/ Al_2O_3/ Nb	NbN/ MgO/ NbN	NbNNb/ MgO/ NbNbN
C_s {μF/cm²}	3	24	6	8	8
j_c {A/cm²}	21	168	42	56	56
$\lambda_{L1}/\lambda_{L2}$ {μm/μm}	150/150	80/150	80/80	>200/>200	>200 to 80/ >200 to 80
l_{max} {μm}	24	10	23	<13	<13 to 18
w_{max} {μm}	67	27	65	<35	<35 to 51
I_{cmax} (μA)	340	450	630	<250	<250 to 510
ΔI_{7max} {μA}	224	297	416	<165	<165 to 337

data show that of all listed junction types, the $Nb/Al_2O_3/Nb$ junctions provide the largest constant voltage step amplitude essential for a stable bias on the step. This type of junction is unaffected by temperature cycling and storage at room temperature, which does not mean that a complex circuit made of these junctions is also stable (cf. Chapt. 3), but the durability of the individual junction is an essential requirement for general durability. For both these reasons the $Nb/Al_2O_3/Nb$ junctions are the best choice for generating broad and stable zero current steps.

NbN/MgO/NbN junctions are even more durable but because of the very large penetration depth of NbN - for thin films it can be even larger than 200 nm, depending very much on the fabrication process - the maximum step size is very small. To obtain a larger step size, suitably thin NbN films can be backed by Nb films. The penetration depth of the NbN films is then decreased by the superconducting proximity effect but the large energy gap voltage of about twice the lead film value is maintained. A large energy gap voltage is desirable because in principle, zero current steps can reach up to about half of the gap voltage. Unfortunately, by backing the NbN films with Nb, one cannot take advantage of the high critical temperature of NbN (15 K).

The Nb/Nb_2O_5/lead-alloy junctions are a good choice, too. The junctions are durable provided there is dry storage, and the current amplitude of the constant voltage steps is also relatively large. A great advantage is that the junction area can be kept small, and complex arrays with a large number of junctions are about a factor of two smaller than arrays of other junction types. This permits a more homogeneous fabrication process and less microwave attenuation because of shorter microstrip lines.

The lead alloy junctions provide sufficient current step amplitude. Their main advantage is the simple fabrication process (cf. Sect. 3). However the stability is affected by temperature cycling and room temperature storage, which makes them unsatisfactory for use in potentiometer circuits.

I_{cmax} must not be exceeded by the critical current of any junction of a series array, otherwise the circuit cannot provide a stable reference voltage. As there is an upper threshold for the critical currents in a series array, the smallest critical current of the array is determined by the quality of the fabrication process, which always causes a certain spread of the critical currents. Unfortunately, as the smallest critical current of the array determines the resulting constant voltage step size, only with an ideal fabrication process - all junctions of the array providing the same critical current - can the maximum possible current width of the voltage steps ΔI_{nmax} be obtained. An absolute lower limit for the step size results from the requirement that it must be stably biased in the presence of the 4.2 K thermal noise of the cooling system. The minimum step width can be derived from the requirement that the Josephson coupling energy must be larger than kT:

$$kT = h\Delta I_{nmin}/4\pi e \quad . \tag{2.6}$$

With T = 4.2 K this yields the very small value of ΔI_{nmin} = 0.18 μA. For practical use a minimum step amplitude of about 15 μA is required to ensure a stable bias in the presence of external electromagnetic noise. This results in a minimum critical current I_{cmin} of about 23 μA, provided that the junction with the smallest critical current is biased onto the 7th step. The large difference between I_{cmin} and I_{cmax} indicates that for a series array driven in the zero current step mode, a larger parameter spread is tolerable than for an array driven in the normal voltage step mode. This estimate is only correct under conditions of optimum microwave coupling into each junction of the array (real junction data for experimental series arrays are listed in tab. 2). In the case of a real array, it is statistically improbable that the junction with the smallest critical current is operated just at the optimum microwave power. The minimum critical current to reach a step width of about 15 μA should therefore not be smaller than 80 μA in a real experiment. This shows that to maximize the step size, besides a fabrication process which provides a small critical current spread, a microwave circuit

Table 2
Experimentally achieved junction parameters of actually fabricated series arrays for references up to about 1 Volt. f_e: frequency of the external oscillator, l: junction length, w: junction width, I_{cmin}: smallest critical current of the array, I_{cmax} largest critical current of the array, I_{cav}: average critical current, ΔI: constant voltage step amplitude at 1 Volt, R_{sg}: subgap resistance, C: junction capacitance.

Junction type / Junction parameter	PbInAu/ PbIn oxide/ PbAu		Nb/ Nb_2O_5/ PbInAu	Nb/ Al_2O_3/ Nb		NbNNb/ MgO/ NbNbN
f_e {GHz}	70	90	96	69	100	69
C {pF}	35	35	108	58	58	51
R_{sg} {Ω}	70	70	100	150	150	100
l {μm}	25	25	15	22	22	17.4
w {μm}	46	46	30	44	44	36.4
I_{cmax} {μA)	250	250	340	380	380	125
I_{cmin} {μA}	150	150	270	240	240	60
I_{cav} {μA}	200	200	300	310	310	93
ΔI_n {μA}	45	70	150	60	50	6

that at least approaches a homogeneous rf distribution over all junctions of the array, is needed.

b) Microwave Circuit

The maximum reference voltage based on zero current voltage steps that can be obtained from a single junction is about half of the gap voltage $2\Delta/e$. For all junction types in Tab. 1 this is about 1 mV. The only exception with $2\Delta/e = 5$ mV is the NbN-junction type. In this case the maximum zero current step might reach up to 2.5 mV. To directly operate a 1 volt reference, series arrays of more than 1000 tunnel junctions must be used for most of the material combinations, and the large number of junctions must be integrated into a suitable microwave distribution circuit. This problem could be solved by arranging the chain of series connected junctions as the top layer of a microstripline [10]. This layer is separated from a superconducting ground plane by an evaporated dielectric layer (Fig. 3). Following [16], the main features of the stripline can be calculated from the simplified model described in Fig. 3. The admittance of the Josephson tunnel junction

$$Y_J = 3/R_{sg} + j\omega_e C \tag{2.7}$$

shows that the rf current through the junction is mainly capacitive: $3/R_{sg} << \omega_e C$. R_{sg} is the subgap resistance of these junctions, $\omega_e = 2\pi f_e$ and $C = \varepsilon_0 \varepsilon_r lw/s$ denotes the junction capacitance. ε_r is the dielectric constant of the

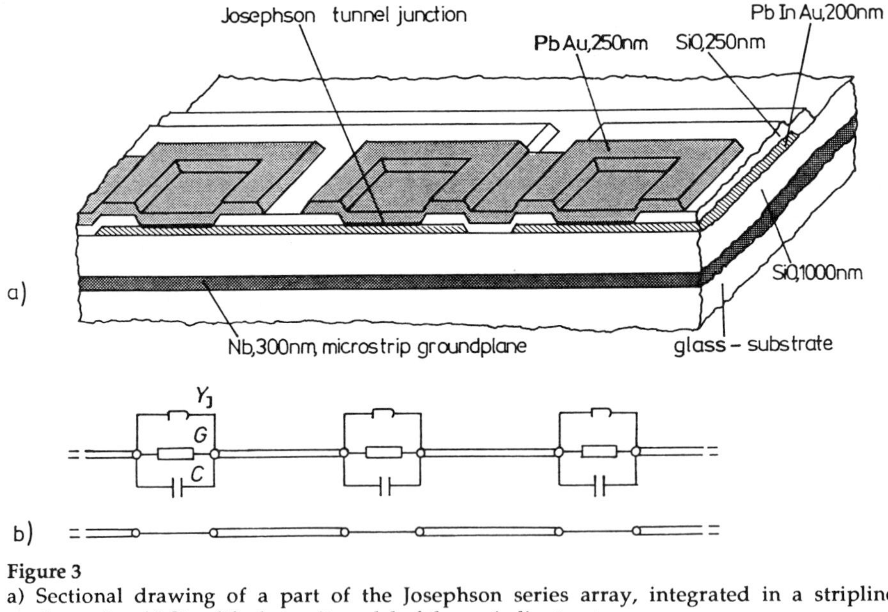

Figure 3
a) Sectional drawing of a part of the Josephson series array, integrated in a stripline configuration. b) Simplified circuit model of the periodic structure.

barrier material. Numerical values are listed in tables 1 and 2. The lower cut off frequency of the structure is

$$f_l = (1/2\pi)\,(Z_o l_k \sqrt{\mu_o \varepsilon_o \varepsilon_{rs}}\,\{C/2\})^{-1/2} \tag{2.8}$$

$$Z_o = (\mu_o/\varepsilon_o \varepsilon_{rs})^{1/2} d_s/w_s \tag{2.9}$$

ε_{rs} is the dielectric constant of the stripline insulator, d_s the thickness of the stripline insulator, w_s is the width of the stripline, and l_k the periodicity length of the array. The wave impedance Z_k of the junction array stripline is

$$Z_k = Z_o\,(1-\{f_l/f_e\}^2)^{1/2} \approx Z_o\,. \tag{2.10}$$

For the lead alloy circuit $Z_o = 2.26\ \Omega$, $f_l = 24$ GHz, and $Z_k = 2.12\ \Omega$ can be derived with $\varepsilon_{rs} = 5.7$, $d_s = 1\ \mu m$, $w_s = 70\ \mu m$, $l_k = 100\ \mu m$, $C = 48$ pF, $f_e = 70$ GHz. The junctions stripline is terminated by a matched lossy stripline. The whole stripline circuit is matched to an E-band waveguide by a fin line antenna which is inserted into a slot at the end of the waveguide. It could be shown experimentally [10] that this design produces reference voltages of up to 1 mV per junction for series arrays of up to at least 400 junctions. For an array of about 1590 junctions, the simple circuit with all junctions arranged in a long stripline did not provide sufficient stability although quantized voltages of up to 1 V has been reached. The attenuation in the long stripline is too large to provide a sufficiently homogeneous rf bias for all junctions. Provided the microwave power is adjusted to the optimum value for the first junction of the array (maximum step amplitude at about 1 mV), the last junctions obtain about 6 dB less power than the first junction in the case of the lead alloy design. If there are some junctions at the end of the array with critical currents close to the upper limit for the critical current density j_c, the smaller amount of microwave power would induce large but unstable low order voltage steps. Because of the relatively low rf power, the phase lock between the incident microwave and the junction oscillators is not tight enough to prevent chaotic behaviour in the high-current density junctions at the end of the long series array. This is evident in Fig. 4. It shows selected constant voltage steps in a 70 GHz and 90 GHz array. Both arrays could be fabricated with the same critical current distribution and the same value of the average critical current. The maximum critical currents reach the upper limit for the critical current densities, at least for a drive frequency of 70 GHz. No stable steps could be found at low voltages. To reach a low voltage step the incident microwave power must be adjusted to a relativly low value which leads to a weak phase lock between the junction oscillators and the external oscillator. It can therefore be expected that low reference voltages from the series array are unstable. For the 70 GHz design the range of instability - from - 800 mV to + 800 mV - is larger than for the 90 GHz design from - 300 mV to + 300 mV,

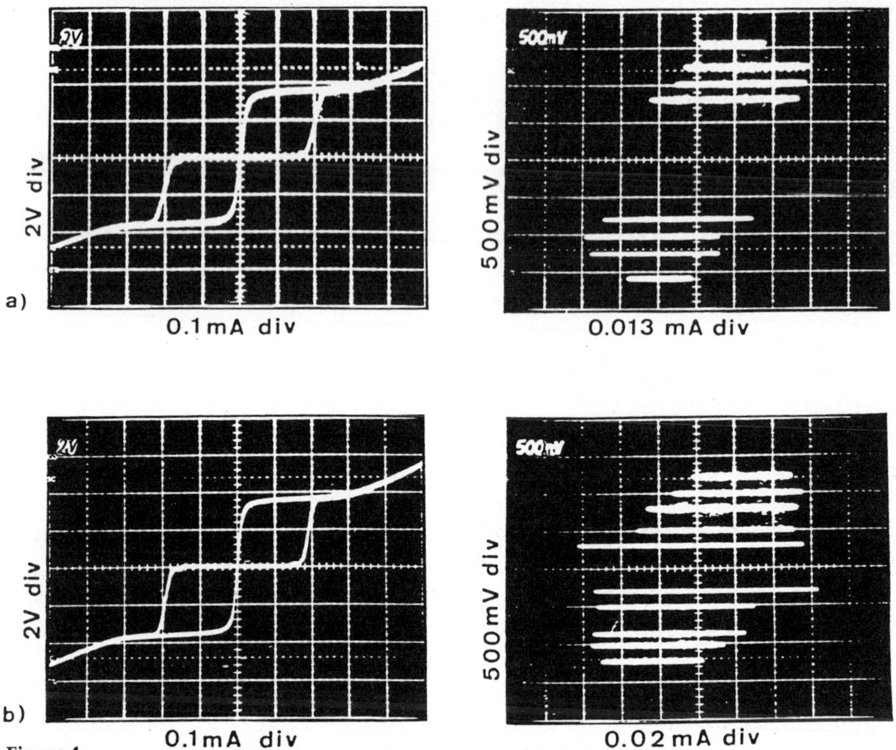

Figure 4
I-V characteristics and multiple exposures of selected constant voltage steps of 1440 junction arrays. a) 70 GHz version b) 90 GHz version. For each constant voltage step the incident microwave power was adjusted to obtain the maximum current width.

because for f_e = 90 GHz the difference between the external frequency and the plasma frequency is larger than for f_e = 70 GHz (cf Eq. (2.1)). This leads to a higher value of j_{cmax} (Eq. (2.2)) for the 90 GHz circuit, and in the case of f_e = 90 GHz, a more stable phase lock can be obtained even at low rf power. In a long stripline of more than 1000 junctions, these effects are so disturbing that stability can be reached only for minutes. For this reason, the junctions stripline for 1440 junctions has been divided into four shorter pieces connected in parallel to the microwave supply via three microwave dividers and two dc blocks [17]. Each of the four parts of the series array is again terminated by a lossy stripline. The dielectric layer for the dc blocks is formed by the insulating window layer which also defines the junction areas. Groundplane, antenna, and waveguide coupling are designed in the same way as for the single stripline circuit (Fig. 5). A detailed calculation of the microwave features of the complete circuit is given in [14]. The design described provided constant voltage steps of up to more than 1 V (Fig. 4). If the average critical current is not larger than 150 μA, the minimum reference voltage of this array is about 100 mV at f_e = 70 GHz. If a reference

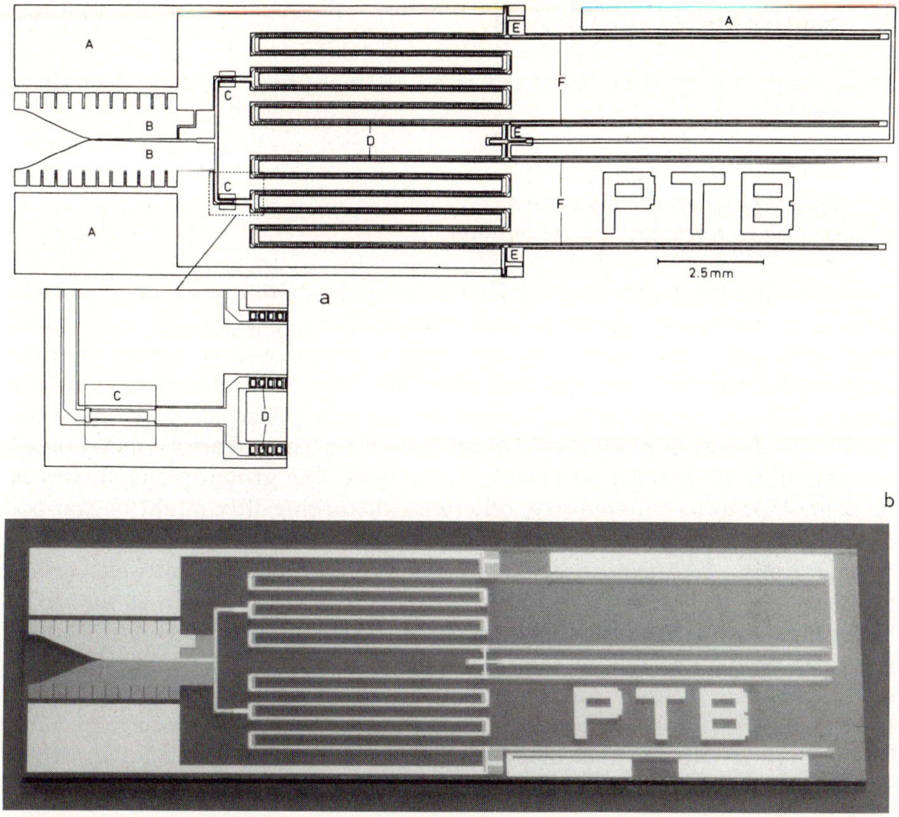

Figure 5
a) 70 GHz version of a series array of 1440 Josephson tunnel junctions. A: dc voltage electrodes; B: fin line antenna; C: dc blocks; D: series arrays; E: filters; F: matched lossy lines. b) Photograph of the standard voltage chip for reference voltages between 0.1 V and 1 V.

voltage of more than 1 V is desired, a correspondingly larger number of junctions can be used. If a stripline exceeds a certain length, the circuit must be divided into a larger number of microwave paths connecte in parallel to the rf supply. In a single path, a total of 1000 junctions should not be exceeded. For a reference voltage of 10 V an $Nb/Nb_2O_5/PbInAu$ circuit with 18992 junctions in series, divided into 16 parallel microstrip paths, has been developed [18]. It could be stably operated at 10 V.

3. Fabrication of the Circuits

Three basic fabrication lines are used to manufacture large Josephson circuits:

- complete lead alloy circuits
- Nb/Nb_2O_5/ lead alloy circuit
- complete refractory metal circuits.

The lead alloy layers can be produced by simple evaporation from a boat; the refractory metals are sputtered from a dc magnetron cathode or evaporated from an e-gun. For patterning the single layers of the circuits, normal photolithography is used, as has been developed for the semiconductor integrated circuits production. The patterning of the lead alloy layers is performed by the usual photoresist lift off processes. The refractory metal layers are patterned by reactive ion-etching processes. The groundplane should be thick enough to be pinhole-free, otherwise disturbing flux might be trapped in the holes during the cooling down of the sample to superconducting temperatures. A thickness of 300 nm is sufficient. If Nb is used for the groundplane, the film can be anodized (Nb_2O_5 layer 20 nm to 40 nm) to be better insulated against the metal layers of the circuit. This is very important for 10 V arrays because of the large active area covered by the extended circuit (see Fig. 6, [19]) and because the voltage applied for recording the characteristics of a 15000 junction array reaches values of more than 50 V. The additional Nb_2O_5 anodization layer prevents the following 1 μm thick SiO layer which forms the dielectric layer of the stripline circuit, from being

Figure 6
Photograph of a two inch diameter wafer with two series arrays of about 15000 junctions and one series array of about 1500 junctions. The junction sandwich is $Nb/Al_2O_3/Nb$.

shorted by the large voltage applied. The junction sandwich layer, consisting of base electrode layer (200 nm), tunneling barrier (1 nm), and top electrode layer (150 nm), is made on top of the SiO layer. In the case of the lead alloy and the Nb/lead alloy process the barrier is formed from the natural oxide of the base electrode by a plasma oxidation process [20]. Artificial barriers can be used for the all refractory metal process: Al_2O_5 for Nb electrodes, formed by thermal oxidation of a thin sputtered Al film [21], and a sputtered MgO layer [22] for the NbN electrodes.

The junction areas are defined by an SiO-window layer of a thickness of about 400 nm. The microwave circuit is designed for an SiO thickness of 250 nm, but the thinner this layer is the more pinholes or cracks can occur. These defects lead to parallel leakage currents forming small SQUID circuits. When the sample is cooled down, these SQUIDs trapp flux and this reduces the critical currents to small values in some junctions, seriously disturbing the stable operation of an array. Even subharmonics can be performed [23], as could be shown by irradiating a sample containing frozen flux with a 70 GHz microwave (Fig. 7). Because a SQUID is a very sensitive magnetic detector, flux is trapped even if the sample is cooled down in a magnetically shielded room for biomagnetic measurements. The evaporation process for the SiO layer must therefore be performed very carefully to avoid any defects [24]. The sample must be stored in a dry atmosphere because SiO absorbs humidity. After the first fabrication step, the all refractory metal tunneling sandwich covers the whole wafer, therefore the base electrode and junction area patterning is done by reactive ion etching after producing the sandwich. This requires a 20 nm MgO etch stopping layer on top of the 1 µm SiO dielectric layer and an additional wiring layer for connecting the single junctions to a series array. To guarantee a superconducting connection between the top Nb electrode and the wiring layer even for the very

200 µV/div

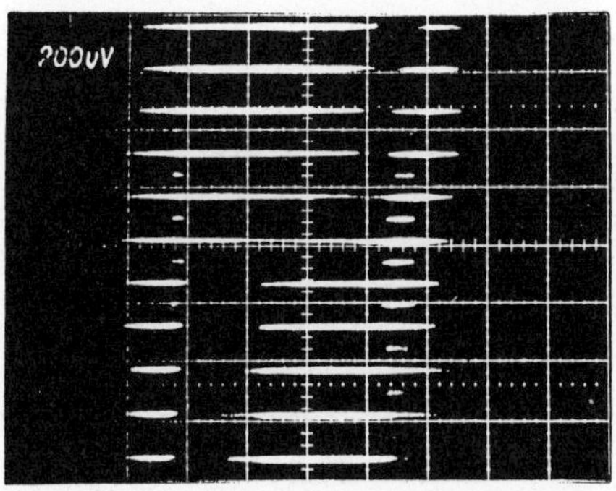

3,5 µA/div

Figure 7
One half subharmonic generation in a series array in the presence of trapped flux. No subharmonics could be found if the flux was removed from the sample by temperature cycling between the normal conducting and the superconducting state.

large rf currents (the rf currents are larger by a factor of 100 than the critical dc currents) the surface of the top electrode must be cleaned by a plasma etching process before the wiring layer is evaporated in situ. The evaporation system should have a starting pressure of 10^{-7} mbar. Tab. 3 shows a comparison of a typical soft metal (lead alloys) and a hard metal ($Nb/Al_2O_5/Nb$) process. For series arrays consisting of 1500 junctions for the critical current a standard deviation of about 5 % in the case of the lead alloy process and of 1 % for the $Nb/Al_2O_5/Nb$ process can be reached (cf Figs. 4

Table 3
Fabrication steps of series arrays with lead alloy- and $Nb/Al_2O_3/Nb$- tunnel junctions.

Circuit layer	Process steps for lead alloys	Schematic representation of the fabrication steps	Process steps for refractory superconductors ($Nb/Al_2O_3/Nb$)
Substrate	Si (highly resistive), glass	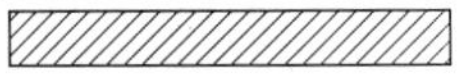	Si (highly resistive)
Groundplane	Photoresisist layer, PbIn(7)Au(2): thermal evaporation, 300 nm, 1 nm/sec, liftoff	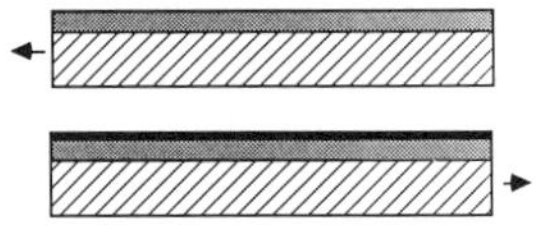	Nb: dc-magnetron sputtering, Ar-plasma, 10 mTorr, 40 sccm, 1 A, 290 V, 300 nm, 1,4 nm/s, anodisation: Nb_2O_5, 40 nm (20 V) photoresist layer, reactive ion etching: CF_4-Plasma, 200 mTorr, 15 sccm, 100 W
Dielectric layer	Photoresist layer SiO: thermal evaporation, 1 µm, 1,3 nm/s, liftoff	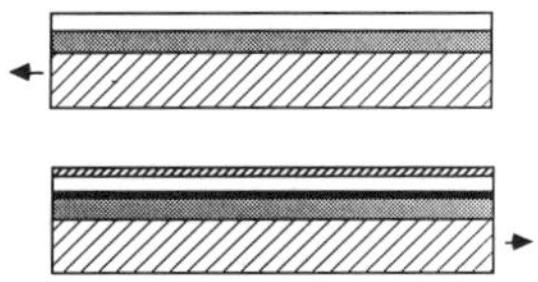	Photoresist layer, rf-sputter-cleaning, O_2-Plasma, 400 V, 40 W, 10 min SiO: thermal evaporation, 1 µm, 1,3 nm/s, liftoff
Etch stopping layer	----------		MgO: electron beam evaporation, 20 nm, 0,3 nm/s
Tunneling sandwich	Base electrode: Photoresist layer, PbInAu, 250 nm, as above liftoff window layer: Photoresist layer SiO: see above, 320 nm, liftoff Greinerprocess (Barrier) Top electrode: Photoresist layer PbIn(7): thermal evaporation, 600 nm, 2 nm/s liftoff	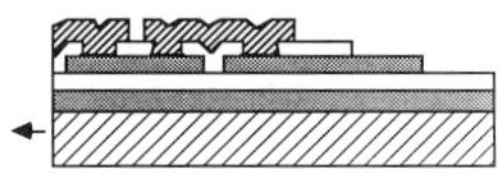 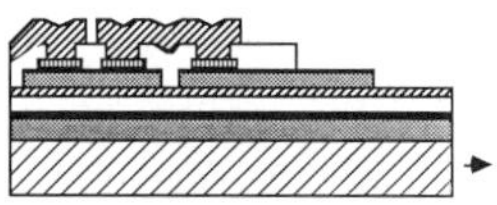	Whole wafer tunneling sandwich Nb: 200 nm, as above, Al: 5 nm: dc-magnetron sputtering as Nb, thermal oxidation of the Al layer, Nb: 150 nm, as above Base electrode: reactive ion etching until the MgO layer is reached, for Nb as above, for Al_2O_3: CF_4, 25 mTorr, 15 sccm, 100 W, 8 min Junction area: reactive ion etching until the tunneling barrier is reached Window layer: photoresiet layer SiO: thermal evaporation, 320 nm, liftoff Photoresist layer, sputter cleaning: Ar, 10 mTorr,10 W, 60 V PbIn(7), thermal evaporation, 600nm liftoff
wiring layer	-----------		or: Nb, dc-magnetron sputtering res. electronbeam evaporation, photoresist layer, reactive ion etching as above
resistive layer	Photoresist layer InAu(40): thermal evaporation, 245 nm, liftoff	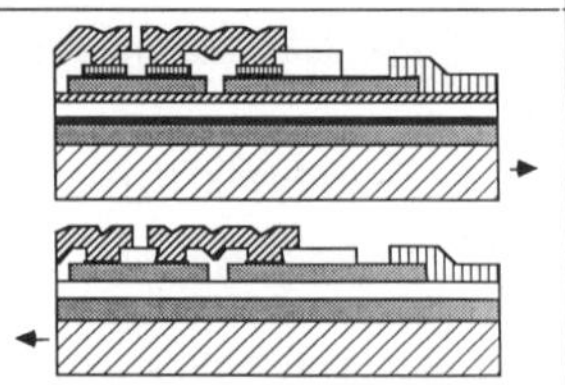	Photoresist layer, InAu(40): thermal evaporation, 245 nm, liftoff or: Mo, dc magnetron sputtering res. electronbeam evaporation, photoresist layer, reactive ion etching as for Nb

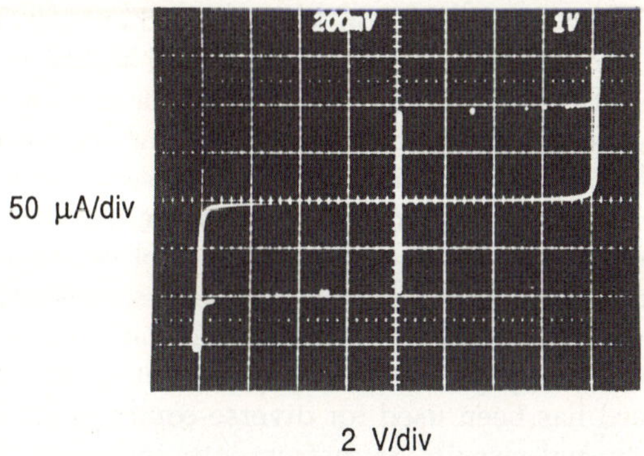

Figure 8
I-V characteristic of an $Nb/Al_2O_3/Nb$ series array of about 1500 junctions. The barrier was formed at 10 mbar O_2-pressure and an oxidation temperature of 22 ^oC.

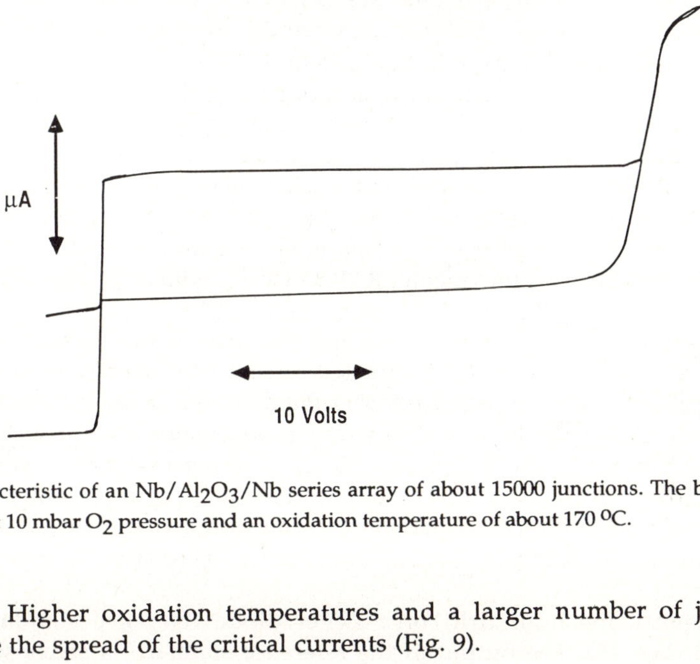

Figure 9
I-V characteristic of an $Nb/Al_2O_3/Nb$ series array of about 15000 junctions. The barrier was formed at 10 mbar O_2 pressure and an oxidation temperature of about 170 ^oC.

and 8). Higher oxidation temperatures and a larger number of junctions increase the spread of the critical currents (Fig. 9).

4. Measuring System and Precision of the Standard Instrument

In principle, the uncertainty of a Josephson reference voltage is limited only by the frequency stabilization of the external oscillator that induces the constant voltage steps on the I-V characteristic. One would therefore expect an estimated uncertainty of about one part in 10^{12}, which is the present short-term stability of atomic clocks, if the external oscillator is controlled by one of these clocks. The Josephson voltage to frequency relation is experimentally proved with very high precision [3] and furthermore, no current dependence of a series array constant voltage step could be found within the measurement uncertainty. The relative uncertainty in a comparison of the reference voltage of two series arrays driven by the same microwave source was a few parts in 10^{13} [25]. Because there is no voltage source of comparable stability available, the decisive experimental test of the ultimate reproducibility would be the comparison of the output voltages of two independently driven series arrays, an experiment which has not yet been performed.

A Josephson standard to be used for calibration and measuring purposes must be connected to equipment or test objects at room temperature. The series array standard has been used for diverse common applications. It was shown that the external circuits or instruments, including the voltage leads to the array and the switches for connecting test objects to the array, determine the reproducibility of the measurements. The uncertainties derived from various experiments are essentially uncertainties of the test objects or circuit elements but not uncertainties of the reference voltage at the dc pads of the series array in liquid helium. This is demonstrated by the experiments subsequently described in this paper. The most accurate room temperature measurement was a comparison of the output of a temperature controlled Weston cell with that of a 1 volt reference voltage from a series array [17]. The low thermal emf switches which were part of the set-up were not actuated in this experiment. It has been shown that a 1 volt reference voltage can be reproduced over a period of several minutes within a relative standard deviation of the mean of 3×10^{-10}, which is the combined deviation caused by the thermal noise of the 500 Ω room temperature internal resistance of the Weston cell and the drift of the thermal emf's of the leads. If switches have to be actuated during the measuring procedure, the uncertainty might be increased. The thermal emf changes, arising from actuating low thermal emf switches, are reproducible within a standard deviation of about 4×10^{-10} V (see table 4). Another example in which the internal noise of the test object determines the reproducibility of the measurement is the direct comparison of a one volt series array output with the output of an electronic voltage standard with relatively large output noise:

A typical one hour drift of a Zener diode reference standard was measured (Fig. 10). The output of the reference standard is stable within a standard deviation of 8 nV. The experiments decribed prove that Josephson series arrays can be used for routine calibration purposes as well as for sophisticated precision measurements. The accuracy of all measurements is limited not by the array output itself but by uncertainties arising from the objects to be calibrated. For the most applications a frequency stabilization of about 10^{-10} is therefore sufficient for a handy, practical calibration device such as the one described below. The system consists of the array and three main parts: the coarse and the fine adjusting devices and the calibration circuit (Fig. 11 shows a block diagram and a photograph of the system). The

Table 4
List of uncertainties of a Josephson series array standard.

Source of uncertainty	Relative uncertainty: parts in 10^{10}			
	Weston cell		Electronic reference standard, Fluke 732A/AC	
- Series array Josephson standard at 4,2 K	0,01		0,01	
- frequency stabilisation	1		1	
- Thermal emf drift	5		5	
- Thermal emf of the switches	4		4	
- Offset current drift of the nano-voltmeter N1a	1		2	
- Insulation resistance of the lines and switches	1		2	
- Insulation resistance of the nanovoltmeter N1a	0,1		0,2	
- Noise of the Weston cell (500Ω, 0,9 nV/$\sqrt{0,1}$ Hz)		9		
- Noise of the electronic reference standard (8 nV/$\sqrt{0,1}$ Hz)				80
- Noise of the nanovoltmeter N1a (0,47 nV/$\sqrt{0,1}$ Hz at 500Ω)		4,7		
(0,9 nV/$\sqrt{0,1}$ Hz at 1000Ω)				9
Total relative uncertainty (1 σ) without measuring time dependent noise	6,5		7	
Total relative uncertainty (1 σ) for 0,1 Hz bandwidth		12		81

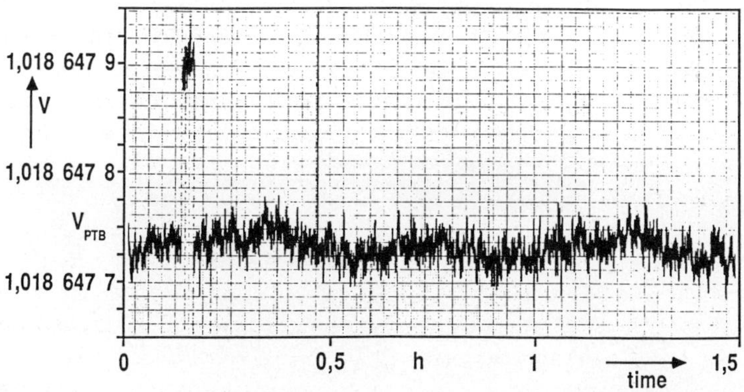

Figure 10
Long-term calibration of a Zener diode reference standard (Fluke 732 A/AC DC Reference Standard, SN 36 75034). The bandwidth of the measuring system is 0.1 Hz and the calibration peak results from tuning the drive frequency by an amount equivalent to 200 nV.

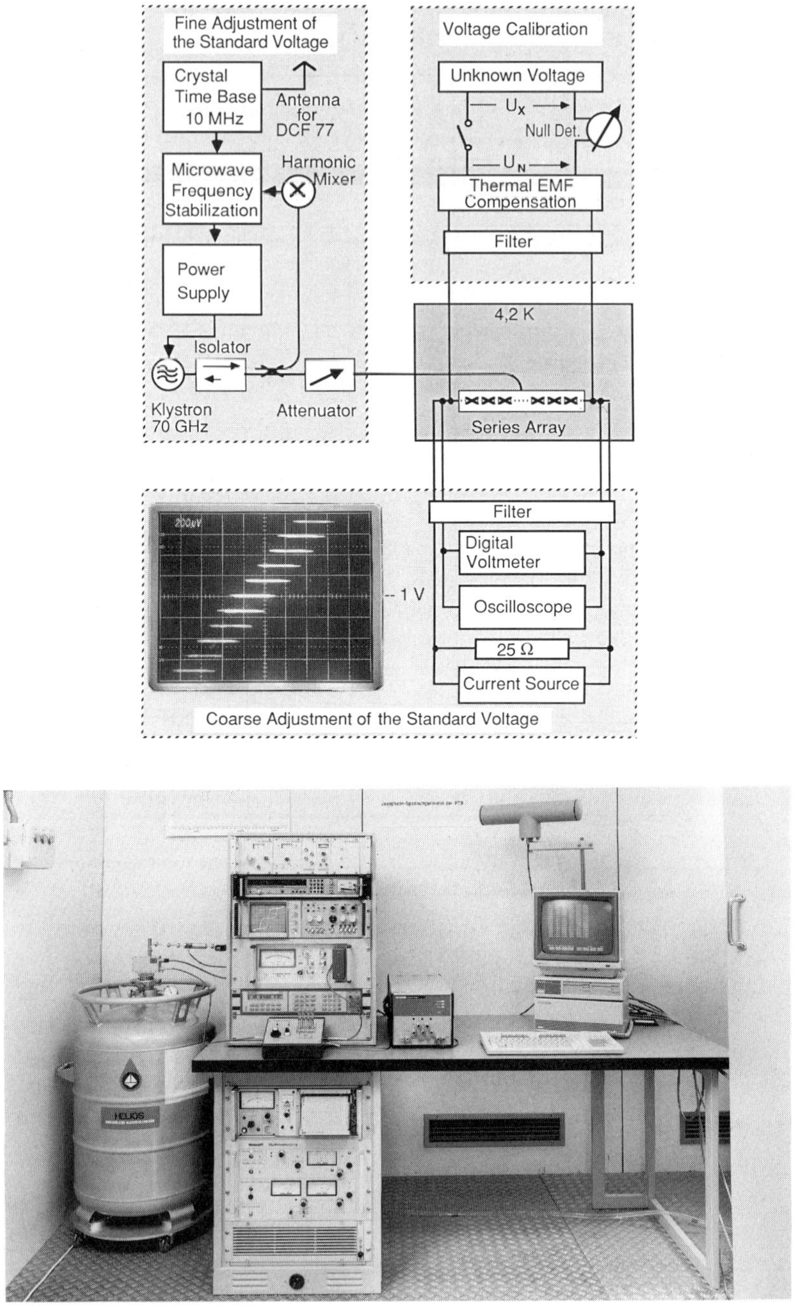

Figure 11
Block diagram and photograph of a transportable Josephson potentiometer. The frequency of the external oscillator is controlled by the radio signal of the long wave transmitter DCF 77 based on the PTB's atomic cesium clock.

coarse adjusting device serves to find a voltage step close to the voltage that has to be calibrated. It contains an oscilloscope - the display of which is showing a few adjacent voltage steps close to 1 V is pictured in the inserted photograph in Fig. 11 - and a digital voltmeter for counting the steps. The 25 Ω parallel resistor is used for a more convenient adjustment of the current bias on the desired constant voltage step. Without the resistor it is difficult to adjust the bias current in the center of a certain voltage step, because all steps more or less totally overlapp without superimposing the parallel resistor charcteristic. The fine adjustment and the frequency stabilization are carried out by means of a source locking microwave counter (EIP, model 578). The frequency of this system is stabilized with an external time base which consists of a crystal oscillator with an excellent short-term stability. The long-term stability of the time base is controlled by the PTB's radio signal which is transmitted via the long wave transmitter DCF 77 that can be received in most European countries. This makes the system easy to use and independent of an extra frequency calibration. The frequency can be tuned electronically over a range large enough to adjust the microwave induced reference voltage quasi continuously between two adjacent constant voltage steps in the voltage range between 0.1 V and higher voltages.

For a single measurement of a Weston cell and an electronic dc reference standard against the series array Josephson standard the estimated partial uncertainties and the total uncertainty are given in table 4 (1 σ values).

The insulation resistance of the lines and switches and of the nanovoltmeter was measured to about 5×10^{12} Ω and 5×10^{13} Ω respectively. The insulation resistance of the nanovoltmeter is the insulation between the shorted input ports and the ground and the measured value is much higher than that garanteed by the manufacturer. The noise of the electronic reference standard and the insulation resistances are measured values. The total relative uncertainties of 12×10^{-10} and 81×10^{-10} are determined with a bandwidth of 0.1 Hz, a room temperature stability of 0.1 K and a total measuring time of less than 6 minutes.

5. Design of an Integrated Potentiometer

The suggested circuit includes two series arrays and two dc SQUIDs as zero instruments, all integrated on a single chip. (Fig. 12) To realize this circuit one of the described all refractory metal fabrication processes is needed. This prevents the tunnel junctions of the series array layer from being destroyed by the following process steps. Two different junction layers are needed because the critical current density for the series array junctions (30 A/cm^2) is much lower than that for the SQUID junctions (1000 A/cm^2). The oxidation temperature for the low current density junctions is 170 °C and for the high current density junctions 25 °C. Therefore The series array sand-

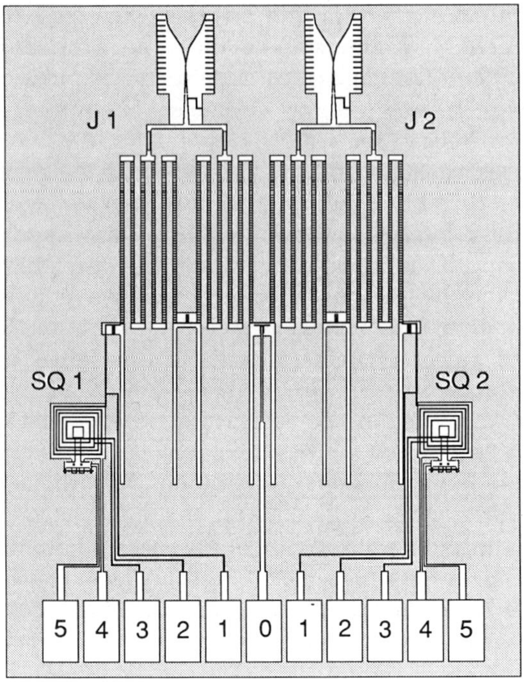

Figure 12
Integrated circuit with two Josephson series arrays, J1 and J2, two SQUIDS, SQ1 and SQ2. Specification of the contact pads: 0; 3: dc current and voltage pads connected to the complete SQUID input coil; 0; 2: dc current and voltage pads connected to a center tapping of the SQUID input coil; 0; 1: dc current and voltage pads connected to the shorted SQUID input coil; 4; 5: SQUID signal output.

wich must therefore be fabricated first, followed by the room temperature process for the SQUID circuit. This ensures that the current density of the first junction layer is not altered by the higher process temperatures of the following steps. In table 5 the steps of the complete fabrication process are listed.

If each series array of the potentiometer contains 8000 junctions distributed over 16 parallel microwave paths, each of these can generate reference voltages of up to 5 V if each is independently driven by a microwave source. Depending on the desired mode of operation, the two circuits can be connected into series or series opposition or operated independently (Fig. 13). The series connection provides reference voltages of up to 10 V for a Zener diode calibration. The series opposition mode formes a difference between two reference voltages which can be adjusted continuously to zero voltage. This is not possible with a single array because of the chaos problem at low rf power (cf. Sect. 2) and because the tuning range of a klystron or Gunn diode is not large enough to cover the voltage range between two adjacent steps continuously at reference voltages smaller than 100 mV. The

Table 5
Fabrication process of an integrated Josephson potentiometer. RIE (Reactive Ion Etching), LO (Lift Off).

Level of the circuit	Patterning
1. SQUID- and series array groundplane, Nb, 400 nm.	RIE
2. Dielectric layer for the series array and insulation of the SQUID against the groundplane, SiO, 1000 nm.	LO
3. Tunneling sandwich, $Nb/Al_2O_3/Nb$, 200/2/150 nm.	RIE
4. Window layer for the array junctions and the input coil, top insulation of the input coil, SiO, 400 nm.	LO
5. Wiring layer for the series array and top leads for the SQUID input coil, Nb, 600 nm.	RIE
6. Resistive layer for the lossy striplines and the SQUID shunts, Pt, 200 nm.	LO
7. Insulation for the top leads of the SQUID input coil, SiO, 500 nm.	LO
8. SQUID tunneling sandwich, $Nb/Al_2O_3/Nb$, 200/2/ 150 nm, SQUID loop Nb 200 nm.	RIE
9. SQUID window layer for the definition of the SQUID junctions and connection of the SQUID shunts SiO, 300 nm.	LO
10. Wiring layer for the SQUID, Nb 500 nm.	RIE

independent mode of operation of the two circuits enables the calibration of two unknown voltages at the same time. This is important for a continuous long term calibration of resistance ratios of common potentiometers. The SQUIDs serve as nullmeters for the comparison of the unknown voltages. If the voltages are not completely balanced, a smaller current signal generates a small magnetic field in the input coil of the SQUID, which is coupled to the SQUID loop. The sensitive area of the loop should be about 50 x 50 μm^2 [26]. The optimal number of turns of the input coil is 10 to 20 [27]. It is possible to use several coil tappings to obtain the optimum fit of the coil to the measuring problem. The flux noise of a SQUID amounts to $10^{-4}\ \phi_0/\sqrt{Hz}$.

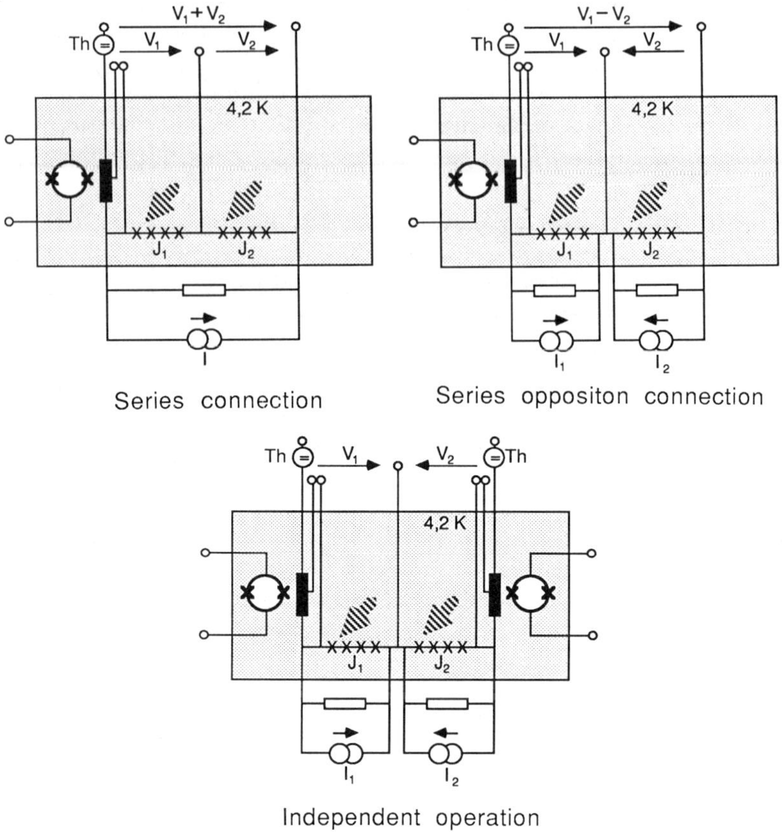

Figure 13
The different modes of operation of the integrated potentiometer. For two 5 V series arrays, the series connection of both circuits provides reference voltages from 0.2 V to 10 V, the series opposition configuration from 0 to 5 V. The third independent mode of operation is useful for voltage ratio calibrations of potentiometers.

$\phi_0 = 10^{-15}$ Vs is the flux quantum. This causes a typical current noise of 10^{-11} A/$\sqrt{\text{Hz}}$ in the input coil. If the test object has an internal resistance of 1 Ω the current noise generates a voltage noise of 10 pV. This shows that a SQUID is an ideal zero instrument if the internal resistance of the test object does not exceed 100 Ω. If the accuracy of the potentiometer is to be fully utilized even at higher internal resistances, the SQUID can be matched by an additional planar transformer to the specific measuring problems [28]. In the normal case, however, the noise of the test object prevents the potentiometer resolution from being fully utilized.

6. The Use of High Critical Temperature Superconductors

Since perovskite-related oxide superconductors with high critical temperatures have been discovered [29, 30], intensive research work has been carried out to characterize and optimize different types of material. The material is strongly anisotropic with a very small coherence length (ξ = 1.6 nm for polycrystalline $YBa_2Cu_3O_7$ [31]). The small coherence length causes even grain boundaries to act like tunneling barriers which limit the superconducting current in polycrystalline samples. The geometrical dimension of the weak connection between the two electrodes of a Josephson junction which can be the thickness of an insulating tunneling barrier, the thickness of a normal layer or any weak link, should be in the order of magnitude of the coherence length. It is therefore, very difficult to fabricate reproducible Josephson junctions from the new high T_c materials. Point contact junctions [32, 33], microbridges where the barrier is formed by grain boundaries inside a constriction of a thin film [34], and superconductor/normal metal/superconductor junctions with the top electrode formed by a conventional superconductor [35] have all been successfully fabricated. Even more complicated will be the preparation of superconductor/insulator/-superconductor junctions - the junction form needed for a series array voltage standard - because to take full advantage of the high critical temperature, the junction top electrode should also be made out of the high T_c material, too. This requires a post annealing and oxidation process which would destroy all artificial insulating barriers known so far.

Provided that, in the face of all these experimental difficulties, it will be possible to prepare reliable and reproducible tunneling junctions with a suitable current density, it must be analysed wether it is possible in principle to operate these junctions in the zero current step mode as described in Sect. 1 and 2. By irradiating adjustable point contact junctions consisting entirely of high T_c material with microwave radiation it could be shown that Josephson oscillations are present at 77 K [32] and attain frequencies of 10 THz at a temperature of 4.2 K [33]. The voltage difference between adjacent constant voltage steps induced on the highly damped, non-hysteretic characteristic is described by Eq. (1.1). The power dependence of the step amplitude is well described by the Bessel function dependence (Eq. (1.3), [33], Fig. 14). Josephson effects can therefore be expected from high T_c tunneling junctions as well as from conventional superconductor tunneling junctions. The considerable effort necessary to produce tunneling junctions for a voltage standard circuit would only be effective, if a voltage standard fabricated from the high T_c material could be stably operated in the presence of the thermal noise at an operating temperature of 77 K. This would allow full advantage to be taken of the simple cooling procedure at liquid nitrogen temperature. The temperature dependence of the stability time of the phase lock between the Josephson oscillator and the external oscillator of a single junction is described by the relation [9]:

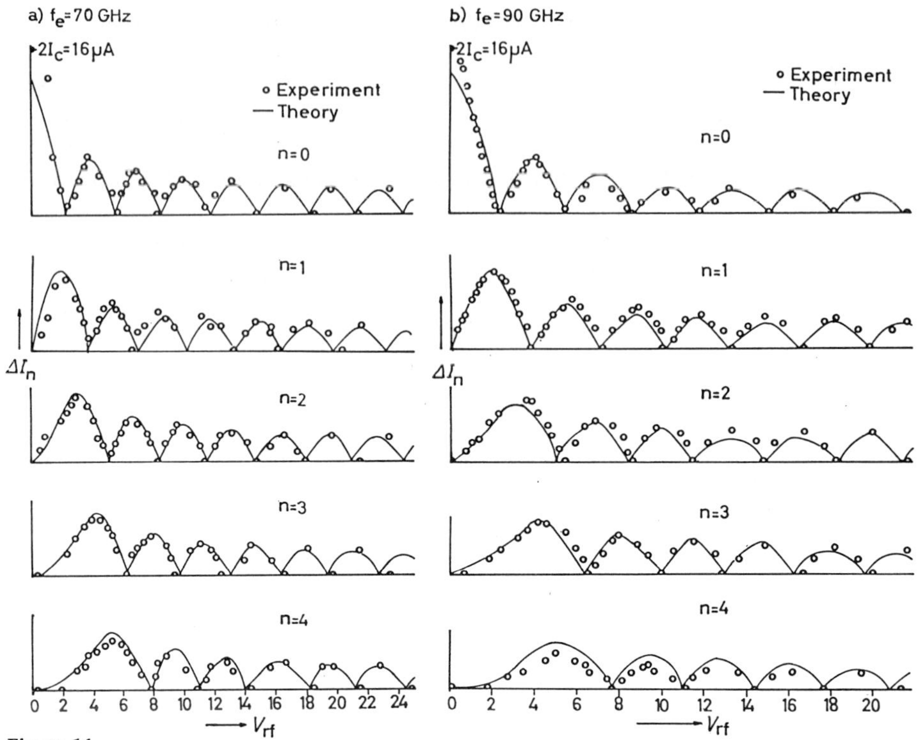

Figure 14
The dependence of the current amplitude ΔI_n of the microwave induced voltage steps from the rf voltage V_{rf} across an $YBa_2Cu_3O_7$ point contact junction. The rf voltage is expressed in units of $hf_e/2e$. The theoretical curve is derived from the expression $\Delta I_n = 2I_c | Jn(2eV_{rf}/hf_e)|$. All experimental data are fitted only at two points to the theoretical curves. a) 70 GHz. b) 90 GHz.

$$\tau = \tau_o \exp(E/kT) \text{ with } E = h\Delta I_n/4\pi e,\ \tau_o = \pi (2I_c/\Delta I_n)^{1/2}/\omega_p \qquad (6.1)$$

For the penetration depth λ_L of $YBa_2Cu_3O_7$, values between 100 nm and 200 nm have been derived [36]. On the assumption that Al_2O_3 might be a stable tunneling barrier between two high T_c superconducting electrodes - at least substrates of Al_2O_3 can be used for preparing thin $YBa_2Cu_3O_7$ films - the drive frequency f_e = 70 GHz, $\omega_p = 2\pi f_p = 2\pi f_e/3$, and the junction should be operated at the 7th constant voltage step only. The following data can be calculated for a single junction:

C_s = 6 μF/cm^2, j_{cmax} = 42 A/cm^2, l_{max} = 17 μm, w_{max} = 48 μm, I_{cmax} = 342 μA, ΔI_{7max} = 226 μA, $\tau_o = 3.7 \times 10^{-11}$ s

From Eq. (6.1) the lifetime of the phase lock of $\tau = 10^{20}$ s can be derived for a single junction operated at 77 K. This result shows that on the above assumptions, a single junction could be stably operated in the zero current step mode at 77 K.

In a series array the experimentally achieved step amplitude which results is much smaller than the optimum value for a single junction. In case of 1500 junctions, a realistic value is ΔI_n = 60 µA. This results in a stability time of only $\tau = 10^{-3}$ s at T = 77 K. To achieve stability at this temperature, the array fabrication process and the microwave distribution circuit must therefore be even better optimized for the high T_c superconductors than for the conventional superconductors.

References

[1] A. Barone, G. Paterno, Physics and Application of the Josephson Effect, John Wiley & Sons, New York, 1982

[2] K.K. Likharev, Dynamics of Josephson Junctions and Circuits, Gordon and Breach Science Publishers, New York, P.O. Box 786, 1984

[3] J.S. Tsai,A.K. Jain, J.E. Lukens , Phys. Rev. Lett. **51** (1983) p. 316

[4] S. Shapiro, A.R. Janus, S. Holly , Rev. Mod. Phys. **36** (1964) p. 223

[5] V. Kose, IEEE Trans. Instrum. & Meas. **IM-25** (1976) p. 483

[6] T.F. Finnegan, A. Denenstein, D.N. Langenberg, Phys. Rev. **B4** (1971) p. 1487

[7] T. Endo, M. Koyanagi, A. Nakamura, IEEE Trans. Instrum. Meas. **IM-32** (1983) p. 267

[8] M.T. Levinsen, R.V. Chiao, M.J. Feldmann, B.A. Tucker, Appl. Phys. Lett. **31** (1977) p. 776

[9] R.L. Kautz, J. Appl. Phys. **52** (1981) p. 6241

[10] J. Niemeyer, J.H. Hinken, R.L. Kautz, Appl. Phys. Lett. **45** (1984) p. 478

[11] C.A. Hamilton, R.C. Kautz, F.L. Lloyd, R.L. Steiner, B.F.Field, IEEE Trans. Instr. Measm. **IM-36** (1987) p. 258

[12] J. Niemeyer, Y. Sakamoto, E.Vollmer, J.H. Hinken, A. Shoji, H. Nagagawa, S. Takada, S. Kosaka, Jap. J. Appl. Phys. **25** (1986) L 343

[13] Y. Sakamoto, T. Endo, Y. Muragama, T. Sakuraba, M. Nakanishi, M. Koyanagi, M. Aoyaki, 1987, Int. Superconductivity Electronics Conference (ISEC), 84

[14] E. Vollmer, Diploma work, Institut für Hochfrequenztechnik der TU Braunschweig (1984)

[15] C.A. Hamilton, R.L. Kautz, R.L. Steiner, F.L. Lloyd, IEEE Electr. Dev. Lett. **EDL-6** (1985) p. 623

[16] J.H. Hinken, Supraleiter Elektronik, Springer Verlag, Berlin, Heidelberg New York (1988) ISBN 3-540-18720-0

[17] J. Niemeyer, L. Grimm, W. Meier, J.H. Hinken, E. Vollmer, Appl. Phys. Lett. **47** (1985) p. 1222

[18] L. Lloyd. C.A. Hamilton, J.A. Beall, D.Go, R.H. Ono,E.Harris, IEEE Elec. Dev. Lett. **EDL-8** (1987) p. 449 and C.A. Hamilton, F.L.Lloyd, K. Chieh, W. Goeke, CPEM 88, to be published.

[19] The $Nb/Al_2O_3/Nb$ circuits have been fabricated in cooperation with the Josephson computer group and the quantum metrology group of the Electrotechnical Laboratory, Tsukuba, Japan.

[20] J.H. Greiner et al., IBM J. Res. & Dev. **24** (1980). p. 195

[21] H. Nakagawa, K. Nakaya, I. Kurosawa, S. Takada, H. Hayakawa, Jpn. J. Appl. Phys. **25** (1986) L 70

[22] A. Shoji, M. Aoyagi, S. Kosaka, F. Shinoki, H. Hayakawa, Appl. Phys. Lett **46** (1985) p. 1098

[23] S. Naito, J. J. Appl. Phys. **24** (1985) p. 449 and C. Vanneste , C.C. Chi, W. J. Gallagher, A.W. Kleinsasser, S.I. Raider, R.L. Sandstrom, to be published
[24] S. Shibayama, S. Hasno, T. Yamaoka, Appl. Phys. Lett. **47** (1985) p. 429
[25] J. Niemeyer, L. Grimm, C.A. Hamilton, R.L. Steiner, IEEE Electron. Device Lett. **EDL -7** (1986) p. 44
[26] J.M. Jaycox, M.B. Ketchen, IEEE Trans. Magn. **MAG-17** (1981) p. 400
[27] M.W. Cromar, P. Carelli, Appl. Phys. Lett. **38** (1981) p. 723
[28] B. Muhlfelder, W. Johnson, M.W. Cromar, IEEE Trans. **MAG- 19** (1983) p. 303
[29] J.G. Bednorz, K.A. Müller, Z. Phys. **B 64** (1986) p.189
[30] M.K. Wu, J.R. Ashburn, C.J. Torng, P.H. Hor, R.L. Meng, L. Gao, Z.J. Huang, Y.Q. Wang, C.W. Chu, Phys. Rev. Lett. **58** (1987) p. 908
[31] S.N. Song, S.J. Hwu, K. Poeppelmeier, T.O. Mason, J.B. Ketterson, J.J. Appl. Phys. **S26** (1987) p. 1039
[32] J. S. Tsai, Y. Kubo, J. Tabuchi, Jpn. Appl. Phys **26** (1987) C 701
[33] J. Niemeyer, N.D. Kataria, M.R. Dietrich, C. Politis, H. Koch, R. Schöllhorn, H. Eickenbusch, Z. Phys. B- Cond. Matt. **69** (1987) p. 1
[34] R.H. Koch, C.P. Umbach, G.J. Clark, P. Chaudari, R.B. Laibowitz, Appl. Phys. Lett. **51** (1987) p. 200
[35] H. Akoh, F. Shinoki, M. Takahashi, S. Takada, to be published in J. J. Appl. Phys. (1988)
[36] A.J. Panson, A.I. Braginski, J.R. Gavaler, J.K. Hulm, M.A. Janocko, H.C.Pohl, A.M.M. Stewart, J. Talvacchio, G. R. Wagner, Phys. Rev. **B 32** (1987) p. 8774

Cryogenic Current Comparator Metrology

P. Gutmann and H. Bachmair
Physikalisch-Technische Bundesanstalt, 3300 Braunschweig

1. Introduction

Current comparators, whether designed for operation at room temperature or cryogenic temperatures, are based on Ampere's law

$$\oint_{(A)} \mathbf{H}\, \mathrm{d}s = \int_{(A)} \int J_n \, \mathrm{d}A = \Sigma I \tag{1.1}$$

where the line integral of the magnetic field strength **H** around a closed path which forms the boundary line of the surface A is equal to the total current crossing this surface. In an ideal current comparator, this total current is carried by two windings and the ampere turns of one winding are equal to and opposite those of the other winding; thus

$$\Sigma I = N_1 I_1 - N_2 I_2 = 0. \tag{1.2}$$

The current ratio of the ideal comparator is therefore equal to the inverse of the turns' ratio:

$$I_1/I_2 = N_2/N_1. \tag{1.3}$$

In practice, this ideal case can only be approximated, which means that the current ratio differs from the turns' ratio by a small amount, E,

$$I_1/I_2 = N_2/N_1 \cdot (1-E) \tag{1.4}$$

To obtain high ratio accuracies, the ampere-turn balance between the two ratio windings must be determined with a detector whose output responds to the line integral in Eq. 1.1. This is realized in practice by a detector sensitive only to the mutual fluxes which link both windings at the same time.

Current comparators - for dc as well as ac applications - operating at room temperature use high permeability toroidal cores, magnetic and eddy current shields, and careful winding techniques [1] to achieve high accuracies. Cryogenic comparators [2, 3, 4] eliminate the need for iron cores as well as magnetic and eddy current shields by using superconducting materials to achieve this ideal coupling and to shield the instrument against external magnetic fields. In addition, a SQUID (Superconducting QUantum Interference Device) magnetometer [5] which has a sensitivity several orders of magnitude better than that of room temperature detectors, is used as a flux detector.

2. Theory and Operation of Cryogenic Current Comparators

2.1 Basic Principle

Besides its infinite dc conductivity, a bulk superconductor is a perfect diamagnetic material, meaning that magnetic fields can neither exist in nor penetrate into a superconductor. In the presence of an outer magnetic field, screening currents flow on the surface producing a counter field, so that the inside of the superconductor is free of any magnetic field neglecting the small penetration depth λ.

The two characteristics - infinite dc conductivity and perfect diamagnetism - make a superconductor an almost ideal ampere-turn balance device for dc current comparator applications. Such a cryogenic current comparator (CCC) is formed by surrounding two current-carrying wires (or windings) with a superconducting shield as it is shown in principle in Fig. 1. Because of the shielding effect, the two currents I_1 and I_2 result in identical counter currents on the inner surface of the shield which close along a path on the outer surface in the direction of the original currents. In general, the resultant current along the outer surface is equal to the summation of all currents within the shield, ΣI. This resultant current generates a magnetic flux equal to the summation of all mutual fluxes within the shield. Thus there is an almost perfect coupling condition between the ratio windings inside the shield and a magnetic flux detector outside the shield which makes the ratio error of a CCC almost independent of the position of the ratio windings.

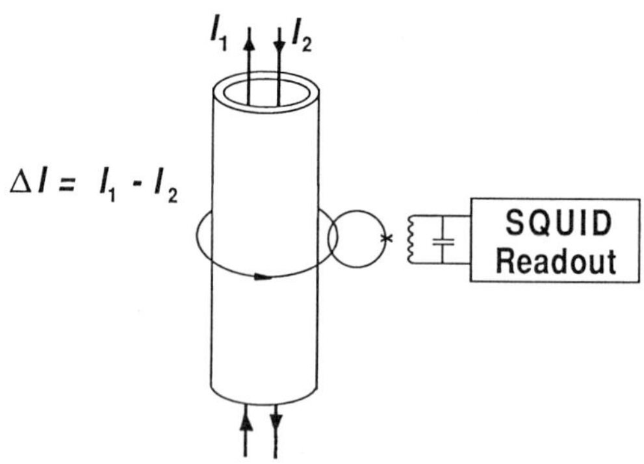

Figure 1
Construction principle of a cryogenic current comparator.

An ideal coupling is only obtained for infinitely long shield cylinders. In practice, edge effects due to the finite length of the cylinder will cause magnetic stray fields influencing the detector and therefore disturbing the ampere-turn balance. To overcome this problem, two different shield arrangements have been developed allowing current comparators to be constructed with very small ratio errors.

The first arrangement is achieved by inverting the shield cylinder in a way which is demonstrated in Fig. 2 [6]. In its final form, the shield forms a toroid which surrounds the detector winding. In order to prevent a shorted turn, the shield is equipped with a thin isolating gap only the principle of which is shown in Fig. 2. The ratio windings are wound outside the shield. Their positions are almost independent of each other as the detector winding only responds to the mutual flux, and stray fluxes are suppressed by the superconducting shield. The resulting flux is measured by a SQUID, either by directly inserting it into the toroid or by coupling it to the current comparator by means of a superconducting flux transformer.

For current ratios other than unity, the necessary number of unit windings must be enclosed in the superconducting shield if the design principle in Fig. 1 is followed. As it is not practical to construct such a cable wound comparator for large current ratios, a shield arrangement with an overlapped tube was developed [7] which permits continuous windings and thus significantly reduces the number of series connections. The shield configuration is shown in Fig. 3. It is isolated at the overlap so that it appears like a snake swallowing its tail. Care must be taken to prevent any continuous superconducting connection around the tubular shield, as this would short the windings. As before, a SQUID sensor is used to detect the ampere-turn balance.

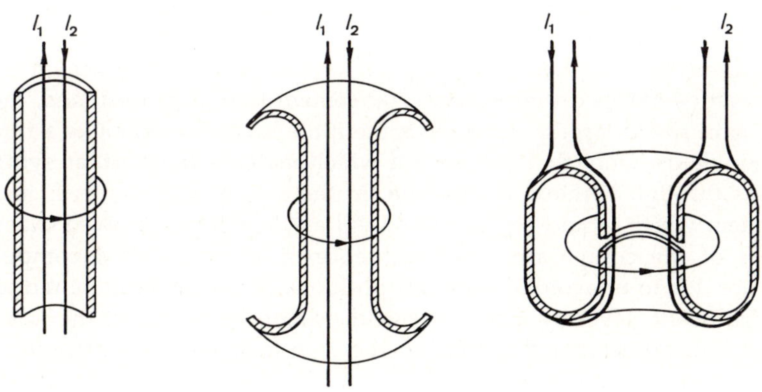

Figure 2
Cryogenic current comparator with a toroidal shield arrangement.

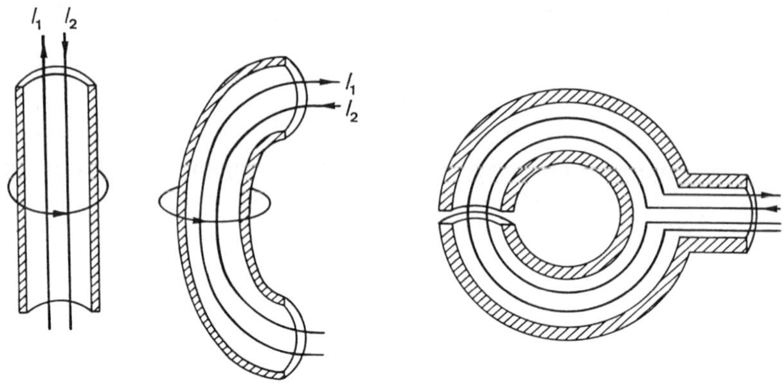

Figure 3
Cryogenic current comparator with an overlapped-tube shield arrangement.

2.3 Ratio Error of a Cryogenic Current Comparator

Ratio errors of a comparator can result from leakage currents between windings, inadequate shielding of the ratio windings or the detector (winding), and trapped fluxes within the shield or in the shield itself. The influence of leakage currents can be greatly reduced by the use of superconducting windings. If necessary, the insulation resistance between the windings can be measured to detect this possible source of error.

Errors resulting from trapped fluxes within the shield or in the shield itself can be eliminated by reducing the ambient field as low as possible before the comparator is cooled down to cryogenic temperatures. In addition, the comparator should be cooled slowly through its transition temperature to prevent flux trapping. The use of a main superconducting shield surrounding the entire comparator further reduces the possibility of external fields coupling to the detector. The effectivness of these shields can be checked by monitoring the detector output while subjecting the comparator to larger external fields.

Respectless, whether a CCC with a toroidal shield arrangement (Fig. 2) or with an overlapped-tube shield arrangement (Fig. 3) is designed, the gap in the main shield largely determines the ratio error. It works like a filter for magnetic fields. Only that component which matches the circular symmetry can pass through without attenuation. In the case of Fig. 2, this is just that component which would be present only if all the currents flowed exactly on the axis of the comparator. The influence of all other field components which are due to unavoidable off-axis positions of currents can be reduced to an insignificant level by a proper shield design. In principle, the following applies: The ratio error depends on the length of the current path of the shield currents as well as the gap width. A decreasing gap width and an increasing gap length result in decreasing errors. A more detailed investigation [8, 9] shows that the error of a CCC is chiefly determined by the

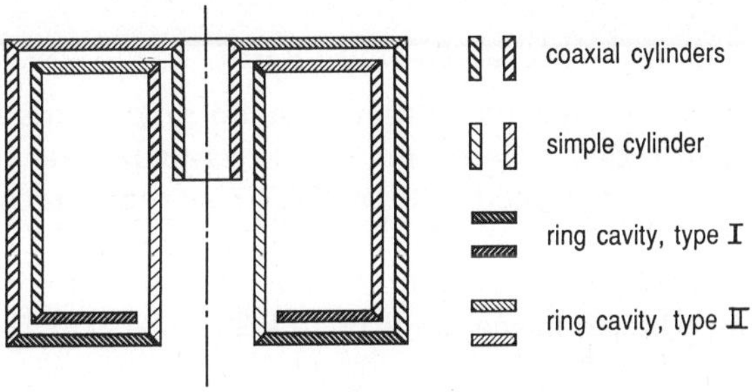

Figure 4
Shield arrangement with coaxial and ring cavities (in addition see table 1).

dipolar component of a transverse magnetic field and that higher order (quadrupolar, hexapolar etc.) components as well as longitudinal magnetic fields are more strongly attenuated. For the various gap geometries shown in Fig. 4, the following relations for the attenuation in terms of the gap dimensions are obtained as given in table 1:

Table 1
Attenuation of radially asymmetric magnetic fields in terms of the gap dimensions
r_i, r_o inner and outer radius for coaxial cylinders and ring cavities of type I
r_o radius of the cylinder for simple cylinders
l length of the gap

Type	Attenuation
Coaxial cylinders	$A_{CC} = \exp\left\{-\left(\frac{2}{1+r_i/r_o}\right)\cdot\frac{l}{r_o}\right\};\ r_o/r_i \le 1.5$
Simple cylinder	$A_{SC} = \exp\left\{-\ 1.84\cdot\frac{l}{r_o}\right\}$
Ring cavity, type I	$A_{RC1} = \left(\frac{r_i}{r_o}\right)^2$
Ring cavity, type II	$A_{RC2} = 1$

A cylindrical or coaxial cavity exponentially attenuates any field not contributing to the net current linkage, and for a ring cavity the attenuation depends on the type of cavity. While for a ring cavity of type I the field is attenuated corresponding to the square of the ratio of the radii, a ring cavity of type II does not contribute to a reduction in the error of a CCC.

The total relative error is proportional to the product of the attenuations as given in table 1:

$$E \sim \Pi A = A_{CC}\cdot A_{SC}\cdot A_{RC1}\cdot A_{RC2}. \tag{2.1}$$

In practical realizations, A_{CC} and A_{SC} play the dominant role, and ratio errors in the order of 1×10^{-12} can be achieved for arbitrary ratios.

2.4 Optimization of the Signal-To-Noise Ratio

The construction of CCC's with extremely small ratio errors needs sensitive detector systems in order to detect the small residual fluxes. Up to now, the best resolution can be obtained with a SQUID system which is matched to the CCC by means of a flux transformer [10]. With an appropriate design and a ferromagnetic core, equivalent noise temperatures below 1 K even at higher source impedances in the range of 10 kΩ can be achieved [11].

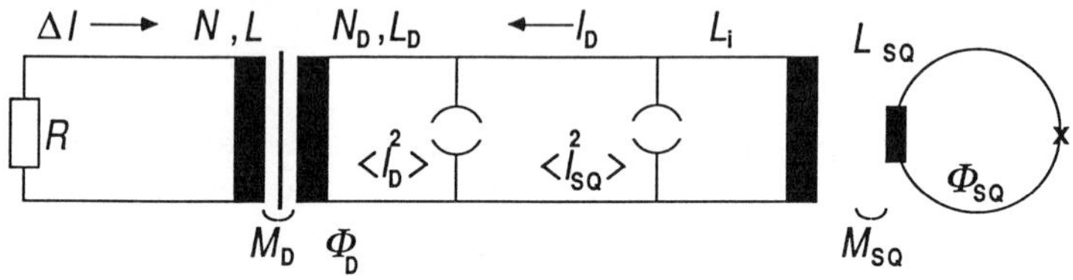

Figure 5
Matching a SQUID instrument to a cryogenic current comparator. R input resistance; I, I_D currents; $<I_D{}^2>$, $<I_{SQ}{}^2>$ noise currents; N, N_D number of turns; L, L_D, L_i, L_{SQ} self inductances; M_D, M_{SQ} mutual inductances; Φ_D, Φ_{SQ} magnetic fluxes (index D: detector winding; index SQ: SQUID).

An ampere-turn difference between the primary and secondary winding $\Delta\Theta = N \cdot \Delta I$ (see Fig. 5) causes a magnetic flux $\Phi_D = \Delta I \cdot M_D$ in the detector winding and due to flux conservation a current $I_D = \Phi_D/(L_D + L_i)$ flows in this winding. This current produces a flux

$$\Phi_{SQ} = I_D \cdot M_{SQ} = \frac{1}{N_D} \cdot \frac{M_{SQ}}{(1 + L_i/L_D)} \cdot \Delta\Theta \tag{2.2}$$

which is detected by the SQUID instrument. In order to match the SQUID as well as possible to the CCC, which means to make the flux Φ_{SQ} as large as possible for a given ampere-turn difference $\Delta\Theta$, N_D should be as small as possible, $N_{Dmin} = 1$, and $L_D \gg L_i$. As $L_D = N_D{}^2 \Lambda_D$ (Λ_D magnetic conductance of the detector winding), Λ_D must be increased (and not N_D!) in order to fulfil the second requirement. This can be done by using a ferromagnetic core.

On the other hand, using a ferromagnetic core causes additional noise, so a compromise has to be found between an optimum signal-to-noise ratio and a best possible matching of the SQUID instrument. The core flux noise $<\Phi_D{}^2>$ generates a noise current $<I_D{}^2>$ in the superconducting loop, and because of flux conservation one obtains

$$<I_D{}^2> = \frac{<\Phi_D{}^2>}{(L_D + L_i)^2}\,. \tag{2.3}$$

The other noise current originates from the SQUID instrument and is given by

$$<I_{SQ}^2> = \frac{2\,W_{SQ}}{L_i}\,, \tag{2.4}$$

where W_{SQ} is the energy resolution of the SQUID instrument. The total noise current

$$<I_S^2> = \frac{2\,W_{SQ}}{L_i} + \frac{<\Phi_D^2>}{(L_D + L_i)^2} \tag{2.5}$$

is transformed to the input by

$$<I_S^2> \cdot (L_D + L_i)^2 = <I^2>\,M_D^2\,, \tag{2.6}$$

and can be represented by an equivalent noise temperature

$$T_{equ} = \frac{R \cdot <I^2>}{4\,kB} = \frac{R}{4\,kB} \cdot \frac{(L_D + L_i)^2}{M_D^2}\left(\frac{2\,W_{SQ}}{L_i} + \frac{<\Phi_D^2>}{(L_D + L_i)^2}\right) \tag{2.7}$$

where R is the input resistance (the white noise of the resistor is neglected), k the Boltzmann constant and B the bandwidth. Using $M_D \approx L_D\,(N/N_D)$, we have

$$T_{equ} = \frac{R}{4\,kB} \cdot \frac{N_D^2}{N^2} \cdot \left[\frac{2\,W_{SQ}}{L_i}\left(1 + \frac{L_i}{L_D}\right)^2 + \frac{<\Phi_D^2>}{L_D^2}\right]. \tag{2.8}$$

As can be seen from Eq. (2.8), the same conditions leading to an optimum matching of the SQUID instrument, namely $N_{Dmin} = 1$ and $L_D \gg L_i$, also minimize the equivalent noise temperature (for more details, see [11]).

3. Resistance Ratio Measurements

Cryogenic current comparators can be used in various applications: as a current detector which is in principle a combination of a SQUID instrument and a flux transformer, and for current as well as resistance ratio measurements, the latter having attained the greatest importance. The first published literature on a CCC [2] already describes its application for resistance ratio measurements.

3.1 Basic Principle

A resistance is defined by the quotient of the potential difference across the terminals of the resistor and the current passing through it. In general, therefore, a resistance ratio is determined by measuring a voltage as well as a current ratio:

$$R_X = U_X/I_X;\; R_N = U_N/I_N; \qquad R_X/R_N = (U_X/U_N) \cdot (I_N/I_X). \tag{3.1}$$

If the same current passes through both resistors, only a voltage ratio has to

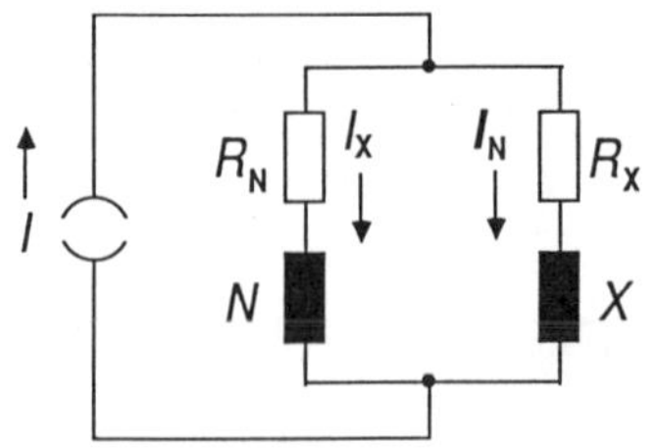

Figure 6
Basic circuit diagram for resistance measurements with a cryogenic current comparator.

be measured, the measuring method being called a potentiometric method. On the other hand, the potential drop across the resistors may be equal and the current ratio is measured by means of a current comparator.

Fig. 6 shows the basic circuit diagram of resistance measurements with a CCC. If the windings of the CCC and the connecting leads are superconducting, the current I splits into the currents I_N and I_X correnssponding to the resistance ratio. On the other hand, the current ratio is equal to the inverse turns ratio, if the CCC is balanced:

$$R_X/R_N = I_N/I_X; \quad I_N/I_X = X/N; \qquad R_X/R_N = X/N. \tag{3.2}$$

For a fixed resistance ratio, the circuit is balanced by supplying a known trim current in an additional winding of the comparator. Alternatively, one resistor could be trimmed to establish a resistance ratio equal to the turns ratio.

3.2 Practical Realizations

There are two different methods of almost equal total uncertainty in practical use: a balance method [12, 13, 14] which uses a conventional null detector in conjunction with a CCC, and a deflection method [15] which makes use only of a CCC. This advantage is offset by the need for a preliminary adjustment and a rather more complicated measuring procedure, a difficulty which, however, can easily be overcome if it is computer controlled.

3.2.1 Deflection Method

Fig. 7 shows the basic circuit diagram of the deflection method with R_X and R_N the resistors to be compared. R_{N1}, R_{N2} and R_{X1}, R_{X2} are symmetrizing resistors which reduce the influence of the lead resistances r_2 and r_3 if they are properly designed. With $r_2/R_{N1} \leq 10^{-5}$ and $R_{X1}/R_{N1} = X/N\ (1+\Delta)$, $\Delta \leq 10^{-4}$, the influence of r_2 on the resistance ratio R_X/R_N is smaller than 10^{-9}. A corresponding condition is valid for R_N, R_X and r_3.

The whole measuring procedure is split into two parts, an auxiliary balance and a main balance. With the auxiliary balance, the switch S_H is

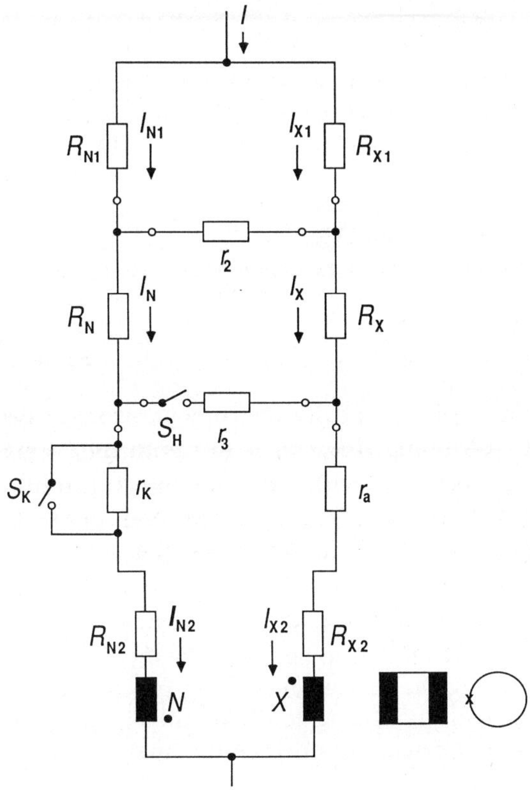

Figure 7
Comparison of two four-terminal resistors with a CCC bridge by means of a deflection method.

closed. At the output of the SQUID instrument two voltages U_1 and U_2 are obtained, according to whether switch S_K is open or closed. The sensitivity of the auxiliary or main bridge is determined by operating this switch. With the main balance, the switch S_H is open and the output voltages U_3 an U_4 are obtained, according to the setting of S_K. When the output voltage U of the SQUID instrument is related to U_o, a voltage which would appear if the differential ampere-turns were equal to the total ampere-turns of one winding, $I \cdot N \cdot X \;/\; (N+X)$, we have

$$\frac{R_X}{R_N} = \frac{X}{N}\left[1 + \frac{\delta_{U2} \cdot \delta_{U3}\left(1 + \delta_{U1} + \delta_{U4}\right) - \delta_{U1} \cdot \delta_{U4}\left(1 + \delta_{U2} + \delta_{U3}\right)}{\left(\delta_{U2} + \delta_{U3} + \delta_{U2} \cdot \delta_{U3}\right) - \left(\delta_{U1} + \delta_{U4} + \delta_{U1} \cdot \delta_{U4}\right)}\right] \tag{3.3}$$

with $\delta_{ui} = U_i / U_o$. If $\delta_{ui}{}^2 \leq 10^{-8}$, the following simplified equation is valid:

$$\frac{R_X}{R_N} = \frac{X}{N}\left[1 + \frac{\delta_{U2} \cdot \delta_{U3} - \delta_{U1} \cdot \delta_{U4}}{\left(\delta_{U2} + \delta_{U3}\right) - \left(\delta_{U1} + \delta_{U4}\right)}\right]. \tag{3.4}$$

The overall relative uncertainty of the method is determined by

o	Linearity and drift of the SQUID instrument	6×10^{-9}
o	Lead resistances of the potential leads (r_2, r_3)	1×10^{-9}
o	Thermoelectric voltages	0.3×10^{-9}
o	Isolation resistance	1×10^{-9}
o	Stability of the components	negligible
o	Ratio error of the CCC	0.1×10^{-9}
o	Scattering of the measurement results	7×10^{-9}

	RSST:	9×10^{-9}

The linearity of the SQUID instrument strongly depends on the electronic circuit. Assuming $\Delta\Phi_0 / \Phi_0 \leq 10^{-4}$ with $\Delta\Phi_0$ = maximum deviation from a linear characteristic within the flux interval from $-\Phi_0$ to $+\Phi_0$ and Φ_0 = magnetic flux quantum ($h/2e \approx 2 \times 10^{-15}$ Vs), the linearity error is $\Delta\Phi_0 / (k_N \cdot \Phi_0)$, where k_N is the number of the flux quanta corresponding to the total ampere turns $N \cdot I_{N2}$. With $\Delta\Phi_0 / \Phi_0 \leq 10^{-4}$ and $k_N \geq 1.5 \times 10^4$, this uncertainty amounts to $\leq 6 \times 10^{-9}$. The drift of the SQUID instrument is almost linear and was found to be $\leq 6 \times 10^{-5}\ \Phi_0/\mathrm{s}$. While a linear drift can be eliminated by using the proper measuring procedure, the non-linear part of the drift, which is about one tenth of the linear part, enters the measurement result and causes an uncertainty smaller than 3×10^{-10} for a measuring time of 10 s.

It has already been mentioned, that the influence of the lead resistances of the potential leads is negligible, if the measuring circuit is properly designed, which means if $r_2 / R_N \leq 10^{-5}$ and $R_{X1} / R_{N1} = X/N \cdot (1+\Delta)$, $\Delta \leq 10^{-4}$.

Thermoelectric voltages act in such a manner that they drive a current $U_{Th} \cdot \Sigma R$, $\Sigma R = R_N + R_{N1} + R_{N2} + R_X + R_{X1} + R_{X2}$ which flows through the windings of the comparator in the opposite directions, causing error ampere-turns. As far as the thermoelectric voltages are constant, their influence can be eliminated by doing two measurements with opposite polarity. This is not true, if the voltage drifts with time. For a typical drift of ≤ 2 nV/min, a measuring time of 10 s and a resistance $\Sigma R = 50\ \mathrm{k\Omega}$, the uncertainty is less than 3×10^{-10}.

If we compare 10 kΩ resistors, an isolation resistor of $10^{12}\ \Omega$ in parallel with one of the 10 kΩ resistors causes a relative error of 10^{-8}. A very critical point is the feedthrough of the cryostat, because besides a high isolation resistance, the feedthrough must also be helium-proof. This problem can be solved by guarding the leads. In this way the parallel resistance of the standards can be increased to more than $10^{13}\ \Omega$, with a negligible influence on the measurement result.

During the measurement, all components must be absolutely stable. In order to fulfil this requirement, only selected components are used, and the measuring circuit must be temperature controlled. The contact resistances of the switches are negligibly small.

The ratio error of the CCC is much smaller than the other error components previously discussed. The scattering of the measurement results depends on the current through the windings and the measuring time: $s = 1.2 \times 10^{-10} / (I/\mathrm{mA}\cdot\sqrt{T/\mathrm{h}})$. With a current of 100 µA and a measuring time of 10 s, the standard deviation amounts to $s = 2.3 \times 10^{-8}$. The main source of the scattering is the noise due to the ferromagnetic core of the CCC. This can be further reduced by increasing the measuring time. With $T = 100$ s we have $s = 7.2 \times 10^{-9}$, a value in the order of the larger systematic uncertainties.

3.2.2 Balance Method

The balance method was introduced by DELAHAYE [12, 13]. Fig. 8 shows the main circuit diagram. The circuit operates on the master/slave principle which has already been used by KUSTERS for a conventional current comparator [1]. The primary current source P supplies a current I_N which flows through the standard resistor R_N and the corresponding winding N of the CCC. The secondary current source S is controlled by a SQUID instrument which is coupled to the CCC by a flux transformer and detects the differential ampere turns of the windings N, X and K. The secondary current I_X flows through the standard resistor R_X and the winding X of the CCC. It is adjusted so that an ampere-turns balance occurs: $I_N \cdot N = I_X \cdot X - I_K \cdot K$. The CCC bridge is balanced by means of the current divider R_T which, together with the compensation resistor R_K, feeds a compensation current I_K into the winding K of the CCC. If the detector D is balanced, the resistance ratio is given by

$$\frac{R_X}{R_N} = \frac{X}{N}\left[1 + \frac{K}{X} \cdot \frac{R_T}{R_T + R_K} \cdot \alpha\right]. \tag{3.5}$$

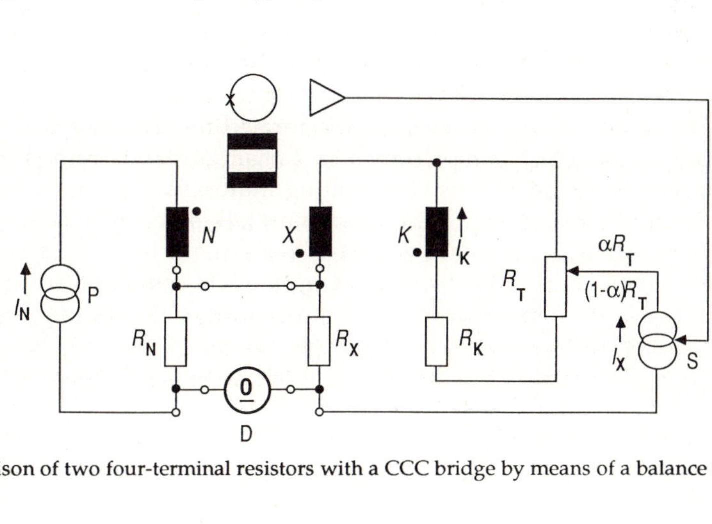

Figure 8
Comparison of two four-terminal resistors with a CCC bridge by means of a balance method.

The range of adjustment can be changed by choosing suitable values for the ratios K/X, $R_T / (R_T + R_K)$ and α. A negative sign in Eq. 3.5 can be obtained by changing the polarity of the compensation winding K. This is necessary for $R_X < R_N$. Due to the automatic tracking of the ampere-turn balance, only one adjustment is necessary.

The measuring uncertainty mainly depends on the following items:

o	Stability and noise of the SQUID instrument	2×10^{-9}
o	Stability and noise of the null detector	1×10^{-9}
o	Leakage current between the primary and the secondary current source	5×10^{-9}
o	Circuit for the bridge balance adjustment	6×10^{-9}

	RSST:	8×10^{-9}

The drift of the SQUID instrument amounts to $5 \times 10^{-6}\ \Phi_0/\text{min}$, the output noise is $8 \times 10^{-5}\ \Phi_0/\sqrt{\text{Hz}}$. For comparison, $8 \times 10^{-4}\ \Phi_0$ corresponds to a relative uncertainty of 1×10^{-8} measuring 1 Ω or 100 Ω resistances. The output noise of the null detector ($0.6\ \text{nV}/\sqrt{\text{Hz}}$) is 1.2×10^{-8} times the voltage drop across a 1 Ω resistor measured with a current of 50 mA. The leakage current between the primary and secondary current source causes a voltage drop along the potential leads of the resistors and therewith an imbalance of the bridge. Assuming a resistance of the potential leads of 1 Ω, the measured leakage current of 1 nA causes an error of 2×10^{-8}. Limited resolution and imperfect calibration as well as linearity of the circuit for the adjustment of the bridge produce a further uncertainty. If all these sources of error are minimized, 1 Ω resistors can be compared with a relative uncertainty smaller than 1×10^{-8}.

3.3 Measurement of Quantized Hall Resistances and Establishment of a Resistance Scale

The calibration of resistors is normally performed by reducing their values to a resistance scale which is represented by a chain of standard resistors with nominal values in decadic steps representing multiples and submultiples of the unit. Since the discovery of the quantum Hall effect by K. v. KLITZING [16], the measurement of quantized Hall resistances has made so much progress that the quantum Hall effect has meanwhile come to form the basis of the most accurate reproduction and maintenance of the unit of resistance ohm. In October 1988, a numerical value for the von Klitzing constant was fixed, $R_{K-90} = 25\ 812.807\ \Omega$. This will be internationally introduced on the first of January 1990 and will lead to worldwide uniformity in resistance calibration from that date [17]. The von Klitzing constant R_K is identical with the quantized Hall resistance on the first plateau, R_{K-90} is the best known value for R_K at this time, and it is believed to represent the SI-value with a relative uncertainty of 2×10^{-7} ($1\ \sigma$). The index "90" denotes the date of introduction.

The establishment of a resistance scale and the measurement of quantized Hall resistances require very accurate decadic and non-decadic ratio devices. Besides conventional devices like Hamon resistors, room-temperature current comparators and bridges, cryogenic methods are gaining increasingly in importance, above all Josephson potentiometry [18] and cryogenic current comparator bridges. These bridges can be operated in a resistance range from $1\ \Omega$ to $12.9\ k\Omega$, and allow a resistance scale to be established with a total relative uncertainty of 1.5×10^{-8} [19]. Besides this, this measuring method is best suited to monitor the drift of standard resistors with time.

4. Future Developments

Cryogenic current comparators are cryoelectronic devices which allow very accurate current ratios to be realized. Resistance ratios can be determined, using these current ratios as a basis.

Further improvements will be possible by

- o replacing conventional null detectors with SQUID instruments, which have a higher sensitivity and smaller noise
- o increasing the resolution of the cryogenic current comparators using an optimized design
- o improving the measuring circuit: increase of the isolation resistance, decrease of thermoelectric emf's, etc.

Furthermore, cryogenic resistors have already been used with great success [20]. These resistors are as stable as the best room-temperature resistors if they are stored at helium temperatures, and they exhibit a white noise which is smaller by a factor of eight than that of room-temperature resistors.

Since the invention of high T_c superconductors, laboratories all over the world have been working to develop these materials in order to obtain even higher transition temperatures and to make them available for technical applications. Even though it is too early to consider these materials for practical use, it is worth-while contemplating their application to cryogenic current comparators. In the first place, the reduced amount of cooling which allows CCC's to be used in a wide range of applications must be mentioned. On the other hand, the higher temperatures cause four times as much noise in resistors as there is at helium temperatures. Up to now, no investigations into the noise mechanisms of SQUID's based on high T_c superconductors have been carried out, even if there are some papers describing first attempts with these materials. For technical applications, the $1/f$ noise of the electronic circuitry dominates, and this will not be influenced when the SQUID is operated at liquid nitrogen temperatures instead of liquid helium temperatures. The characteristics of the core material will improve with higher temperatures: while the permeability increases (μ_r = 3 000 at 4.2 K, μ_r = 500 000 at 300 K) the core noise decreases with decreasing hysteresis losses and magneto-striction effects.

Besides its application in a resistance bridge, a CCC can also be used as a current standard [21]. In this system a persistent supercurrent is used as a stable reference and a current comparator compares an external current with the persistent one. A SQUID senses the net flux difference between them and allows the external current to be servoed to be equal and opposite to the stable supercurrent.

References

[1] N.L. Kusters, IEEE Trans. Instr. Meas. **IM-13** (1964) pp. 197-209

[2] J.K. Harvey, Rev. Sci. Instr. **43** (1972) pp. 1626-1629

[3] K. Grohmann, H.D. Hahlbohm, H. Lübbig, H. Ramin, PTB-Mitt. **83** (1973) pp. 313-318

[4] D.B. Sullivan, R.F. Dziuba, Rev. Sci. Instr. **45** (1974) pp. 517-519

[5] J.E. Zimmerman, P. Thiene, J.T. Harding, J. Appl. Phys. **41** (1970) pp. 1572-1580

[6] K. Grohmann, H.D. Hahlbohm,H. Lübbig, H. Ramin, IEEE Trans. Instr. Meas. **IM-23** (1974) pp. 261-263

[7] R.F. Dziuba, D.B. Sullivan, IEEE Trans. Mag. **MAG-11** (1975) pp. 716-719

[8] K. Grohmann, H.D. Hahlbohm, D. Hechtfischer, H. Lübbig, Cryogenics **16** (1976) pp. 423-429 and pp. 601-605

[9] K. Grohmann, D. Hechtfischer, Cryogenics **17** (1977) pp. 579-581

[10] J. Clarke, E. William, D. Woody, Journ. Appl. Phys. **42** (1971) pp. 3859-3865

[11] P. Gutmann, V. Kose, IEEE Trans. Instr. Meas. **IM-36** (1987) pp. 267-270

[12] F. Delahaye, IEEE Trans. Instr. Meas. **IM-27** (1978) pp. 426-429

[13] F. Delahaye, D. Reymann, IEEE Trans. Instr. Meas. IM-34 (1985) pp. 316-319

[14] A. Hartland, NBS Spec. Publ. **617** (1984) pp. 543-548

[15] P. Gutmann, to be published

[16] K. von Klitzing, G. Dorda, M. Pepper, Phys. Rev. Lett. **45** (1980) pp. 494-497

[17] T.J. Quinn, Metrologia **26** (1989) pp. 69-74

[18] P. Warnecke, J. Niemeyer, F. Dünschede, L. Grimm, G. Weimann, W. Schlapp, IEEE Trans. Instr. Meas. **IM-36** (1987) pp. 249-251

[19] F. Delahaye, A. Satrapinsky, T. Witt, IEEE Trans. Instr. Meas. **IM-38** (1989) pp. 256-259

[20] P. Warnecke, E. Braun, IEEE Trans. Instr. Meas. **IM-38** (1989) pp. 293-295

[21] J.C. Gallop, SQUID, Berlin, New York: Walter de Gruyter (1977) pp. 267-276

Fast SQUID Pseudo Random Generators

M. Albrecht and W. Kessel
Physikalisch-Technische Bundesanstalt, D-3300 Braunschweig

1. Introduction

Digital and counting techniques provide very comprehensive measuring methods. Even in the field of noise measurements which seems to be the domain of analogous signals digital methods become more interesting which try to replace the conventional methods both in generating and detecting noise signals. The generation of random signals by fast digital methods is the subject which will be considered here. In the field of noise measurement, the generation of well defined noise power within the rf and hf range has become an important task. A number of applications in engineering, such as long-distance communication, navigation, localization, remote sensing and others need random electromagnetic wave generators in this frequency range. Even in meteorological and medical investigations and not forgetting radioastronomical observations, the measurement of noise spectra provides new ways of gathering information about systems. Almost all of these applications rely on the fact that electromagnetic noise waves have a power spectrum which is widely distributed in frequency. Since noise measurement is constituted by the intercomparison of two noise powers, noise sources of high accuracy are needed, the power of which can be exactly calculated. The convenient method adapted in the RF frequency range is to generate thermal noise in a dissipative element. Here, the noise power density is given via Nyquist's theorem by the temperature. However, these thermal systems suffer from the fact that on account of thermal effects, particulary the heating, they are not reliable over long time periods. As will be shown in this chapter, pseudo random generators based on fast superconducting switches present a new solution to this problem of generating noise power spectra. They allow the output noise power to be defined in terms of the measurements of a voltage, a resistance and a frequency. The work described was sponsored by the "Stiftung Volkswagenwerk".

2. Principles of the Generation of Digital Random Noise

To generate random noise signals by digital techniques, various methods have been proposed [1]. They all use a generator for random waveforms which creates the desired random signal by precise processes and they are based on random telegraph signals. This waveform was described in great detail in the early literature on random fluctuations [2].

2.1 The Continuous Process of Generation

A typical random telegraph wave $u(t)$ is shown in Fig. 1. At arbitrary points in time the signal value switches from $+U$ to $-U$ or vice versa and stays at that value for the next time interval. If the switching occurs without any correlation or memory the probability that there will be k transitions within an arbitrary time interval from t_1 to t_2 is given by Poisson's distribution formula

$$P(k, t_2 - t_1) = \frac{[\rho(t_2 - t_1)]^k}{k!} \exp[-\rho(t_2 - t_1)] \tag{2.1}$$

Here

$$\rho = \frac{\langle k \rangle}{t_2 - t_1} = \frac{1}{t_2 - t_1} \sum_{k=0}^{k} k P(k, t_2 - t_1) \tag{2.2}$$

is the mean number of transitions per time interval. The brackets <..> mean the average over the probability distribution specified. The signal value at time t_2 is simply related to the signal value at time t_1 by the number k of transitions which have occurred within the time interval from t_1 to t_2:

$$u(t_2) = u(t_1)(-1)^k \tag{2.3}$$

This equation together with the probability distribution (2.1) gives the autocorrelation function of the random telegraph signal

$$C(t_2 - t_1) = \langle u(t_2) u(t_1) \rangle = U^2 \exp(-2\rho |t_2 - t_1|) \tag{2.4}$$

For noise source application the accuracy of its spectral intensity is the most important quantity to know. It is derived from the autocorrelation function via the Wiener Chintchine theorem:

$$G(\omega) = 4 \int_0^\infty C(t) \cos \omega t \, dt \qquad C(t) = \frac{1}{2\pi} \int_0^\infty G(\omega) \cos \omega t \, d\omega \tag{2.5}$$

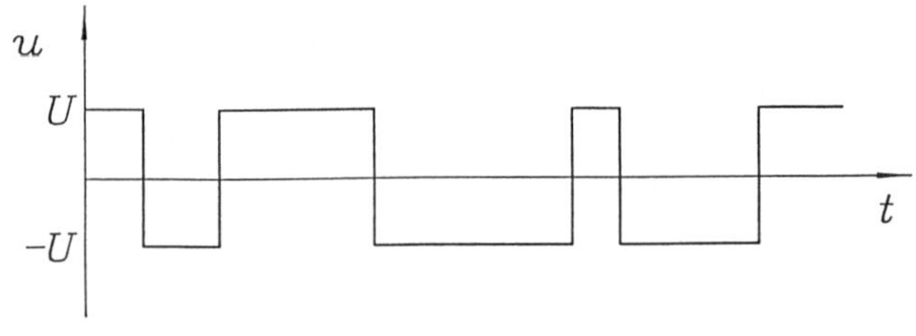

Figure 1
Random telegraph wave of voltage u as a function of time generated by a continuous statistical process.

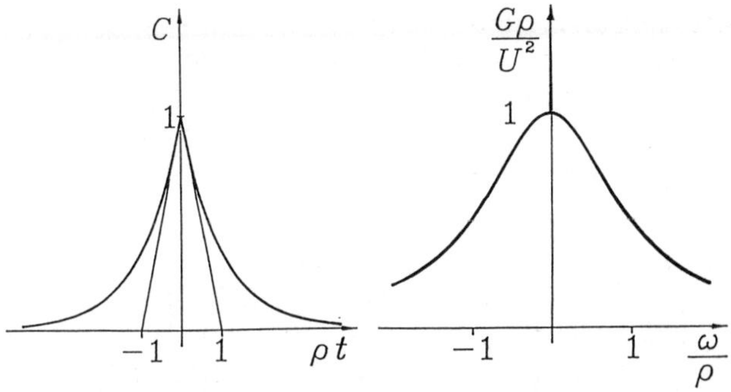

Figure 2
Autocorrelation function C (left) and spectral density G (right) for the continuous process of generation C is plotted as a function of the product of the mean number of transitions per time and the ellapsed time t. G normalized to its maximum value U^2/ρ with U equal to the value of the switched voltage is plotted as a function of the quotient of the angular frequency and the mean number of transitions per time.

which states that the spectral density function is the Fourier transformed function of the autocorrelation function. It should be noted that both functions are symmetrical with respect to their arguments. In the case of random telegraph signals described above, this results in a simple Lorentzian shaped frequency dependence of the spectral density

$$G(\omega) = \frac{8\rho U^2}{(2\rho)^2 + \omega^2} \tag{2.6}$$

which is schematically shown together with the corresponding correlation function in Fig. 2.

A noise generator constructed on the basis of this principle of generation and working at frequencies below 100 kHz has been presented by WONHAM and FULLER [3]. They used a radioactive source and a Geiger-Müller counter to obtain the Poisson distributed transition points. Because of the difficulty in invoking the necessary Poisson process by simple electronic methods, it only has a small field of application and will not be described in more detail.

2.2 The Discrete Process of Generation

For application in modern digital circuits, the discrete process of generating random telegraph waveforms is more suitable. It is easily related to the continuous process described above. Its difference consists in the fact that the transition from $+U$ to $-U$ does not occur at random time points but can happen only on the arrival of clock pulses at fixed intervals T. The switching itself is still random (cf. Fig. 3). This process is described by a binomial distribution

$$P(n,k) = \binom{n}{k} p^k (1-p)^{n-k} \tag{2.7}$$

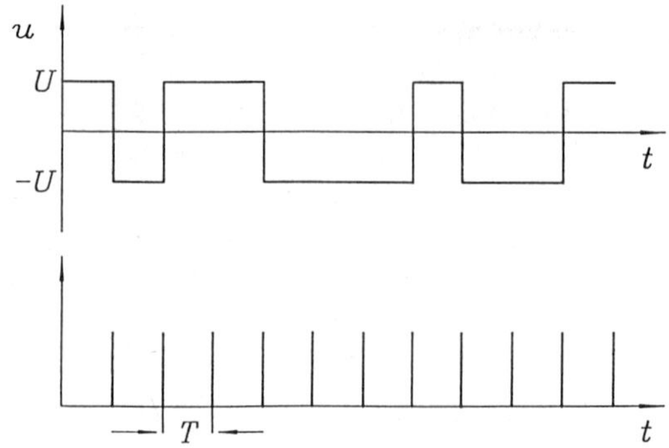

Figure 3
Random telegraph wave of voltage u as a function of time t (above) generated by a discrete statistical process induced by clock pulses (below) at fixed time intervals T.

It gives the probability for the event that within the time interval of n clock pulses, k transitions have taken place. Here p is the probability that at the edge of a clock pulse switching will occur. This probability normally has the value of 1/2, which means that it is equally probable that switching will or will not occur. On the same lines as in the case of continuous generation process, the autocorrelation function of the discrete random telegraph signal is derived. It takes the simple triangular form:

$$C(t_1\text{-}t_2) = \begin{cases} U^2\left(1-\dfrac{|t_2-t_1|}{T}\right) & |t_2-t_1| < T \\ 0 & |t_2-t_1| > T \end{cases} \tag{2.8}$$

Its graph is shown in Fig. 4. The corresponding spectral density is given by the formula (see also Fig. 4)

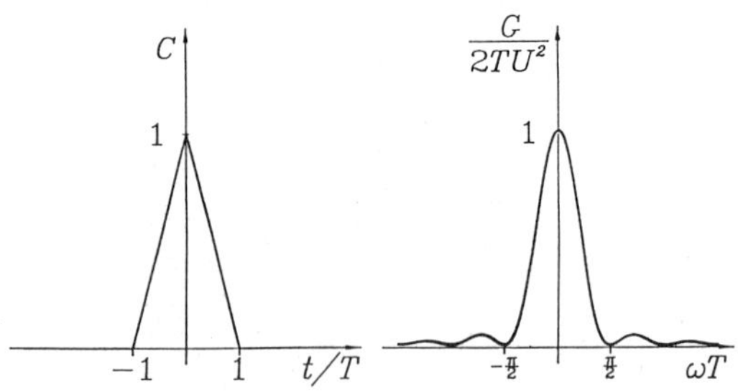

Figure 4
Autocorrelation function C (left) and spectral density G (right) for the discrete process of generation. C is plotted as a function of the time t normalized to the repetition time T of the clock pp.ulses. G normalized to its maximum value $2TU^2$ with U equal to the value of the switched voltage is plotted as a function of the product of the angular frequency and the repetition time T.

$$G(\omega) = 2TU^2\left(\frac{\sin\frac{1}{2}\,\omega T}{\frac{1}{2}\,\omega T}\right) \tag{2.9}$$

which shows that for frequencies small compared with the repetition frequency $\omega_0 = 2\pi/T$ of the clock pulses the noise spectrum will be almost white, with the density $2TU^2$. In this expression U is the switched value of the random wave $u(t)$ which can normally be measured with great precision. T can also be determined very accurately, since it is the period of the regular pulse train. The remaining feature open to question is the method of realizing the random choice for the transitions at the time points set by the clock pulses. In the case of low frequency, SUTCLIFFE and KNOTT [1] developed a method which uses a wideband RF noise source together with a bistable circuit to generate the switching process with the probability p = 1/2. The bistable circuit is switched to its 'high-state' whenever the output signal of the RF noise source exceeds a fixed value or to its 'low-state' if the signal remains below this fixed value. The state signal of the bistable circuit is fed to a pulse-forming device which gives the desired random waveform. In order to get good switching conditions, the lower limit of the frequency range of the RF noise source must be a factor of ten to hundred higher than the highest frequency needed in the spectral density function of the random wave. For noise sources designed to operate at radio frequencies this method is no longer applicable because a wideband UHF noise source is needed to generate the statistics.

2.3 Pseudo-Random Noise

Stimulated by the development of random number generators in digital computers, a method for generating the transition probabilities p in Eq. (2.7) has been used which produces pseudo-random binary signals as 'm-sequences' or sequences of maximum length [4]. The basis for their generation process is a shift register of N stages which can hold the information of binary states as shown in Fig. 5. The information contained

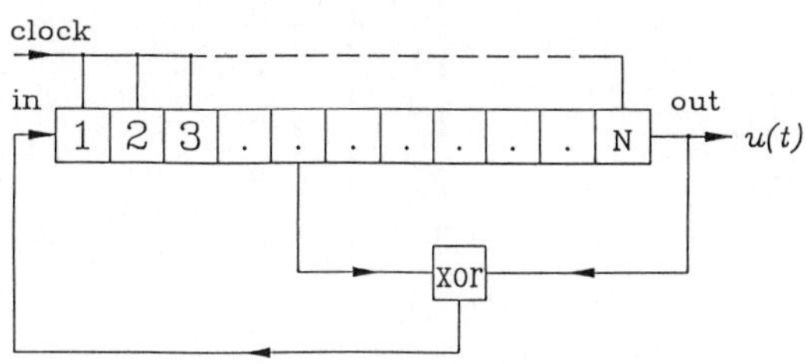

Figure 5
Schematical diagram of a shift register with 'xor' feedback for generation of pseudo random pulse trains. The feedback signal derived from the signal $u(t)$ at the output and the state of an intermediate storage cell is supplied to the input and shifted through the register cell by cell at every clock pulse.

in the cells of the register is shifted by the edge of the clock pulses simultaneously from the left to the right, cell by cell. In this process the binary state of the last cell on the right is given as an output, whereas the binary state at the input is taken into the first cell on the left. At every instant the output waveform $u(t)$ reflects the information which was contained in cell i at N-i clock cycles before. The pulse train at the output therefore gives the binary state pattern which was fed into the input at the left, but with a delay in time of N clock cycles. A random appearance of the output signal is attained by a feedback loop which generates a random pulse train at the input. This is achieved by a digital circuit which at every moment derives the new input signal from the actual state of the shift register. The usual way of doing this is to combine the signal at the output and the information of an intermediate cell via an 'exclusive-or' gate which gives a binary one if one and only one of its inputs is set to binary zero.

The possible states of a shift register with this feedback consist of all distributions of binary ones and zeros in its cells. If the register has the length of N cells, there are 2^N different states. The pulse sequences of maximum length mentioned above as 'm-sequences' are generated if all possible states of the shift register are run through with the exception of the zero state (which means that all cells contain a binary zero). This state must be excluded because it is self-reproducing. 'm-sequences' will be obtained by a proper choice of the register length and the intermediate cell which is used for the feedback. The mathematical theory for the generation of 'm-sequences' has been given by ELSPAS [5] together with detailed tables. Cases of special interest are reproduced in table 1 which shows the total number N of cells in the shift register and the index n of the intermediate cell to be used in an 'exclusive-or' feedback. Only those combinations are given which produce distributions of transitions with good uniformity. A more detailed explanation of this term is given later in this section.

Table 1
Some ideal combinations of total number N of cells in a shift register and the index n of the intermediate cell used for the feedback loop.

N	n
15	7
17	5
23	9
31	6

The determination of the autocorrelation function now follows the same lines as for the discrete process of generation in Sect. 2.2. However, two differences must be taken into account. First, it should be noted that the pulse train at the output has only a pseudo-random character due to the fact that the output signal certainly arises from a large but limited number of states of the shift register. This is reflected in an overall periodicity of the output waveform with the period equal to the time which is necessary to go through all possible states once

$$T_L = (2^N - 1)\, T \tag{2.10}$$

This means that the system does not exhibit random behaviour at long time distances. However for time distances smaller than the periode T_L the pulse train is random in character. The process is therefore uniform if the mean number of transitions per time derived from time intervals which are short compared with the periodicity time T_L is a constant parameter of the process and does not depend on time.

The second deviation from the discrete process results from the fact, that because of the self-reproducing property of the zero-state, the averages over the different pulse series at the output have to be taken without this zero-state. Consequently, the autocorrelation function derived from Eq. (2.7) is a periodic function with period T_L. It is determined within a periodicity cell by

$$C(t) = U^2 \times \begin{cases} 1 - \dfrac{|t|}{T} - \dfrac{1}{T_L} & |t| < T \\ -\dfrac{1}{T_L} & T < |t| < T_L \end{cases} \tag{2.11}$$

Values outside are obtained by analytical continuation. The Fourier transformation involved in the calculation of the spectral density produces discrete lines with the line spacing given by

$$\Delta\omega = \frac{2\pi}{T_L} \tag{2.12}$$

The envelope of their spectral amplitude is the spectral density function of Eq. (2.9).

2.4 Synthetic Noise Sources for RF Frequencies

Noise sources which operate according to the principles described in the last section present a nice possibility of defining random signals on electrical magnitudes only. The background for this is Eq. (2.9) which determines the spectral density of the noise power emitted. This density can be exactly calculated if the parameters T and U are known. The first one is the repetition time T of the clock pulses. It is the reciprocal of the clock frequency

$$f = \frac{1}{T} \tag{2.13}$$

which can be measured with high precision. The second parameter is the value of the voltage U involved in the switching process. It can be also determined with high precision.

The fact that the signal generated by the switching process has a random distribution over the two values $+U$ and $-U$ only is no real hindrance. In all measurements, especially in the high frequency regime which we have in mind, low-pass filtering by the signal lines and the measuring instruments takes place. This cuts off the high frequency tails of the spectral density and a normal distribution of the random signal values is produced. This statement is justified considering the response of the low-pass filter in the time domain when the input is wide-band in the frequency domain. The filter output at any instant may be regarded as the sum of the input values before that instant, weighted with the filter response function which extends to a time interval with the length of the reciprocal cut-off frequency. According to the central limit theorem summing the random pulse trains in this way results in a normal distribution of the voltage values. This qualitative argumentation has been elaborated in detail by WONHAM and FULLER [3] for their simple RC low-pass filter.

To produce switching circuits for the generation of random waves two techniques come under consideration: conventional semiconducting devices and circuits containing superconducting Josephson elements. If operation frequencies up to 10 GHz are desired the necessary repetition time for switching lies below 100 ps and the switching time itself must be smaller than 10 ps. Whereas for superconducting elements switching times below 5 ps have been reported in the literature semiconducting devices just start to step into the field of fast switching below 50 ps.

What also has to be considered is the power density level which is generated by these two techniques. The noise power density is the energy emitted per second by the noise source in a frequency interval of 1 Hz width. It is derived by dividing the spectral density function $G(\omega)$ in Eq. (2.5) by the value of the resistance R of operation.

$$S(\omega) = \frac{G(\omega)}{R} \tag{2.14}$$

From this, the noise temperature T_R of the source can be calculated which is the temperature of a thermal noise source of equal noise power density. The noise temperature of the pseudo-random generator of Sect. 2.3 becomes frequency dependent with a frequency relation analogous to Eq. (2.9) (see also Fig. 4). Its maximum value is given by

$$T_R = \frac{2U^2}{k_B R f} \tag{2.15}$$

where k_B is Boltzmann's constant ($k_B = 1{,}38 \times 10^{-23}\,\mathrm{JK^{-1}}$).

For semiconductors, typical values of the switching voltage U lie in the range of the threshold voltage of the semiconducting elements and are about 0.7 V for Si-transistors or diodes, for example. With the resistance values of 50 Ω which can be realized in semiconducting circuits and which correspond to the impedance of coaxial lines, the typical noise temperature becomes $T_R = 1.42 \times 10^{11}$ K. In the case of superconducting Josephson elements, on the other hand, the switching voltage values are of the order of the gap voltage which is about 2 mV for PbInAu-superconductors, and the resistance takes values of about 5 Ω. From these values a noise temperature of $T_R = 1{,}16 \times 10^6$ K results. Because typical noise temperatures in technical applications lie in the range of 2.000 to 20.000 K superconducting switching elements are appropriate devices and will be considered in greater detail.

3. Superconducting Shift Registers for the Generation of Random Noise

To realize fast shift registers with clock rates up to 10 GHz high speed switches are needed. Josephson elements are suitable for this purpose because they are well understood and give excellent switching performance. Storage cells using Josephson elements are SQUIDs, which are interconnected to arrays of larger storage area.

3.1 Dynamics of a Single Josephson Element

We will start with the features of the single Josephson element, the current-voltage characteristic of which is shown in Fig. 6 for low damping (see later in this section). Through an ideal junction in the superconducting state a current I, which is lower than the critical current I_0, can flow without any voltage drop across the element ('S-state' as mentioned in ref. [6]). If the current I increases above the threshold value I_0, a voltage $u(t)$ appears across

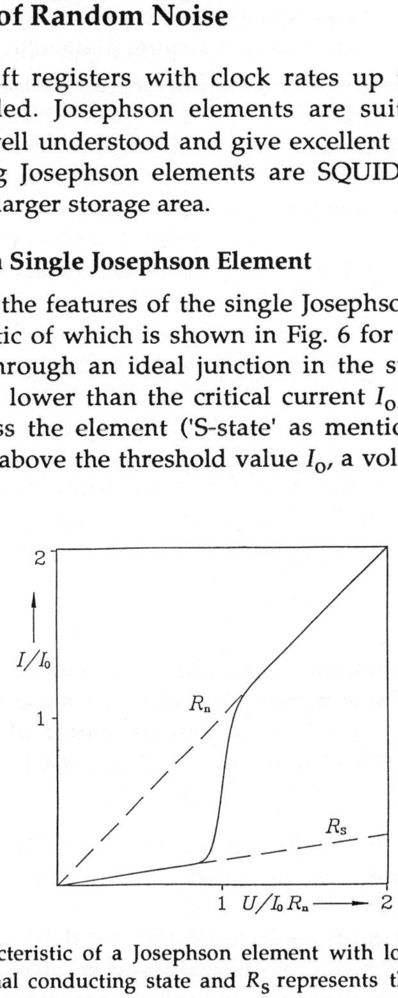

Figure 6
Current-voltage characteristic of a Josephson element with low damping. R_n shows the resistance in the normal conducting state and R_s represents the quasi-particle resistance which shows up in the subgap regime. Also shown is the supercurrent I in the 'S state' (bold line at $U = 0$) which can exist for currents lower than the critical current I_0.

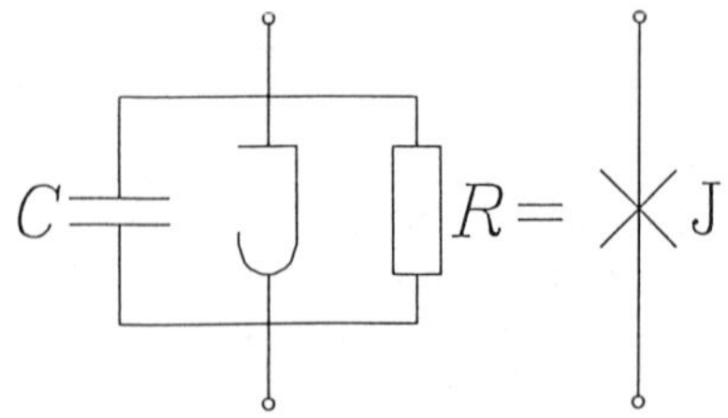

Figure 7
Quasi classical circuit diagram of the RSJ model. *R* represents the resistance in the 'R state' and *C* the capacitance of the Josephson element.

the Josephson element - the element switches to the 'R-state'. In this state the current through the Josephson element contains an ac component with the frequency $f = 2eu/h$. Here e stands for the elementary charge and h for Planck's constant.

The dynamics of a single small Josephson element has been described in a form suitable for almost all purposes by McCUMBER [7], STEWART [8] and JOHNSON [9]. Their calculations are based on a model conveniently called the <u>R</u>esistively <u>S</u>hunted <u>J</u>unction model (RSJ model). This model can be understood from the quasi classical circuit diagram with three circuit elements shown in Fig. 7. The non-classical Josephson behaviour is represented by the symbol J, which stands for the Josephson equations [10]

$$I = I_o \sin\varphi \tag{3.1}$$

$$\frac{d\varphi}{dt} = \frac{\hbar u}{2e} \tag{3.2}$$

They relate the voltage drop $u(t)$ across the element and the current I to the difference of the macroscopic phase difference φ across the contact. The Josephson junction is shunted by the resistance R which describes the damping within the element, and the Josephson contact is connected in parallel to the capacitor C, which reflects the effects of the electrical charges in the contact area. The equation resulting from summing the currents through these circuit elements and equating them to the supply current of an external source has the structure

$$\hat{D}\varphi = c_2 \frac{d^2\varphi}{d\tau^2} + c_1 \frac{d\varphi}{d\tau} + c_o \sin\varphi = \alpha \tag{3.3}$$

where the coefficients c_i (i = 0...2) and the normalized time τ depend on the normalizations and time scales used. Depending on the particular application they are chosen to facilitate the development of suitable approximations. The notations which are used in the literature are summarized in table 2. This gives the values of the three coefficients c_2, c_1 and c_o of Eq. (3.3) and the normalized current α together with their convenient names in terms of the parameters of the model.

Table 2
Parameters of the various normalizations of Eq. (3.3) in the RSJ-model which are used in the literature. The different formulations are developed for different cases of interest: McCumber's notation is suitable for high damping, Stewart's for the treatment of oscillations and Johnson's in the case of low damping. The constant of time τ_o used in all three notations is the time constant of the *RC* part in the Josephson circuit.

	Notation of		
parameter	Mc CUMBER [7]	STEWART [8]	JOHNSON [9]
c_2	$\beta_o = \omega_m \tau_o$	1	1
c_1	1	$1/\tau_o$	$\beta_J = 1/(\omega_J \tau_o)$
c_o	1	ω_o^2	1
α	I/I_o	I/I_o	I/I_o
τ	$\omega_m t$	t	$\omega_J t$
characteristic frequency	$\omega_m = 2eRI_o/\hbar$	$\omega_o^2 = 2eI_o/\hbar C$	$\omega_J^2 = 2eI_o/\hbar C$

The three columns show the notations for the three cases of interest as stated in the description of table 2: Josephson elements with high damping, Josephson elements which show strong oscillations and Josephson elements with low damping.

Even though the Johnson formulation is the normalization suitable for low damping it will be used in the further dicussion although quite different damping factors have to be examined. Because the numerical calculations are accomplished without any approximation, details of the normalization are not important and the choice can be made according to practical convenience.

When the nonlinearities of the resistance of the Josephson element are taken into consideration, the discription can be extended to the so called RSJN model, where N characterizes the replacement of the linear resistance relation by a nonlinear one. In the calculations the damping parameter β is replaced by the expression:

$$\beta = \frac{1}{R_n}\left[\frac{R_n}{R_s} + \left(1 - \frac{R_n}{R_s}\right)\frac{\left(\frac{U}{U_g}\right)^n}{1 + \left(\frac{U}{U_g}\right)^n}\right] \qquad (3.4)$$

Here, R_N is the resistance in 'R-state', R_S the resistance in the 'S-state' and U_g the gap voltage. The result of simulations made on this basis for a Josephson element with the Johnson damping factor $\beta_J = 1$ is shown in Fig. 8. At the point X the Josephson element is switched to the 'R-state' by a very small increment in the external current I which raises this current to a value just above the critical current I_o. After the turn on time this transition

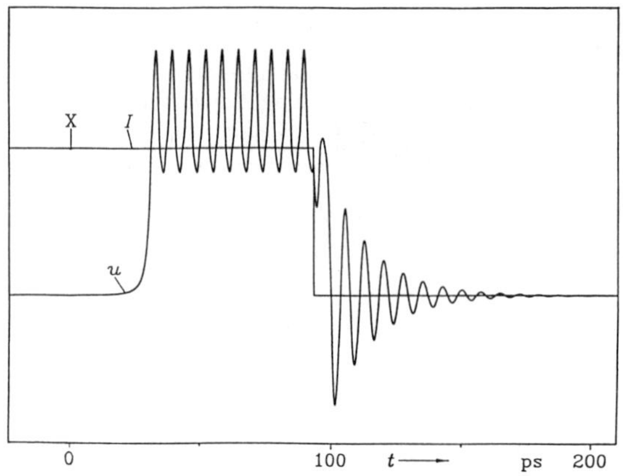

Figure 8
Switching behaviour of a Josephson element based on the RSJN model. I is the current through the element, u the voltage across the element. At the point X the current is raised just above the critical current I_0 (from the value 0.999 I_0 to the value 1.001 I_0). After a turn on time of about 30 ps the transition to the 'R-state' occurs and Josephson oscillations appear. The switching off at about 95 ps causes damped plasma oscillations. The simulations were done with a value of 1 for the Johnson damping parameter β_J.

to the 'R-state' can be detected by the voltage which rises sharply and shows very pronounced Josephson oscillations. When the external current I is switched off the voltage breaks down showing the phenomena of damped plasma oscillations. This damping is determined by the leakage current of the nonlinear resistance relation in the subgap region.

The influence of nonlinearity has recently been checked in detail by BRUNK [11], LÜBBIG [12] and JUTZI [13]. They show independently that for most cases of practical application with elements which are highly damped by shunt resistors nonlinearities can be neglected. They give significant effects only near the gap voltage which is not important in the case of the switching applications considered here.

3.2 Switching Performance of a SQUID

As has been mentioned, a cell convenient for the storage of information in Josephson techniques is the SQUID, the circuit diagram of which is shown in Fig. 9. It consists of two Josephson elements J_1 and J_2 linked together in a loop with an inductance L, which is split in the two parts L_1 and L_2

$$L = L_1 + L_2 \tag{3.5}$$

in order to feed in the so called gate current I_g. The magnetic flux stored in this inductive loop is controlled by the current I_c. This can be done by inductive coupling as suggested in Fig. 9 or by direct galvanic coupling. Because of the quantization of magnetic flux in multiple connected

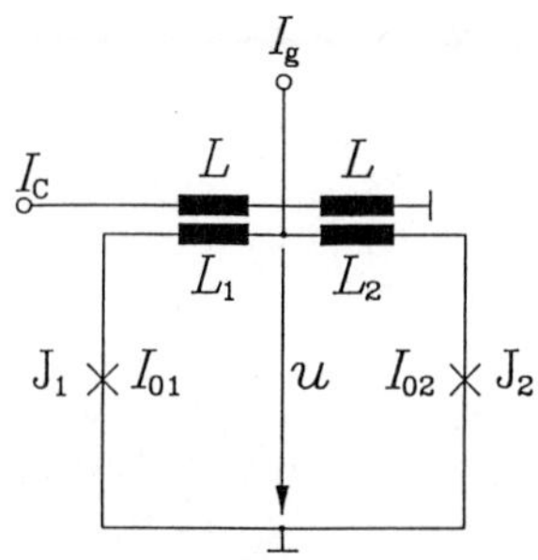

Figure 9
Circuit diagram of the SQUID model. I_g is the gate current, I_c the control current. I_{01} and I_{02} are the critical currents of the Josephson elements J_1 and J_2 which are coupled by the loop inductances L_1 and L_2. u represents the voltage across the SQUID.

superconducting systems it is possible to store information via these flux quanta.

The static characteristic gives the maximum gate current I_g which can be fed into the system at a prefixed value of the control current without destroying the 'S-state' of any of the Josephson elements. The mathematical procedure to obtain the characteristic consists in determining the extremal value of the gate current which flows through the Josephson elements in the 'S-state'

$$I_g = I_{01} \sin\varphi_1 + I_{02} \sin\varphi_2 \tag{3.6}$$

with respect to variations in the phase differences φ_1 and φ_2 in the single elements. These variations are to be submitted to the secondary condition

$$\Phi = \Phi_{ext} + \frac{1}{2\pi}\left(\lambda_1 \sin \varphi_1 - \lambda_2 \sin \varphi_2\right) \tag{3.7}$$

which results from the inductive coupling of the two Josephson elements in the storage loop. Here, Φ_{ext} represents the magnetic flux externally coupled into the loop and the λ_i are related to the inductances L_i by the equation

$$\lambda_i = \frac{L_i I_{oi}}{\Phi_o} \qquad (i = 1,2) \tag{3.8}$$

The process of variation finally leads to the equation

$$I_{01}\cos \varphi_1 + I_{02}\cos \varphi_2 + \left(\lambda_2 I_{01} + \lambda_1 I_{02}\right) \cos \varphi_1 \cos \varphi_2 = 0 \tag{3.9}$$

which relates the phase differences of the two Josephson elements at the transition point of the SQUID. The characteristic of a system with equal critical currents of the single Josephson elements $I_{01} = I_{02}$ and an asym-

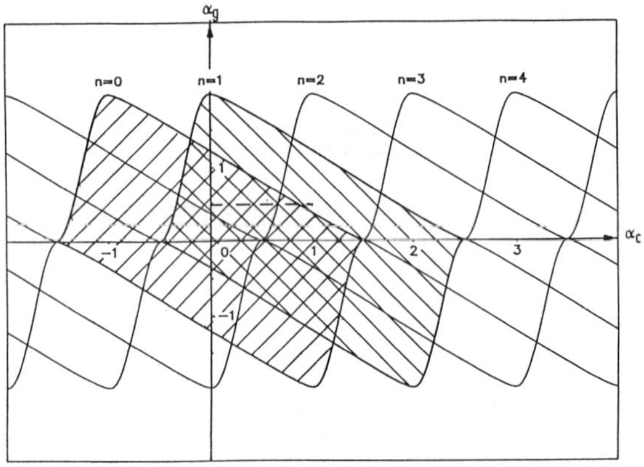

Figure 10
The characteristic of an asymmetrically-fed SQUID with equal critical currents of the single Josephson elements showing the areas of existence of the flux-quantum states. The curves represent the normalized maximum gate currents α_g of the SQUID for given normalized control currents α_c. The shaded areas mark the areas of the flux-quantum states n = 0 and n = 1 which overlap. This is important in digital applications. The dashed line marks the path of the control current used for the simulation of the switching process in Fig. 12.

metrically fed-in gate current I_g that gives $\lambda_1 = 0$ and $\lambda_2 = 2\pi$ is shown in Fig. 10. The closed curves reflect the various flux-quantum states with different numbers n of flux quanta enclosed.

To obtain the equations for the dynamics of a SQUID the static current through the Josephson elements is replaced by the dynamic current given in Eq. (3.3). This leads to (Johnson formulation)

$$\widehat{D}_i\varphi_i = \frac{d^2\varphi_i}{d\tau^2} + \beta_{Ji}\frac{d\varphi_i}{d\tau} + \sin\varphi_i = a_i\alpha_i \quad (i = 1,2) \tag{3.10}$$

with the stimulation terms

$$\alpha_1 = p_2\,\alpha_g + \frac{1}{\lambda}\left[\varphi_2 - \varphi_1 + 2\pi\left(n - \Phi_{ext}\right)\right] \tag{3.11}$$

$$\alpha_2 = p_1\,\alpha_g - \frac{1}{\lambda}\left[\varphi_2 - \varphi_1 + 2\pi\left(n - \Phi_{ext}\right)\right] \tag{3.12}$$

Here, Φ'_{ext} is the external magnetic flux normalized with the flux quantum Φ_o and the set of parameters is given by the equations:

$$I_o = \sqrt{I_{01}I_{02}}\,,\quad \alpha_i = \frac{I_i(t)}{I_o}\,,\quad a_i = \frac{I_{oi}}{I_o}\,,$$
$$p_i = \frac{L_i}{L}\,,\quad \lambda_i = p_i\,a_i\,,\quad (i = 1,2),\quad \lambda = \lambda_1 + \lambda_2 \tag{3.13}$$

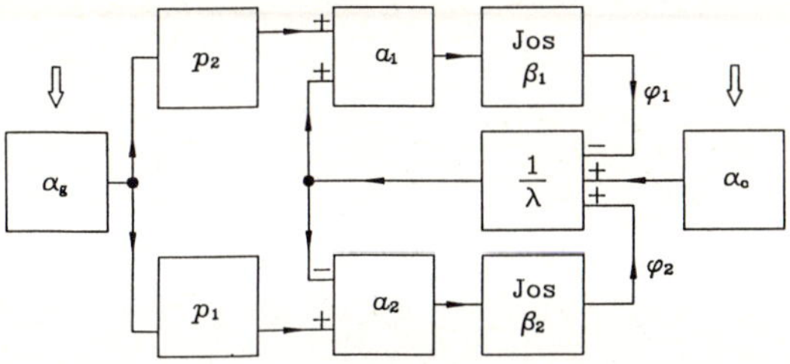

Figure 11
Mathematical model for the quasi-analog simulation of the behaviour of a SQUID. The graphical structure reflects the relations between the phase differences φ_1 and φ_2, the normalized currents α_g and α_c and the parameters of the model which are given by the Eq. (3.10), (3.13). The arrows point out the sequence of the calculations and the flow of data: Jos means the operation for solving the differential Eq. (3.10) for a given current and p_1, p_2, a_1, a_2, and $1/\lambda$ the multiplication of the input values with the magnitudes specified. α_g and α_c represent the supply with external currents. If a box has more than one input the data values are summed before executing the operation.

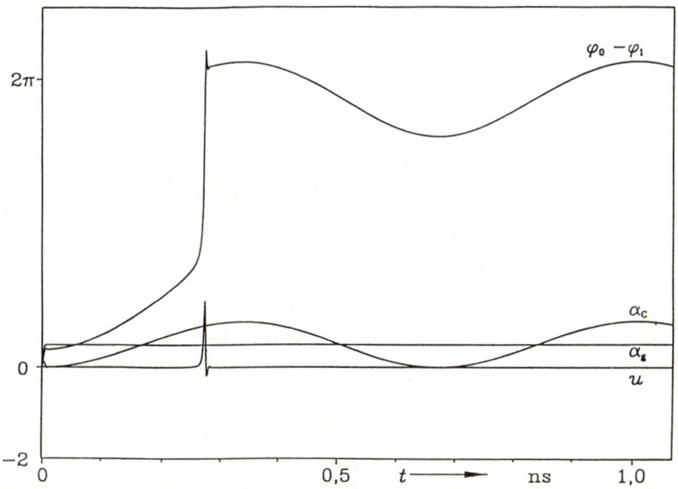

Figure 12
Numerical simulation of the storage of a flux quantum in a SQUID. The difference of the phase difference φ_1 - φ_2 represents the magnetic flux showing the transition from flux-quantum-state n = 0 to n = 1. α_g is the normalized gate current, α_c the normalized control current and u the voltage across the interferometer. The critical currents of both Josephson elements are equal. The inductance is chosen in order to give $\lambda = 2\pi$. The Johnson damping parameter β_J has the value 1.

For numerical evaluations these equations have been translated into a model the structure of which is shown in Fig. 11 in terms of analog computer boxes.

Some of the results of the numerical calculations are presented here. Fig. 12 shows the process of storage of one flux quantum in a SQUID driven

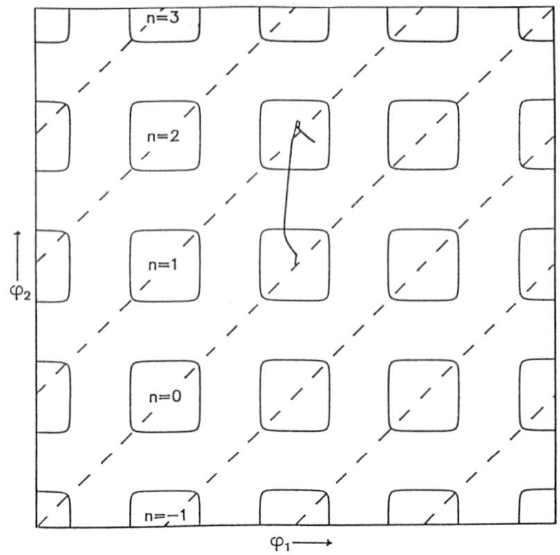

Figure 13
Existence areas of 'S-states' of a SQUID for diverse flux-quantum-numbers n in the plane of the two phase differences φ_1 and φ_1. The areas with equal numbers of flux quanta are connected via diagonals. Also shown is the path of the flux quantum change from n = 0 to n = 1 which has been simulated in Fig. 12.

with a constant gate current I_g by means of a sinusoidal control current I_c. If the control current (marked in Fig. 10) exceeds a threshold value on the characteristics which depends on the gate current I_g, the storage cell jumps from the flux-quantum-state with n = 0 to the state with n = 1. If the decreasing sinusoidal control current I_c does not cross the characteristic of the flux-quantum state with n = 1 on the left (see Fig. 10), the so-called interferometer stays at n = 1. Fig. 13 shows the path of the storing process in the plane of the two phase differences φ_1 and φ_2. The jump into the other flux-quantum state can be clearly seen. The current distribution in the Josephson elements during the storing process is presented in Fig. 14. Knowledge of this is of great importance in the calculations of arrays of Josephson elements in which a large number of storage loops are coupled to a long chain.

The results of the simulations clearly show, that with interferometers which are highly damped the switching performance of a SQUID is well defined. Fig. 12 distinctly illustrates that due to damping of plasma oscillations the switching performance of a SQUID is much better than that of a single Josephson element with the same parameters. The switching time is shorter because the Josephson element which stays in the 'S-state' during the whole process shunts the other which has been switched. This causes the leakage current to rise and the plasma oscillations are less pronounced than in Fig. 8. Moreover, it is not necessary to take the nonlinearity of the resistance of the Josephson elements into account. To

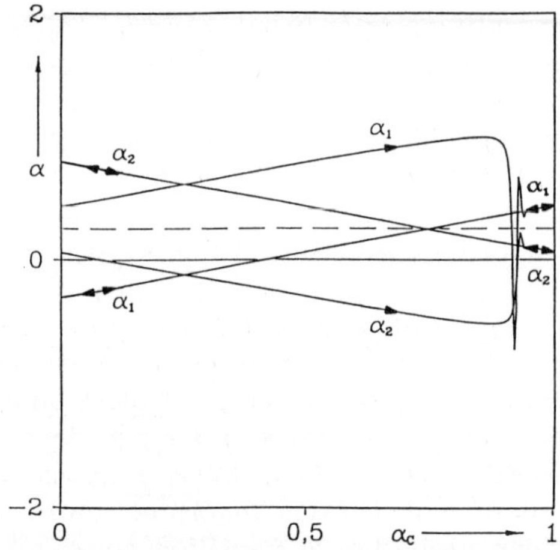

Figure 14
The current distribution in the SQUID during the switching process illustrated in Fig. 12. α_1 and α_2 are the normalized currents through the two Josephson elements and α_c is the normalized control current. Also shown is the arithmetic mean of α_1 and α_2 (dashed line) which is equal to half the value of the biasing gate current α_g. The arrows indicate the temporal evolution. α_1 i.e. rises from the starting value along the upper curve until switching takes place. Thereafter it stays on the lower curve.

confirm this, some numerical calculations with nonlinear characteristics of the Josephson elements were done on the basis of the RSJN model mentioned above [6].The simulations carried out with this model at high damping rates ($\beta_J > 1$) show only minor differences compared with the results obtained with the RSJ model which do not affect the switching performance.

3.3 The Flux Shuttle

There are several means of realizing a shift register in Josephson techniques. FULTON, DYNES and ANDERSON [14] proposed an array of directly coupled SQUIDs which they called the flux shuttle. They calculated maximum information transfer rates of 40 ps at critical current densities of 200 A/cm^2. HERRELL [15] presented another directly coupled shift register with only one control current which consists of very short pulses. The shift register is build up of SQUIDs which are driven as galvanically coupled flip-flops. This has never been realized, but it is an interesting alternative because it allows the DATA and NOT DATA signals to be obtained. JUTZI, CROCOLL, HERWIG, KRATZ, SADORF and WUNSCH [16] constructed a prototype of a flux shuttle consisting of four storage loops with one nondestructive readout interferometer which was driven by three pulsed control currents. It worked up to 250 MHz and was designed in PbInAu technique with critical current densities of about 4.3 kA/cm^2. NAKANISHI and YOSHIKIYO [17] constructed an ultra high-speed shift register from resistor loaded three-junction interferometers which were coupled magnetically and driven by only two phase-shifted sinusoidal control currents. The computer simulations they made suggested good performance for frequencies up to 12 GHz. FUJIMAKI, IMAMURA and HASUO [18] realized a 9-bit shift register in Modified Variable Threshold Logic, with the shifting process controlled by three phase-shifted control currents. Their prototype set-up in the $NbAl_xO_yNb$ technique was tested up to 2.2 GHz.

In order to build up a shift register for clock rates up to 10 GHz, it is necessary to have a device with a performance which does not critically depend on the shape of the control currents. At these high frequencies it is not possible to generate clock pulse trains of distinct shape nor to form them later. The SQUID as described in Sect. 3.2 is a device which meets the requirements. Simulations with various shapes of control currents confirm the same good switching properties at transition points which depend only on the level of the gate current.

On the other hand taking the application as a pseudo-random generator into account, the switching edges must be as sharp as possible both for storing a binary one and for dropping it (storing a binary zero). Here the SQUID is also the favoured device because it shows very sharp edges of transition and the plasma oscillations are strongly damped out as long as the storage loop is used in such a way that both of its Josephson elements are never driven into the 'R-state'.

A shift register consists of a number of storage cells linked together in an array to transfer the binary state from one cell to the other. The coupling between the individual cells must also fulfill the criteria stated above for the storage cells. Because the inductive coupling provides these conditions, the flux shuttle is the design with the optimum performance. But it is not as simple as the version which has been proposed by FULTON [14]. Every storage cell consists of a storage loop which may be called the master loop and which holds the information, and a coupling loop called the slave loop which performs the transfer to the next cell. To build a shift register with N stages an array of 2N SQUID loops must be produced.

The determination of the characteristics of a flux shuttle with N storage cells is a very complex and tedious problem. SCHULZ-DUBOIS and WOLF [19] however, have reported methods of approximation which can be used in computer calculations.

The dynamic description of a flux shuttle is a direct generalization of the equations of a single SQUID loop. The model used is shown in Fig. 15. A chain of N storage loops consists of a linear array of N + 1 Josephson elements which are coupled via the inductances L_i to their nearest neighbours. When Kirchhoff's sum rule for every node is taken into account the equation for the currents through the Josephson elements reads:

$$\hat{D}_i\varphi_i = \frac{d^2\varphi_i}{d\tau^2} + \beta_{Ji}\frac{d\varphi_i}{d\tau} + \sin\varphi_i = \alpha_{gi} - \alpha_{gi-1} \qquad (i = 0 \ldots n) \tag{3.14}$$

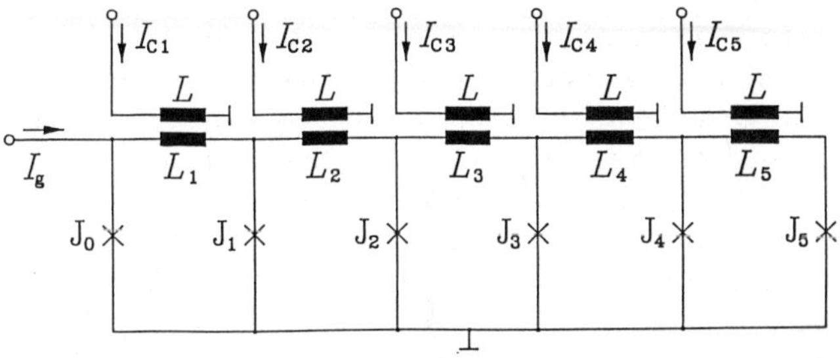

Figure 15
The flux shuttle as a shift register. I_{ci} (i = 1...5) are the control currents for the shifting of the flux which are phase shifted 180^0 against each other. I_g is the gate current which sets the bias point on the haracteristic and L_i (1 = 1...5) are the coupling inductances. J_i (i = 0...5) are the Josephson elements of the various storage loops. The control currents are coupled into the loops via the inductances L.

Here $\hat{D}_i$ is the differential operator (Eq. (3.3)) of the i-th element and the α_{gi} are the normalized electrical currents flowing through the inductivities L_i. For the first and the last element of the array appropriate boundary conditions must be applied:

$$\alpha_{g0} = \alpha_g, \qquad \alpha_{gn+1} = 0 \tag{3.15}$$

The magnetic coupling of neighbouring Josephson elements in the model of Fig. 15 is the same as in a single SQUID (nearest neighbour approximation). For the various loops one obtains

$$\varphi_{i-1} - \varphi_i = \lambda_i (\alpha_{gi} + \alpha_{ci}) + 2\pi n_i \qquad (i = 1...n) \tag{3.16}$$

where the quantities λ_i and the α's are defined in the same way as in Eq. (3.13). In the development above, the implicit assumption has been made that the Josephson elements of the array have an equal contact area and equal critical currents. The nearest neighbour approximation used in the derivation above and the assumption that the boundaries besides the galvanic condition of Eq. (3.15) have no serious effect on the characteristics of the storage loops, have both been proved by BEHA et al. [20] to be valid for shift registers with more than three cells.

Fig. 16 shows the computer simulation of the shifting of one quantum of flux by means of two sinusoidal control currents in a shift register with three loops. The curves of the phase differences show, besides sharp switching edges, a structure which is attributed to the shapes of the control and gate currents. By a proper choice of the amplitude of the control current and the characteristics of the simple storage loop, this influence can be reduced.

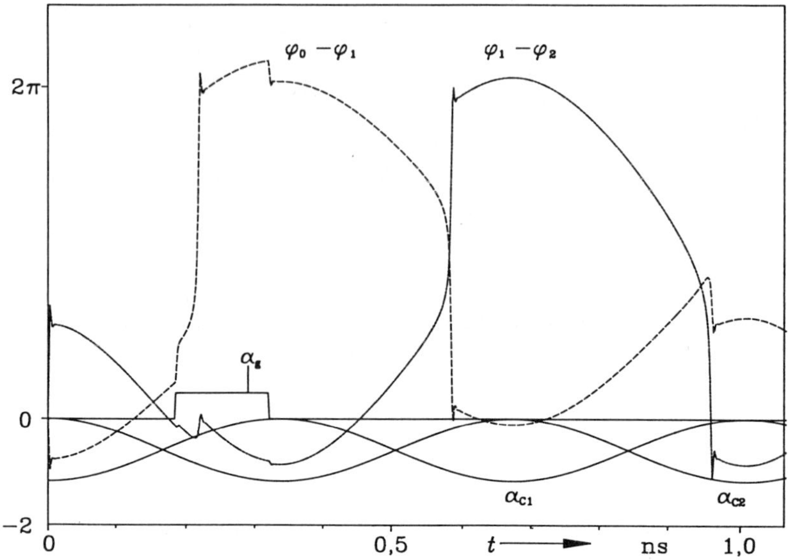

Figure 16
Simulation of shifting of a flux quantum through a shift register by means of two normalized sinusoidal control currents α_{c1} and α_{c2}. The differences of the phase differences $\varphi_0 - \varphi_1$ and $\varphi_1 - \varphi_2$ show the flux-quantum-states of the storage loops. α_g is the normalized gate current which is raised for a short time to store one flux quantum in the first loop. When α_{c1} increases and α_{c2} decreases simultaneously the stored flux is shifted to the next loop. This can be seen by the crossing of the curves $\varphi_0 - \varphi_1$ and $\varphi_1 - \varphi_2$ at about 0.6 ns.

3.4 The Feedback Logic

As explained in Sect. 2.3 the pseudo random generator needs a feedback loop which combines the signal from the last and one of the previous storage cells and feeds it back to the input of the first stage. The 'exclusive-or' gate which is needed for this purpose usually is built up from an 'or' gate and a 'nand' gate linked via an 'and' gate. There are other possibilities of building up the 'exclusive-or' function but the main requirement is the realization of an inverting gate which is included in the 'nand' gate of the example mentioned. This problem can easily be solved in the traditional transistor logic which is an inherent inverting, voltage controlled logic.

In contrast to the case of the semiconductors the Josephson logic is a genuine non-inverting, current-controlled logic. The control currents are used to reduce the critical currents of the switching device which is either a single Josephson element or a SQUID or another. By means of this reduction, the device is switched to the 'R-state' and the main part of the supply current is diverted to the output. Switching the control current to the 'high state' therefore switches the output to the 'high state', too. The logic circuit families of the first generation were based on single Josephson elements of large area. HERRELL [21] and CHAN and VAN DUZER [22] described

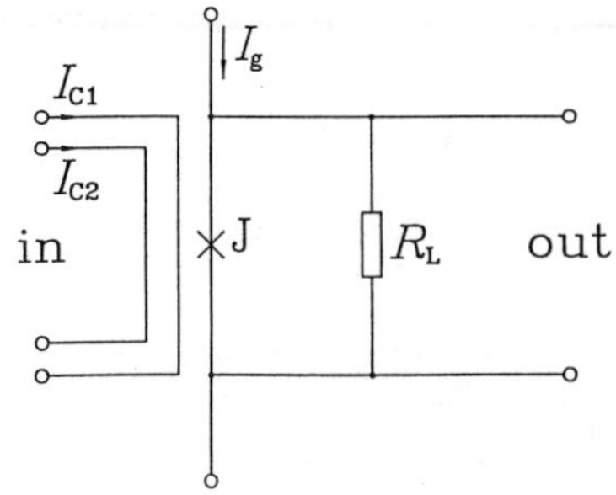

Figure 17
A logic cell in HERRELL's Josephson Tunneling Logic. I_g is the gate current which biases the element. I_{c1} and I_{c2} are control currents which change the critical current of the Josephson element J. R_L is the load resistance.

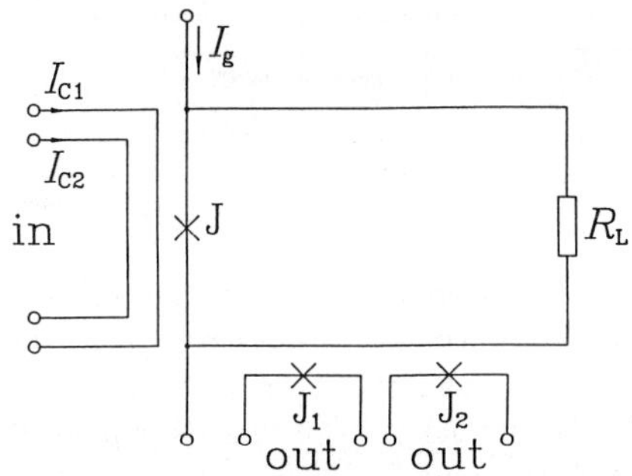

Figure 18
A logic gate in CHAN and VAN DUZER's Nonlatching Josephson Tunneling Logic. The symbols are the same as in Fig. 17. J_1 and J_2 represent the magnetically controlled output gates.

various means of realization. In the work of Herrell a magnetic field is coupled into a Josephson junction via control currents I_{ci} which affect the critical current I_o (see Fig. 17). Depending on the bias current I_g and the direction of the control currents I_{ci} an 'and', 'or' or 'not' gates can be realized. The advantage of this set up on one side is the high fan-out of about 10 mA to 30 mA for loading the following logic elements and the simple design with only one Josephson element. This is important if techniques with relative low reproducibility and long time stability of the Josephson junctions - such as the PbInAu technique - are used. A severe disadvantage is the large contact area. It leads to time delays of about 100 ps per gate and small margins in the production process. A basic problem of this logic family is the fact that the circuits are latching, and therefore have to be reset after each operation cycle by reducing the currents to zero. The last point has been improved by Chan and van Duzer who developed a logic circuit family based on two magnetically controlled Josephson elements as described above in each logic gate (see Fig. 18). One is used for the switching, the other for supplying the output. The switching circuit is thus decoupled from the output and its parameters can be chosen that the circuit becomes

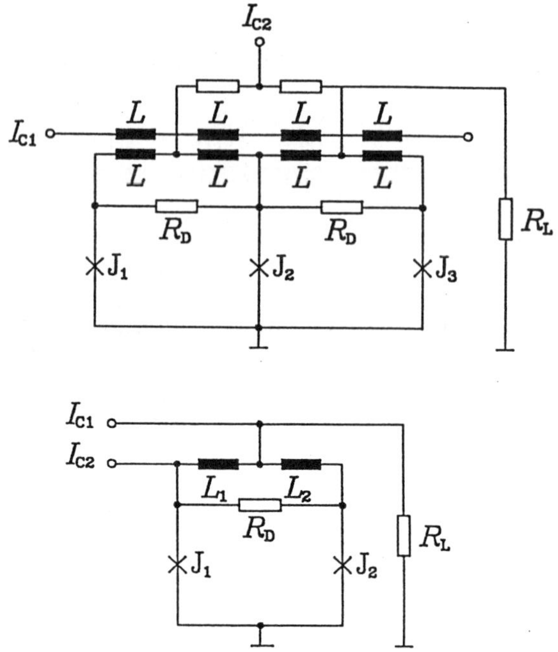

Figure 19
Basic gates of GHEEWALA's Current Injection Logic. I_{c1} and I_{c2} are the control currents, L the coupling inductances, R_D are damping resistances and R_L is the load resistance. The 'or' gate (above) is built from a three-junction interferometer. The 'and' gate (below) is an asymmetrically-fed SQUID.

self-resetting. Besides this, their design makes it possible to control more than one output gate.

More recent techniques use SQUIDs instead of simple Josephson elements. Different critical currents of the Josephson elements are taken to create distinct characteristics. GHEEWALA [23] introduced the current injection logic (CIL). In this logic family the various gates are coupled galvanically. The 'or' gate (Fig. 19) is realized by a 3 junction interferometer with the distribution of the critical currents given by the relation 1:2:1. The 'and' gate consists of an asymmetrically-fed SQUID. To obtain a 'not' gate, a timed inverter has been constructed. This logic family has been realized in the PbInAu technique, which has the drawback of relatively low reproducibility of the critical currents. Nevertheless, the performance was good enough to construct some basic computer circuits.

TAKADA, KOSAKA and HAYAKAWA [24] suggested a directly coupled logic family with four Josephson elements in a 'non-interferometer' device. The advantage of this logic family is the small area needed because of the absence of inductances. This is important in computer applications. Another advantage is that it is possible to model almost all logic gates. Using this

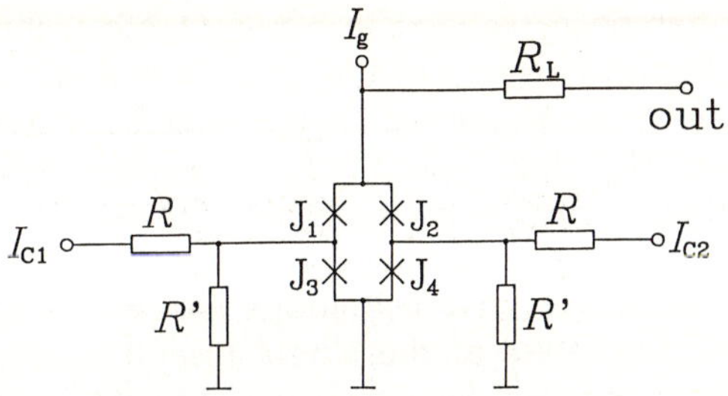

Figure 20
An 'exclusive-or' gate in the Direct Coupled Josephson Logic. The four Josephson elements J_i (i = 1...4) are galvanically coupled without any inductance loop. I_g is the clocked gate current, I_{c1} and I_{c2} are the control currents, R' are shunt resistances and R are input resistances. R_L is the load resistance.

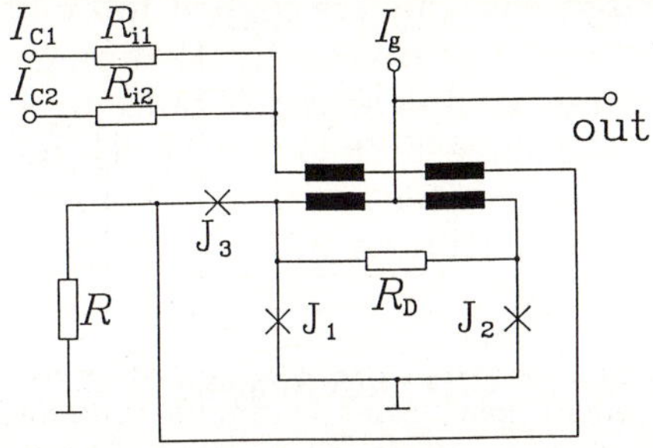

Figure 21
An 'or' gate in the Modified Variable Threshold Logic. I_g is the clocked gate current, I_{c1} and I_{c2} are the control currents, R_D is a damping resistance. The third Josephson element J_3 and the resistance R are used to decouple the inputs from the output.

techniques NITTA, HOHKAWA and ISHIDA [25] developed the 'exclusive-or' gate shown in Fig. 20. This logic family offers the advantage of high fan out and wide operation margins. The disadvantage is the relatively long transfer time up to 80 ps per gate [26]. This results from the fact that both circuit branches are driven to the 'R-state' one after the other. During the subsequent resetting cycle, the plasma oscillations described above, which are chiefly responsible for limiting the transfer time, are initiated.

FUJIMAKI, IMAMURA and HASUO created flip-flops in modified variable threshold logic [18], [27]. The basic gate they use consists of an interferometer with two Josephson elements J_1 and J_2, as can be seen in Fig. 21 and a further element J_3 which acts as a decoupling circuit between the input and the output. The performance, whether it works as an 'and' or an 'or' gate, is adjusted by choosing the inputs for the signal currents. Fig. 21 shows the wiring of an 'or' gate. The advantage of this technique is the fact that it offers small dimensions and wide margins for control currents in the class of interferometer devices. Moreover, a good insulation between input and output is achieved and the same basic cell is used for all types of logic gates.

This logic family has been realized in $NbAl_xO_yNb$ technique to keep the deviations of the critical currents of the single junctions small. The authors have measured gate delay times not longer than 9 ps. Even smaller values have been reported. The overall switching properties, however, are limited by the plasma oscillations as has been explained above. No details on this are given in the literature.

In order to construct feedback loops for a fast shift register, the uniformity of the parameters of the single Josephson elements is not the most serious problem because the area covered by the whole device is very small. A simple design with short transmission lines is necessary to avoid additional signal time delays in the feedback loop which are long compared with the period of a clock pulse. Although the direct coupled Josephson logic [24] gives overall delay times of 80 ps it is this logic which is most promising for the feedback loop in a shift register. It admits clock rates up to 12.5 GHz and only four Josephson elements are needed to build up an 'exclusive-or' gate (see Fig. 20) in one logic gate which works during one clock cycle. Other logic families use three complete logic gates or more which achieve 'exclusive-or' output after three clock cycles. With a more effective damping of plasma oscillations, the overall delay time of the direct coupled logic gates may be further reduced, shifting the possible clock rates to higher frequencies.

3.5 Realization of a Fast SQUID Shift Register

To realize the shift register with the feedback loop described above the conditions for producing the various parts must be chosen carefully with respect to the techniques which are available for the production of logic circuits with Josephson elements. The two techniques commonly used are the PbInAu technique developed in the IBM research group and the $NbAl_xO_yNb$ technique which has been promoted mainly in Japan. Whereas the latter technique produces Josephson circuits of long lifetimes and small scattering of the parameters, and allows very fine structuring, it requires a sophisticated control of the production process. It is used for the generation of high element integration on chips which are intended for computer applications. In this field, uniformity on one chip and between different chips is a crucial requirement. For devices which do not need such high element integration, the PbInAu technique offering a wider spread of the parameters of the various elements may be considered. This is not important if the whole circuit only consists of a few hundred of elements or so. The process of production is well known and relatively easy to handle. It allows a good fine structuring which is suitable for most of the digital applications.

The calculations for the shift register are based on Josephson elements with a density of critical current of 1 kA/cm^2. This value can be achieved with both techniques and as will be shown shortly gives fast switching parameters. With contact areas of 5 µm x 5 µm the critical current has a value of 250 μA. A Josephson contact of this area has a capacitance of about 0.9 pF (PbInAu technique). From these values a plasma frequency of 147 GHz is derived (table 2: characteristic frequency ω_J, column 3) which guarantees good switching up to frequencies higher than 10 GHz.

As has been discussed in detail in section 3.1, the Josephson elements should be strongly damped. The simulations shown above have been done with the value of the Johnson damping parameter of $\beta_J = 1$. The Josephson elements used for switching must therefore be shunted by resistances of about 1.6 Ω. The values used for the layout are compiled in table 3.

Table 3
Parameters used in the layout of the flux shuttle shift register.

parameter	symbol	value
density of critical current	j_o	1 kA/cm^2
contact area	F	25 μm^2
critical current	I_o	250 μA
capacitance	C	0.9 pF
plasma frequency	f_J	147 GHz
Johnson damping parameter	β_J	1
shunt resistance	R	1.6 Ω
critical phase	λ	≥ 2 π
linking inductance	L	8 pH

This table also gives the inductance of the storage loops of the shift register, referred to as linking inductance. Its value is chosen that each loop can hold one quantum of flux.

A section of the layout is shown in Fig. 22. It presents the basic cell of the shift register which consists of two storage loops. Three Josephson contacts with two linking inductivities which form the meandering top electrode are shown. Each bend is covered by one of the control line inductivities which are extricated alternating to the two control lines extending along the sides of the shift register chip. The large quadratic patches at the resistors and the control electrodes are contact areas which connect the resistors to the top electrode, and the control electrodes to the ground electrode via Josephson contacts of large area (20 μm x 20 μm). These Josephson elements are not shown. The meandering form of the top and the control electrode has been chosen in order to have the necessary inductance of 8 pH with small overall extension.

A chip with the parameters described above is produced by running through the following procedure. First a niobium shielding 150 nm thick is sputtered on the substrat. It serves as guard against scattering RF signals, gives higher inductance values (electromagnetic mirror) and at the same time functions as an adhesive ground plate. This ground plate is electrically

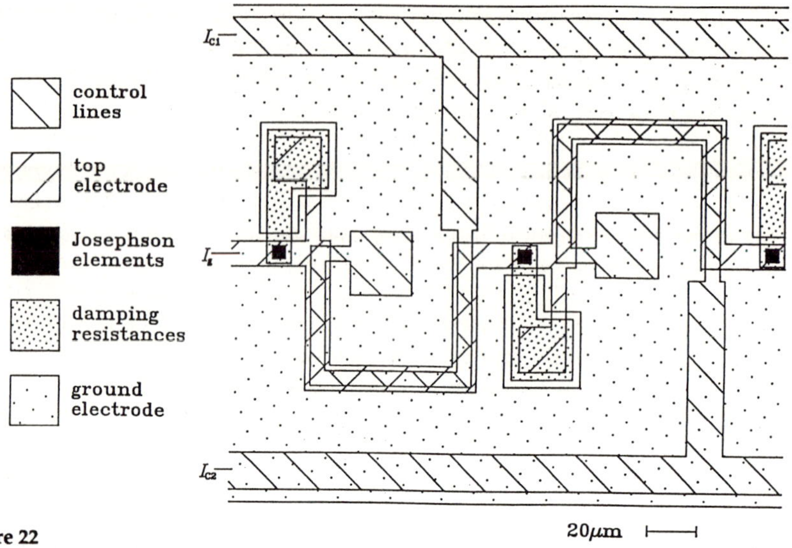

Figure 22
Section of the layout of the flux shuttle shift register showing the consecutive storage loops which represent one storage cell of the register. I_g, I_{c1} and I_{c2} mark the lines for the gate and control currents respectively (for further details see text).

insulated from the circuit by first oxidizing and then depositing a SiO layer. The following layer contains the resistors which are composed of AuIn. The resistance is controlled by the thickness and the ratio of the length to the width. A value of 1.6 Ω is obtained with an aspect ratio of 2 and a layer thickness of about 50 nm. The next layer of 200 nm PbInAu is the ground electrode covered with an insulation layer of 300 nm SiO which leaves windows for the Josephson contacts. The Josephson barriers are created by controlled oxidizing of the ground electrode within the uncovered areas (Greiner process). This process is stopped at the desired thickness of the barrier by evaporating of the top electrode which is formed by a PbInAu-layer of 400 nm thick. The last electrode is the control line (PbInAu layer 600 nm thick) which is insulated from the top electrode by a SiO layer 500 nm thick. Finally, the whole chip is covered with a protective SiO layer 700 nm thick.

The structuring of each layer is done by a lift-of process. The thickness of the layers increases from the bottom to the top in order to guarantee the covering of edges in all planes.

3.6 Modifications Using High T_c Superconductors

One serious problem connected with the application of the shift register in the high frequency range is the difference in temperature between the noise generating element and the measuring equipment. The superconducting shift register must be cooled down to about 5 K, whereas the noise power must be available at room temperature. The transmission line which transfers the noise power from the low temperature end to room temperature causes losses which depend on the distribution of temperature on the line. Because of the nonuniformity of this distribution, these absorption losses are neither easy to calculate nor to measure, thus giving rise to additional uncertainties in the noise power delivered to the output at room temperature. The use of high T_c superconductors would reduce this problem because of the reduction of the difference between the operating temperature of the shift register and the output at room temperature.

On the other hand, the higher operation temperature gives rise to stronger influences of the inherent noise on the characteristics of the switching elements. This means that the scattering of the parameters becomes more important and narrower margins must be chosen.

In accordance with microscopic theory, high T_c superconductors with transition temperatures of about 100 K exhibit gap voltages of about 20 mV. Taking Eq. (2.9) into account, this results in an increase of the output noise power by a factor of 100 compared with conventional superconducting materials. The desired noise power level is available only after appropriate attenuation. With high T_c superconductors which have low gap voltages in the range of 1 mV, shift registers could be developed which can be used in noise measurement applications.

The design of the shift register discussed here is based on small plane contacts with a thin insulating barrier. The switching properties have been deduced from the quasi classical RSJ circuit shown in Fig. 7. The underlying theory presumes that the coherence length of the superconducting materials involved is larger than the thickness of the barrier. For the high T_c superconductors known, the coherence length is very short and it is not possible to make conventional barriers. Other principles of switching and storing information must be developed, which will surely lead to a different design. In conclusion, high T_c superconductors can be employed in advantage in the design proposed if materials with a high transition temperature but a low energy gap and long coherence length become available.

References

[1] H. Suttcliff and K.F. Knott, Radio and Electron. Engineer **40** (1970) pp. 132-135

[2] S.O. Rice, Bell Sept. Techn. J. **23** (1944) pp. 282-332 and **24** (1945) pp. 46-156

[3] W.M. Wonham and A.T. Fuller, J. Electronics and Control **4** (1958) pp. 568-576

[4] P.D. Roberts, B. Eng and R.H. Davies, Proc. IEE **113** (1966) pp. 191-196

[5] B. Elspas, IRE Trans. **CT-6** (1959) pp. 45-60

[6] K.K.Likharev, Dynamics of Josephson Junctions and Circuits, Gordon and Branch Science Publishers, p. 48 (Amsterdam 1986)

[7] D.E. McCumber J. Appl. Phys. **39** (1968) pp. 3113-3118

[8] W.C. Stewart, Appl. Phys. Lett. **12** (1968) pp. 277-280

[9] W.J. Johnson, Ph.D. Thesis, University of Wisconsin, Madison (1968)

[10] B.D. Josephson, Phys. Lett. **1** p. 251 (1962)

[11] G. Brunk, SQUID'85, de Gruyter & Co. Berlin New York, (1985) pp. 77-82

[12] H. Lübbig, G. Brunk, Chr. Zurbrügg, PTB Jahresbericht 1985 pp. 214-215
[13] H. Kratz, W. Jutzi, IEEE Trans. Mag. **MAG-23** (1987) pp. 731-734
[14] T.A. Fulton, R.C. Dynes, and P.W. Anderson, Proc. IEEE **61** (1973) pp. 28-35
[15] D.J. Herrell, IBM Tech. Disc. Bull. **NY17** (1975) pp. 3089-3092
[16] W. Jutzi, E. Crocoll, R. Herwig, H. Kratz,M. Neuhaus, H. Sadorf and J. Wunsch, IEEE Electron Devices Letters **EDL-4** (1983) pp. 49-50
[17] T. Nakanishi and H. Yoshikiyo, Trans. IECE Jpn. **E69** (1986) pp. 430-431
[18] N. Fujimaki, T. Imamura, S. Hasuo, Extended Abstract of the Electrochemical Society **87/2**, Fall Meeting, Honolulu, Hawaii, Oct. 18/23 Abstract No.366, Low Temp. Electr. (1987)
[19] E.O. Schulz-DuBois, P. Wolf, Appl. Phys. **16** (1978) pp. 317-338
[20] H. Beha, W. Jutzi and G. Mischke, IEEE Trans. Electron Devices, **ED-27** (1980) pp. 1882-1887
[21] D.J. Herrell, IEEE Journal of Solid-State Circuits **SC-9** (1974) pp. 277-282
[22] H.W.K. Chan, T. van Duzer, IEEE Journal of Solid-State Circuits **SC-12**, (1977) pp. 73-79
[23] T.R. Gheewala, IBM J. Res. Develop. **24** (1980) pp. 130-142
[24] S. Takada, S. Kosaka and H. Hayakawa, SQUID '80 (1980) pp. 713-724
[25] J. Nitta, K. Hohkawa and A. Ishida: Electr. Lett. **19** (1983) pp. 233-234
[26] G. Matheron, Ph. Migny and O. Sie, IEEE Trans. Mag. **MAG-21** (1985) pp. 555-557
[27] N. Fujimaki, S. Kotani, S. Hasuo and T. Yamaoka, Jpn. J. Appl. Phys. **24** (1985) pp. L1-L2

Subject Index